T0190387

'This provocative yet deeply researched collection of essays edited by Alexander Geppert reveals the profound connection between the climacteric of manned spaceflight after Apollo 11 and the onrush of globalization in the 1970s. Pausing after the moon landings in its cosmic quest, humanity, as it were, deepened its global connections; and this book opens up that hitherto unexplored linkage.'

—Charles S. Maier, Harvard University

'For ages, mankind envisioned venturing to the moon. Surprisingly, once that vision was realized, popular fascination with spaceflight vanished quickly. The stars became disenchanted, and spaceship earth began to mirror itself with thousands of satellites instead. From perspectives as diverse as geopolitics, architecture and law, this intriguing book outlines continuities and transformations of astroculture during the post-Apollo era. It offers thought-provoking insights by adding a third dimension to the more than ambivalent 1970s and 1980s.'

—Dirk van Laak, Universität Leipzig

'This is a highly original volume on the surprising lull in space exploration during the crisis-ridden 1970s. The particular value of its multinational chapters lies in their transdisciplinary investigation of how the end of the Apollo moon landings coincided with a growing disillusionment of space imaginaries during the onset of globalization.'

—Konrad H. Jarausch, University of North Carolina, Chapel Hill

'*Limiting Outer Space* illustrates the rich possibilities of seeing spaceflight and astroculture as integral components of the pivotal decade of the 1970s. Representing an array of disciplines and geographies, the authors in this volume collectively complement and amend previous understandings of the cultural and geopolitical transitions of the age. Highly recommended for its broad scope and well crafted essays.'

—Emily S. Rosenberg, University of California, Irvine

Palgrave Studies in the History of Science and Technology

Series Editors
James Rodger Fleming
Colby College
Waterville, ME, USA

Roger D. Launius
Washington, DC, USA

Designed to bridge the gap between the history of science and the history of technology, this series publishes the best new work by promising and accomplished authors in both areas. In particular, it offers historical perspectives on issues of current and ongoing concern, provides international and global perspectives on scientific issues, and encourages productive communication between historians and practicing scientists.

More information about this series at
http://www.palgrave.com/gp/series/14581

Alexander C.T. Geppert
Editor

Limiting Outer Space

Astroculture After Apollo

European Astroculture
Volume 2

palgrave
macmillan

Editor
Alexander C.T. Geppert
New York University Shanghai
Shanghai, China

and

New York University, Center for
   European and Mediterranean Studies
New York, NY, USA

Palgrave Studies in the History of Science and Technology
European Astroculture, Volume 2
ISBN 978-1-349-67660-6   ISBN 978-1-137-36916-1   (eBook)
https://doi.org/10.1057/978-1-137-36916-1

Library of Congress Control Number: 2017951526

Cover image: © Gösta Röver
Cover design by Tom Howey

Printed on acid-free paper

This Palgrave Macmillan imprint is published by Springer Nature
The registered company is Macmillan Publishers Ltd.
The registered company address is: The Campus, 4 Crinan Street, London, N1 9XW, United Kingdom

CONTENTS

ACKNOWLEDGMENTS

The idea of alien invasion is not entirely foreign to scholars of the past. In more than one instance in his voluminous *œuvre,* Eric Hobsbawm – arguably one of the greatest historians of the twentieth century – fantasized about the advent of extraterrestrial colleagues on planet Earth. 'Suppose that one day, after a nuclear war, an intergalactic historian lands on a now dead planet,' begins, for instance, his *Nations and Nationalism,* published in 1992.[1] Little did Hobsbawm know that such an obscure breed of historians from outer space had long touched down and even regularly convened at international conferences which would then, in turn, give rise to books like this one. Early versions of almost all of the 13 chapters gathered here were originally presented at such a symposium, entitled *Envisioning Limits: Outer Space and the End of Utopia* and convened together with Daniel Brandau and William Macauley in Berlin in April 2012. Those who enabled us to host an interplanetary gathering of this magnitude must be thanked first, and I would like to express sincere gratitude to both the Center for International Cooperation at Freie Universität Berlin (FU) and, in particular, the Deutsche Forschungsgemeinschaft (DFG). It is the latter institution, internationally known as the German Research Foundation, which has also been funding the Emmy Noether research group 'The Future in the Stars: European Astroculture and Extraterrestrial Life in the Twentieth Century' at Freie Universität Berlin, which I have had the pleasure to direct from 2010 through 2016.[2]

The Berlin symposium and this ensuing volume are tangible outcomes of that group's work. *Limiting Outer Space: Astroculture After Apollo* pursues some of the problems raised and issues discussed in an earlier anthology, *Imagining Outer Space: European Astroculture in the Twentieth Century,* a companion volume published with Palgrave Macmillan in 2012 and now reissued in paperback.[3] While *Imagining Outer Space* set out to establish and contour a new field of historical inquiry – 'astroculture' –, the scope of the present book is more limited, yet also more narrowly focused. It is more limited because it zooms in on a single decade in the history of imagining, thinking and practicing outer space, the crisis-ridden 1970s. At the same time the volume foregrounds one particular problem, the limits of utopian thought and practice during this aptly called post-Apollo period. What both volumes have in common, however, is a cultural-interpretative approach, a commitment to combining a multiplicity of disciplinary perspectives, and the intention to push space history's geographical focus beyond the borders of the two Cold War superpowers. *Militarizing Outer Space: Astroculture, Dystopia and the Cold War*, a forthcoming third volume in form and format identical with the existing two, will expose the 'dark' side of global astroculture by exploring the militant dimensions of outer space in science fiction and science fact. Concentrating on weapons, warfare and violence, *Militarizing Outer Space* will conclude the unintended 'European Astroculture' trilogy.[4]

Engineering such scholarly large-scale enterprises would not be possible without the help of many. Thanks are due to both the conference speakers whose presentations could, alas, not be included in this volume as well as more than a dozen commentators and discussants. Their insight and criticism shaped the original symposium and, in turn, this volume. These critical interlocutors include Philippe Ailleris, Debbora Battaglia, Peter Becker, Thomas Brandstetter, Ralf Bülow, Matthew Hersch, John Krige, Neil Maher, Patrick McCray, Lisa Messeri, Agnes Meyer-Brandis, Gonzalo Munévar, Virgiliu Pop, Claudia Schmölders, Matthias Schwartz, Helmuth Trischler, Christina Vatsella, Janet Vertesi and Thomas P. Weber. Insisting that the colors of space are black and silver, FU's chief designer Gösta Röver developed our own visual language; her conference posters also formed the basis for the cover illustrations of all three volumes. FU photographer Hubert Graml helped prepare images for publication. Several anonymous reviewers offered invaluable criticism and pointed advice. Kayalvizhi Saravanakumar and her team oversaw the production process with equanimity and punctilious attention to detail. Audrey McClellan created the index with great professionalism and utmost care. I would also like to express my heartfelt gratitude to the contributors themselves, in particular for their patience and willingness to let me subject them to one round of revisions after another. The final word of thanks, however, must go to all members of the 'Future in the Stars' research group at Freie Universität Berlin. They include doctoral students Daniel Brandau and Jana Bruggmann, postdoctoral research associate Tilmann Siebeneichner as well as student assistants Björn Blaß, Ruth Haake, Friederike Mehl, Tom Reichard, Katja Rippert and Magdalena Stotter. Ruth proved particularly indispensable during the final stretches; without her, neither this book nor its editor would have survived the interminable publication process. Once the group has dissolved in the not too distant future, I shall terribly miss working with an entire crew of intergalactic, and indeed stellar, historians. Fortunately, we still have a ways to go before our mission can be declared accomplished.

Shanghai
November 2017

Alexander C.T. Geppert

Notes

1. Eric J. Hobsbawm, *Nations and Nationalism: Programme, Myth, Reality*, 2nd edn, Cambridge: Cambridge University Press, 1992, 1.
2. A detailed conference program can be found at http://limits.geschkult.fu-berlin. de. For comprehensive reports see Friederike Mehl, 'Envisioning Limits: Outer Space and the End of Utopia. 19.–21. April 2012,' *H-Soz-u-Kult* (9 July 2012), online at http://hsozkult.geschichte.hu-berlin.de/tagungsberichte/id=4303; and idem, 'Berlin Symposium on Outer Space and the End of Utopia in the 1970s,' *NASA History News & Notes* 29.2–3 (2012), 1–5. For further information on the Emmy Noether research group 'The Future in the Stars: European Astroculture in the Twentieth Century,' consult http://www.geschkult.fu-berlin.de/astrofuturism (all accessed 1 October 2017).
3. Alexander C.T. Geppert, ed., *Imagining Outer Space: European Astroculture in the Twentieth Century*, Basingstoke: Palgrave Macmillan, 2012; 2nd edn, London: Palgrave Macmillan, 2018 (= *European Astroculture*, vol. 1); and idem, ed., *Astroculture and Technoscience*, London: Routledge, 2012 (= *History and Technology* 28.3).
4. Alexander C.T. Geppert, Daniel Brandau and Tilmann Siebeneichner, eds, *Militarizing Outer Space: Astroculture, Dystopia and the Cold War*, London: Palgrave Macmillan, forthcoming (= *European Astroculture*, vol. 3).

LIST OF FIGURES

The 1969–72 moon landings marked a shift in planetary perspectives. Inspired by iconic NASA photographs Earthrise (1968), Blue Marble (1972) and the first picture of earth and moon captured in a single frame (1977), the cover image positions the post-Apollo spectator beyond the moon, looking back at the distant home planet from outer space. © Gösta Röver, Freie Universität Berlin.

Abbreviations

ABM	Anti-ballistic Missile
AFOL	Adult Fan of Lego
AHR	*American Historical Review*
AIAA	American Institute of Aeronautics and Astronautics
ASAT	Anti-Satellite Weapon
ASE	Association of Space Explorers
ASTP	Apollo-Soyuz Test Project
BBC	British Broadcasting Corporation
BIS	British Interplanetary Society
BRD	Bundesrepublik Deutschland
BUL	Bulgaria
CETS	Conférence Européenne des Télécommunications par Satellites
CFF	Committee for the Future
CNES	Centre National d'Etudes Spatiales
CNN	Cable News Network
COPERS	Commission Préparatoire Européenne de Recherche Spatiale
COPUOS	Committee on the Peaceful Uses of Outer Space
COSPAR	Committee for Space Research
DDR	Deutsche Demokratische Republik
DEFA	Deutsche Film-Aktiengesellschaft
DFG	Deutsche Forschungsgemeinschaft
DFVLR	Deutsche Forschungs-und Versuchsanstalt für Luft- und Raumfahrt
DLR	Deutsches Zentrum für Luft- und Raumfahrt
DoD	Department of Defense
EEC	European Economic Community
ELDO	European Launcher Development Organization
ERNO	Entwicklungsring Nord
ESA	European Space Agency
ESOC	European Space Operations Centre
ESRO	European Space Research Organization
ET	Extraterrestrial

ETI	Extraterrestrial Intelligence
EU	European Union
EVA	Extravehicular Activity
FAZ	*Frankfurter Allgemeine Zeitung*
FR	France
FU	Freie Universität Berlin
GfW	Gesellschaft für Weltraumfahrt
HAEU	Historical Archives of the European Union
HAL	Heuristically Programmed Algorithmic Computer
IAF	International Astronautical Federation
IBMP	Institute of Biomedical Problems
ICBM	Intercontinental Ballistic Missile
IGY	International Geophysical Year
ISS	International Space Station
JPL	Jet Propulsion Laboratory
LGA	LEGO Group Archives
LSD	Lysergic Acid Diethylamide
MGM	Metro-Goldwyn-Mayer
MIRV	Multiple Independently Targetable Re-entry Vehicle
MOU	Memorandum Of Understanding
MOUSE	Minimum Orbital Unmanned Satellite, Earth
MTR	Military-Technical Revolution
n.d.	No date
n.p.	No publisher/pagination
NACA	National Advisory Committee on Aeronautics
NAS	National Academy of Sciences
NASA	National Aeronautics and Space Administration
NASM	National Air and Space Museum
NATO	North Atlantic Treaty Organization
NIEO	New International Economic Order
NRC	National Research Council
OST	Outer Space Treaty
Pan Am	Pan American World Airways
PAP	Post-Apollo Program
PL	Poland
RAE	Royal Aircraft Establishment
SDI	Strategic Defense Initiative
SDS	Students for a Democratic Society
SETI	Search for Extraterrestrial Intelligence
SKA	Stanley Kubrick Archive
SMI²LE	Space Migration, Intelligence Increase, Life Extension
STS	Space Transportation System
SYNCON	Synergistic Convergence
TNA	The National Archives (UK)
UAR	United Arab Republic
UFO	Unidentified Flying Object
UK	United Kingdom
UN	United Nations

USA	United States of America
USAF	United States Air Force
USSR	Union of Soviet Socialist Republics
V-2	Vergeltungswaffe 2
VEB	Volkseigener Betrieb
VfR	Verein für Raumschiffahrt
WDR	Westdeutscher Rundfunk
ZDF	Zweites Deutsches Fernsehen

NOTES ON CONTRIBUTORS

Thore Bjørnvig holds an MA in the History of Religions from the University of Copenhagen. A former associated member of the Emmy Noether research group 'The Future in the Stars: European Astroculture and Extraterrestrial Life in the Twentieth Century' at Freie Universität Berlin, his research focuses on intersections between science, technology and religion, with a particular emphasis on outer space. His publications include 'The Holy Grail of Outer Space: Pluralism, Druidry, and the Religion of Cinema in *The Sky Ship*' (2012); 'Outer Space Religion and the Ambiguous Nature of *Avatar*'s Pandora' (2013); and a special issue of *Astropolitics* on spaceflight and religion (2013, co-ed.). He blogs on astroculture for the Danish popular science news site videnskab.dk and the Nordic popular science news site sciencenordic.com.

Martin Collins is a curator at the Smithsonian National Air and Space Museum in Washington, DC. His research focuses on the history of the United States in the world after 1945, as seen through the history of technology. He recently concluded his tenure as editor of the journal *History and Technology* and is managing editor of the book series *Artefacts: Studies in the History of Science and Technology*. His book *A Telephone for the World: Iridium, Motorola, and the Making of a Global Age* is forthcoming in 2018.

Luca Follis is a lecturer at Lancaster University Law School. He works at the interface of socio-legal studies, historical sociology and political theory. Recently published articles include 'Power in Motion: Tracking Time, Space and Movement in the British Penal Estate' (2015); 'Democratic Punishment and the Archive of Violence: Punishment, Publicity and Corporal Excess in Antebellum New York' (2016); and 'Discipline Unbound: Patuxent, Treatment and the Colonization of Law' (2017).

Alexander C.T. Geppert holds a joint appointment as Associate Professor of History and European Studies and Global Network Associate Professor at New York University Shanghai as well as NYU's Center for European and Mediterranean Studies in Manhattan. From 2010 to 2016 he directed the Emmy Noether research group 'The Future in the Stars: European Astroculture and Extraterrestrial Life in the Twentieth Century' at Freie Universität Berlin. Book publications include *Fleeting Cities: Imperial Expositions in Fin-de-Siècle Europe* (2010, 2013); *Imagining Outer Space: European Astroculture in the Twentieth Century* (2012, 2018, ed.); *Obsession der Gegenwart: Zeit im 20. Jahrhundert* (2015, co-ed.); *Berliner Welträume im frühen 20. Jahrhundert* (2017, co-ed.); and *Militarizing Outer Space: Astroculture, Dystopia and the Cold War* (forthcoming, co-ed.). At present, Alexander Geppert is completing a cultural history of outer space in the European imagination, entitled *The Future in the Stars: Time and Transcendence in the European Space Age, 1942–1972*.

Andrew Jenks is Professor of History at California State University, Long Beach. He has published widely in both traditional and digital media on Russian history, environmental history and the history of science and technology. He is the author of *Russia in a Box: Art and Identity in an Age of Revolution* (2005); *Perils of Progress: Environmental Disasters in the Twentieth Century* (2010); and a biography of the cosmonaut Yuri Gagarin, *The Cosmonaut Who Couldn't Stop Smiling: The Life and Legend of Yuri Gagarin* (2012). Jenks is co-founder of the *Russian History Blog* and editor of *Kritika: Explorations in Russian and Eurasian History*. His current project looks at spaceflight as a key moment in the formation of transnational forms of collaboration and consciousness.

David A. Kirby was an evolutionary geneticist before leaving the laboratory to become Senior Lecturer in Science Communication Studies at the University of Manchester. His experiences as a member of the scientific community informed his internationally recognized studies into the interactions between science, media and cultural meanings. His book *Lab Coats in Hollywood: Science, Scientists and Cinema* (2011) examines collaborations between scientists and the entertainment industry in the production of movies. He received a Wellcome Trust Investigator Award to analyze the interactions among the biosciences, religion and entertainment. His current book project *Indecent Science: Religion, Science and Movie Censorship, 1930–1968* explores how movies served as a battleground over science's role in influencing morality.

Florian Kläger is Professor of English Literature at Universität Bayreuth and the author of *Forgone Nations: Constructions of English National Identity in Elizabethan Historiography and Literature. Stanihurst, Spenser, Shakespeare* (2006). He has published on cultural negotiations of collective identities, the contemporary novel and reflexivity in fiction, and is currently finishing a monograph on 'reading into the stars' in British and Irish narrative fiction. His book publications include *Diasporic Constructions of Home and Belonging* (2015, co-ed.) and *Early Modern Constructions of Europe: Literature, Culture, History* (2016, co-ed.).

Roger D. Launius was Associate Director for Collections and Curatorial Affairs at the Smithsonian National Air and Space Museum in Washington, DC. Between 1990 and 2002 he served as NASA chief historian. A graduate of Graceland College in Lamoni, Iowa, he received his PhD from Louisiana State University, Baton Rouge, in 1982. Roger Launius has written or edited more than thirty books on aerospace history, most recently *Space Shuttle Legacy: How We Did It and What We Learned* (2013, co-ed.); *Exploring the Solar System: The History and Science of Planetary Probes* (2013, ed.); *Historical Analogs for the Stimulation of Space Commerce* (2014); and *NASA Spaceflight: A History of Innovation* (2017, co-ed.)

Doug Millard is Deputy Keeper for Technologies and Engineering at the Science Museum in London. Millard has produced numerous space exhibitions, written articles, papers and books including a history of the Black Arrow satellite launch vehicle and its engines, and lectured and appeared on television and radio. In 2006 he gained his MSc in the history of science, technology and medicine at the University of London. In 2014 he edited the accompanying book to the 'Cosmonauts: Birth of the Space Age' exhibition on display at the Science Museum in 2015–16. His most recent book publication is *Satellite: Innovation in Orbit* (2017).

Regina Peldszus is with the Department of Space Situational Awareness at the German Aerospace Center (DLR) Space Administration, where she works on European governance of space surveillance and tracking infrastructure. From 2013 to 2015 she was a research fellow at the European Space Agency, based in the Studies and Special Projects Division at the European Space Operations Centre (ESOC). Previously, Peldszus contributed to space projects in Europe, Russia and the United States. She holds a PhD in human systems integration for exploration missions from Kingston University, London. Her research interests focus on resilience of large-scale sociotechnical systems in space, and transfer of expertise from polar and nuclear domains.

Robert Poole is a British historian and author of *Earthrise: How Man First Saw the Earth* (2008). He is currently Guild Research Fellow at the University of Central Lancashire and a former associate member of the 'Future in the Stars' Emmy Noether research group at Freie Universität Berlin. Recent publications include 'The Challenge of the Spaceship: Arthur C. Clarke and the History of the Future, 1930–1970' (2012); 'What Was Whole about the Whole Earth? Cold War and Scientific Revolution' (2014); and '*2001: A Space Odyssey* and the Dawn of Man' (2015).

Tilmann Siebeneichner is a research associate at the Zentrum für Zeithistorische Forschung (ZZF) in Potsdam. He holds a degree in philosophy and history from Georg-August-Universität Göttingen, where he completed his doctoral dissertation in 2011. From 2013 to 2016 Tilmann Siebeneichner was a member of the Emmy Noether research group 'The Future in the Stars: European Astroculture and Extraterrestrial Life in the Twentieth Century' at Freie Universität Berlin. Recent publications include *Proletarischer Mythos und realer Sozialismus: Die Kampfgruppen der Arbeiterklasse in der DDR* (2014); 'Europas Griff nach den Sternen: Das Weltraumlabor Spacelab, 1973–1998' (2014); 'Die "Narren von Tegel": Technische Innovation und ihre Inszenierung auf dem Berliner Raketenflugplatz, 1930–1934' (2017); and *Berliner Welträume im frühen 20. Jahrhundert* (2017, co-ed.). His current research focuses on the militarization of outer space in the 1970s.

Peter J. Westwick is Assistant Research Professor in History at the University of Southern California and Director of the Aerospace History Project at the Huntington-USC Institute on California and the West. He is the author of *The National Labs: Science in an American System, 1947–1974* (2002) and *Into the Black: JPL and the American Space Program, 1976–2004* (2006), editor of *Blue Sky Metropolis: The Aerospace Century in Southern California* (2012) and co-author of *The World in the Curl: An Unconventional History of Surfing* (2013). Peter Westwick is currently working on a history of the US National Academy of Sciences and a history of the Strategic Defense Initiative.

Introduction

CHAPTER 1

The Post-Apollo Paradox: Envisioning Limits During the Planetized 1970s

Alexander C.T. Geppert

People aren't interested in the future any more. [...] One could say that the moon landing was the death knell of the future as a moral authority.
J.G. Ballard, 1970

We are now in an interesting transition period when we can compare the realities of space with earlier imaginings of artists.
Arthur C. Clarke, 1972[1]

For much of the twentieth century, human possibilities in outer space seemed endless. Not the skies, but the stars were the limit. During the 1970s this relationship was reversed and outer space reconfigured. After the six moon landings between July 1969 and December 1972 (Figure 1.1), for many the 'unrepeatable spectacle of a lifetime,' disillusionment set in.[2] All successes in planetary exploration by robotic spacecraft were overshadowed by the memory and legacy of the American Apollo program. Machine-generated close-up photographs of Venus, Mars and Jupiter could not outrival a human being walking on earth's closest celestial neighbor. Against the backdrop of the raging Vietnam War and the global oil crisis of 1973/74, imaginary expansion was shrunk, bounded and

Alexander C.T. Geppert (✉)
New York University Shanghai, Shanghai, China
New York University, New York, NY, USA
e-mail: alexander.geppert@nyu.edu

© The Author(s) 2018
Alexander C.T. Geppert (ed.), *Limiting Outer Space*
European Astroculture, vol. 2
https://doi.org/10.1057/978-1-137-36916-1_1

Figure 1.1   Apollo 11 lunar module 'Eagle' as it returned from the surface of the moon on 21 July 1969 to dock with the command module *Columbia*. While a smooth mare area is visible on the moon below, the half-illuminated earth hangs over the horizon in the background. Command module pilot Michael Collins (1930–), the NASA astronaut who took this picture when the lunar module ascent stage was about four meters away, has sometimes been described as 'the only human alive or ever to have lived not contained within the frame of this photo.'
Source: Courtesy of NASA.

grounded. With human spaceflight confined to low-earth orbit ever since the last astronaut returned to earth, the skies once again became the limit. If the Apollo era, in particular the new picture of planet Earth as its key legacy, constituted the apogee of worldwide space enthusiasm and the apex of the global Space Age, how did the latter's demise affect space thought and astroculture? Is the argument correct that it was during this aptly termed 'post-Apollo period' that the long-established link between sociotechnical imaginaries of outer space and phantasmagoric visions of a collective, imminent future in the stars loosened? And that, as a consequence, outer space itself lost much of the political relevance, cultural significance and popular appeal which it had been gaining worldwide since the mid-1920s, in particular after the end of the Second World War?

Limiting Outer Space has a triple focus. First, it zooms in on a particular time period, situated within a specific geographical setting, and foregrounds a clear-cut historical question. Concentrating on the 1970s – according to the late New York University historian Tony Judt the 'most dispiriting decade of the twentieth century' – the book's thirteen chapters examine this now widely debated transition process from expansion to reduction, often considered concomitant with disillusionment and disenchantment, from a multiplicity of disciplinary perspectives. Second, the majority of contributions aim to replace oft-repeated US- and USSR-centric narratives of a bipolar Cold War rivalry and an escalating Space Race between East and West with more nuanced, less formulaic and more comprehensive analyses, integrating and indeed featuring European, if not global views on and contributions to 1970s astroculture. Finally, chapters ask whether the new 1970s sense of 'general space fatigue' marked the end of that hitherto inextricably intertwined nexus between outer space and the quest for utopia, when widespread belief in infinite human expansion was superseded by the discovery of inner space.[3]

I   The growth of limits in the decade of crisis

It has taken historians a while to realize the wide-ranging implications and indeed epochal significance of what Eric Hobsbawm termed the 'crisis decades' or, more drastically: 'the landslide.' With the first oil-price shock of 1973/74, the standard argument now goes, an unprecedented quarter-century-long boom era came to an end in the West. The *trente glorieuses* had been a long period of relative political stability that was characterized by rapid economic growth, material prosperity for larger sections of society than ever before, and a reassuring sense of having successfully overcome two devastating world wars.[4] In March 1972, more than a year prior to the oil crisis, the Club of Rome had published its notorious 600-page *Limits to Growth* study on the 'predicament of mankind.' Translated into 35 languages and selling 9 million copies worldwide, the book's computer-based predictions for the future seemed to be validated by the unfolding course of events.[5] During the following years, a new sense of worldwide interconnectedness and global interdependence found its counterpart in the individualization of society and a withdrawal from the collective to the self. In an oft-cited article, American writer Tom Wolfe (1931–) coined the term 'Me Decade' to portray an ego-centered generation that had replaced 'man's age-old belief in serial immortality' with a narcissistic 'I have only one life to live.' The golden postwar era thus gave way to a less romantic, less optimistic and much more troubled, if not entirely 'lost,' decade, as contemporary observers in both Europe and the United States were quick to point out. 'In the long run,' *Time* magazine forecasted correctly, 'this decade and the next may well constitute an historical era of transition.'[6]

A majority of contemporary historians now echo these contemporaneous readings, impressionistic, unsystematic and incomplete as they may have been both then and now. Hardly surprising, economic and environmental historians were among the first to draw attention to the decade's transformative character. The former declared the 1970s 'of great interest for the economic and social historian,' while the latter pointedly termed the all-encompassing reinterpretation of the man-environment relationship during these years the '1970s diagnosis.'[7] Within the past decade or so, literature on the so-called long 1970s, usually understood as lasting through the conservative turn of the early 1980s, has mushroomed both in European[8] and American historiography.[9] Contrary to usual experience, a rare consensus has eventually emerged among 'general' historians that the 1970s are to be regarded as a key period in the history of the twentieth century. Standing for structural rupture and constituting an epochal caesura, they should be conceptualized as a major turning point. Accordingly, a plethora of competing labels has been created to come to terms with a decade once overhastily described as a time when nothing happened: the 1970s as the 'end of confidence,' 'the age of fracture,' the period 'after the boom,' the 'decade without a name' that nonetheless constituted the 'threshold of change,' or the moment in time when all of a sudden the 'shock of the global' set in, simultaneously limiting and liberating. Others, somewhat predictably, have objected to any such forms of 'decadology,' as if historians were not well aware of their periodizations' artificial character, necessitated by professional pragmatism to come to terms with change over time.[10] There is opportunity in every crisis, goes another trite cliché, and labeling the 1970s as a global crisis consequentially leads to emphasizing their Janusfacedness, as a period of inertia *and* change, when the established post-Second World War consensus was revoked while giving way to the rise of post-industrial society in Europe and the world that dominates today's planetized present.[11]

As consequence and effect of such a structural rupture, not the least in contemporary self-understanding, the future changed its character during these years as well, often considered an unmistakable sign of epochs drawing to a close. 'My children, or today's teenagers, they are not interested in the future,' English novelist J.G. Ballard (1930–2009) deplored in a 1970 interview with British *Penthouse* magazine. 'What you see is the death of outer space, the failure of the moon landing to excite anyone's imagination on a real level, and the discovery of inner space in terms of sex, drugs, meditation, mysticism,' Ballard stated, thus giving expression to a frequently diagnosed assessment of the 1970s as a self-questioning time of troubles that looked neither forward nor outward but backward and inward.[12] Retrospection replaced prospection. Continual progress, exponential growth and outward expansion – previously considered the basis of incessant improvement of the human condition by means of technoscience – went into reverse. Large-scale technology ceased to be the trustworthy engine of societal change and humankind's betterment proved itself a problem, if not indeed its very obstacle.

Images and imaginaries of outer space and spaceflight, vastly popular and usually utopia-saturated in previous decades, changed correspondingly. Three cover images of the West German weekly *Der Spiegel* – published in 1966, 1970 and 1979, respectively – illustrate the shifting space-future nexus over the course of the decade. Quoting at length Arthur C. Clarke (1917–2008), British techno-prophet bar none, the *Spiegel*'s 6 December 1966 issue indulged in 1960s technocratic planning fervor. The future could be forecast because it was man-made and therefore controllable (Figure 1.2). Published

Figure 1.2–4 From planning fervor to threat via irrelevance: changing expectations for the future over the course of the 1970s as illustrated by the West German weekly *Der Spiegel*. The headlines translate as 'Futurology: Man's Future is Being Planned' (1966, left), 'The Seventies: Planless into the Future?' (1970, center) and 'Skylab Falls to Earth: Danger for Mainz?' (1979, right). Mainz, the capital of Rhineland-Palatinate, was the largest German city lying within the forecasted hazardous zone.
Source: Courtesy of *Der Spiegel* 20.53 (6 December 1966); 24.1 (5 January 1970); 33.27 (2 July 1979).

only a couple of years later, the *Spiegel*'s 5 January 1970 issue denounced the formerly utopian ideal of total feasibility not only as outmoded ideology but as the very 'trauma of the modern world' (Figure 1.3). Scenarios of future expansion into outer space were now marginalized; the only mention of spaceflight in this 12-page feature was an image of a moon colony illustrating the article. In a third *Spiegel* cover story published in 1979, another nine years later, space was no longer a futuristic promise nor an irrelevant epiphenomenon but had transformed into an otherworldly threat. Dangerous debris raining down from Skylab (1973–79), the decommissioned and long uninhabited first American space station, might cause considerable damage upon re-entry, the article warned its readers (Figure 1.4).[13]

The same modernist faith in technoscientific rationalism that had propelled the Apollo program into the 1960s skies and beyond was feared to be falling from the heavens at the end of the 1970s. Ballard, commenting in another *Penthouse* interview conducted a decade later, agreed. 'The world of "outer space," which had hitherto been assumed to be limitless, was being revealed as essentially limited, a vast concourse of essentially similar stars and planets whose exploration was likely to be not only extremely difficult, but also perhaps intrinsically disappointing,' the writer pointed out. For him, the Space Age had irrevocably ended in 1974, when the last Skylab mission returned to earth, having long given way to an era of limits in which the future developed in one direction only – toward home. 'The twentieth century began with a futuristic utopia and ended with nostalgia. Optimistic belief in the future was discarded like an outmoded spaceship,' literary scholar Svetlana Boym has summarized this drastic volte-face in hindsight. The turn from a prospective and extroverted to a retrospective and introverted reasoning simultaneously marked the inglorious end of the much celebrated Age of Space.[14]

That outer space, whether imagined, journeyed or feared, should have played a key role in the genesis of the 1970s as a transitional period might surprise middle-of-the-road historians of the twentieth century more than experts in space history.[15] 'Post-Apollo period' – the term suggested here to characterize the decade *succeeding* the classical Space Age, namely the time period from December 1972 until the early 1980s – is an example of how mainstream historiography – in this case 1970s scholarship in particular – and space history can supplement, illuminate and enrich each other.[16] The benefit is mutual: on the one hand, 'post-Apollo' provides students of outer space, spaceflight and astroculture with a broader intellectual and conceptual context, which in turn allows them to situate their analyses within a recognized interpretative framework to which general historians can equally relate. On the other hand, christening the 'decade without a name' the 'post-Apollo period' suggests that the end of the postwar consensus, the widely shared sense of societal crisis, the growth of limits and the oft-noted introspective spirit of the 1970s did not only coincide but also shared a common denominator. It is not by chance that humankind's outward movement correlated with a new sense of planetized globality; the irony is that both only emerged *after* the classical Space Age had drawn to a close.

II   The Post-Apollo paradox

According to contemporaneous experts, the historical assessment would be unambiguous. When asked what the American Apollo missions *meant* for mankind and how their societal impact was to be characterized then and in the future, American, British, French and German historians, anthropologists, philosophers, scientists and public intellectuals all but agreed. According to notables such as Arnold M. Schlesinger Jr., Arnold J. Toynbee, C.P. Snow, Margaret Mead, Claude Lévi-Strauss, Hoimar von Ditfurth and many other *hommes de lettres*, landing a man on the moon was an unprecedented achievement of unforeseen dimensions which later generations would hail as an epoch-making step in human history. 'The twentieth century will be remembered,' historian Schlesinger forecasted in 1972 in a later oft-repeated statement, 'as the century in which man first burst his terrestrial bonds and began the exploration of space.'[17] Yet, as to what characteristics and societal consequences the just-entered Moon Age would entail, the experts were divided. Some reckoned the moon to be a stepping stone toward the discovery of new worlds and their imminent colonization, while others warned of a rise of 'cosmic claustrophobia' should humankind fully comprehend its aloneness throughout the universe. 'Was the voyage of Apollo 11 the noblest expression of a technological age, or the best evidence of its utter insanity?,' wrote Norman Mailer (1923–2007), bringing the dilemma to a head.[18] A third, originally less prominent, reading suggested that the truly alien planet and the only newly discovered frontier was, indeed, planet Earth itself. Bridging unparalleled physical distances and reaching a new vantage point in space made it possible to turn the gaze around, to look back and inward rather than forward and outward. Accordingly, the most precious souvenirs brought along from the journey were neither the pictures of Neil Armstrong's footprints on the moon's gray, dusty surface nor the 382 kilograms of lunar rock the six missions brought back, but rather two unplanned, low-priority by-products of the $20 billion Apollo program, 'Earthrise' (1968) and 'Blue Marble' (1972). Two photographs of the home planet, epitomizing this newly reversed perspective from without, proved the program's inadvertent legacy.

Present-day geographers, historians, art historians and philosophers have readily taken up and now widely echo this third reading, arguably elevating it to one of the few widely accepted standard arguments in space history. Geographer Denis E. Cosgrove has attested to Earthrise and Blue Marble having 'altered the shape of the contemporary geographical imagination,' whereas historians Robert Poole and Benjamin Lazier have, respectively, declared Earthrise as providing the 'defining moment of the twentieth century' which gave rise to an entire 'Earthrise era.' Similarly, art historian Horst Bredekamp has used philosopher Peter Sloterdijk's notion of a 'Copernican revolution of the gaze' to argue that Blue Marble became the image of earth par excellence as it allowed for a complete reversal of viewing directions only possible from an extraterrestrial standpoint. Distance made for a reorientation and complete reversal of perspective, which in turn led literally to a new *Weltanschauung* on earth.[19] Following these and other assessments, the Apollo program did indeed prove epoch-making – albeit hardly for the reasons put forward by the majority of observers, analysts and critics at the time. Apollo was not tantamount to a caesura in human history because it meant twelve men walking on earth's closest celestial body, but because the spacefarers, acting as representatives of all of humankind, returned with portraits of everyone's communal home, the world's first selfie.

The minority of experts who had predicted that jaunting into outer space would, paradoxically, lead to a rediscovery of *inner* space were correct. As some had argued as early as 1965, 'man's thrust into outer space' proved ultimately a return to himself. Correspondingly, when in 1977, five years after the end of the Apollo program, US probe Voyager 1 sent back a color photograph that showed earth and moon floating together in the vast darkness of outer space, public resonance was limited. Lacking the implicit 'human touch' of the earlier souvenirs, the novelty of this machine-generated image was not sufficient to excite the public anew, and neither did it make front-page headlines (Figure 1.5).[20]

How then to connect this new, earth-centered image of outer space featuring *planet* Earth with the transitional 1970s, and why suggest labeling these years the 'post-Apollo period'? Signalling its problematique in its very name, post-Apollo denotes a period, a program and a problem. First, the term obviously refers to the time period *after* the completion of the Apollo missions in 1972.[21] Second, it also stands for NASA's spaceflight program by the same name, first discussed in Congress in August 1965, laid out in a September 1969 report and culminating in President Richard Nixon's announcement on 5 January 1972 in which he committed to build the Space Shuttle. Vehemently debated nationally and internationally, the task was to find an answer to the question of where the American nation would 'go in space in the Post-Apollo period.' As historian John M. Logsdon has argued, the set of decisions made during those three short years defined human spaceflight activities in the United States for the next four decades, until the termination of the Shuttle program in 2011.[22] But in addition to marking a historical time period and denominating a national space policy of long-term impact, post-Apollo also points, third, to a particular historical problem: the Post-Apollo paradox. As the contributions to this book testify, neither spaceflight nor astroculture ceased to exist during the 1970s, even if their already complex relationship further loosened once the future moved elsewhere and enthusiasm began to dwindle all the more.[23] Yet, it was precisely at this moment in time that, by many accounts, the world-encompassing process of international entanglement now usually referred to as globalization finally unfolded with full force. That the term 'global' took on its contemporary theoretical connotations in the early 1970s and turned into the conceptual category so familiar today is not a coincidence but a by-product of the post-Apollo period.[24]

Surprisingly absent from the flourishing historiography is the causal connection between the heyday of space exploration, space thought and astroculture of the 1950s and 1960s, and the sense of crisis and incipient globality of

Figure 1.5 Recorded by NASA's Voyager 1 spacecraft on 18 September 1977, this photograph was the first complete image of earth and moon together in a single frame. The spacecraft had been launched two weeks earlier, on 5 September 1977, and was at this point 11.7 million kilometers away. Voyager 1 passed the boundaries of the solar system in 2012 and continues to travel in interstellar space, making it to date the farthest human-made object away from planet Earth. *Source*: Courtesy of NASA.

the 1970s. 'Achievement of the Apollo goal resulted in a new feeling of "oneness" among men everywhere,' the aforementioned 1969 US report stated, stipulating that any subsequent program would have to continue promoting a similar 'sense of world community.'[25] In addition to such imaginative repercussions, while difficult to distinguish from their propagandistic value, there was also a more tangible technological component behind the globalizing impact of the Space Age whose significance is easily overlooked: the incipient telecommunication satellite revolution. According to contemporaneous estimates no fewer than 7,600 satellites were launched between October 1957 and 1975 alone. Even though the vast majority is no longer operational, together with undersea cables they constituted the key infrastructure for processes of world-encompassing interconnectedness and increasing global entanglement.[26] Thus, the polymorphic and multinational thrust into outer space after the Second World War was a major factor in making our planet as planetized as it is. Indeed, it is arguably the most unintended *and* most far-reaching consequence of the reach for the stars, both realized and imagined.

And this is the very 'Post-Apollo paradox': the full impact of the Space Age only came to the fore when it had by most accounts effectively passed, with 1970s astroculture proving more earth-centered than previous imaginaries. Because the post-Apollo period was characterized by epochal change for which spaceflight proved a central motif, space historian Martin Collins has suggested terming it the 'in-between decade.'[27] Another historian has argued that the 1970s marked as much the end of the *trente glorieuses* as they constituted the beginning of a new epoch of globalization and individualization. If this is accurate, the paradigmatic shift in humankind's self-understanding – caused by a temporary departure from home as well as earth's communicative coalescence based on space-placed infrastructures – was a decisive factor in this transition.[28]

III   Spaceflight after Apollo

On closer inspection, spaceflight during the post-Apollo period was far from being 'marked by matter-of-factness rather than by lofty visions,' but the 1970s proved indeed a time of disenchantment, disillusion and disengagement. Not only in the United States but also in Europe and the Soviet Union, outer space lost much of its capacity to arouse and engage divergent publics.[29] In January 1972, Nixon announced the new Post-Apollo Program which would 'revolutionize transportation into near space by routinizing it' and 'take the astronomical costs out of astronautics.' Featuring a 'space vehicle that can shuttle repeatedly from earth to orbit and back' in addition to a space station, a lunar base and a manned voyage to Mars, the program would take a decade to produce tangible results.[30] Between the Apollo-Soyuz Test Project (ASTP) in July 1975 – the first docking of a US and a Soviet spacecraft in earth orbit that symbolized for many the end of the Space Race – and the maiden Space Shuttle launch in April 1981 there was a gap of almost six years without a manned US mission.

Judged by public memory, this 'post-Apollo, pre-Space Shuttle interregnum' lasting from 1975 through 1981 looks indeed unremarkable.[31] Preceding decades had each been characterized by a single dominating event, each with diverse and wide-ranging repercussions: the late 1920s by the first space 'fads' in the Soviet Union, Europe and the United States; the 1940s by the development of the A4/V-2 and the 'invention' of the flying saucer in 1947; the 1950s by the launch of Sputnik, the first artificial satellite; the 1960s by Yury Gagarin's orbit of planet Earth in 1961 and, of course, the Apollo moon landings. The 1980s, on the other hand, were under the sign of the Space Shuttle, subject to a first catastrophe only five years after its first launch, while the 1990s saw the initial steps toward the assembly of the most cost-intense civilian project ever undertaken, the International Space Station (ISS). But the 1970s? While the classical Space Age came to a symbolic close with astronaut Eugene Cernan (1934–2017) stepping off the moon on 14 December 1972, spaceflight and space exploration continued in a variety of ways.

As a necessary historical grounding of the chapters that follow, four parallel strands of this post-Apollo, pre-Shuttle period need to be sketched: first, the renewed emphasis, both in East and West, on positioning space stations in earth orbit; second, the surprising interdependence between planetary exploration on the one hand, and the new interest in the search for extraterrestrial intelligence (SETI) on the other; third, the sweeping, yet short-lived space colonization fad during the second half of the 1970s; and, fourth, the so-called rebirth of Space Age Europe throughout this decade of transition and reconfiguration.

First, although Apollo did indeed 'end up as a dead end project' at the peak of its acclaim, as historian Logsdon has remarked, its hardware had a material afterlife as it continued to be used for Skylab and the Apollo-Soyuz Test Project.[32] Partially assembled from recycled leftovers and less specialized than later space stations, Skylab was launched in May 1973. It remained circling earth in a low orbit of 480 kilometers for much of the 1970s, even if its so-called Orbital Workshop was only inhabited for a total of 171 days during the first two years of operation. A similar return to a much older itinerary, namely the positioning of a space station in earth orbit rather than directly going to the moon or beyond, occurred simultaneously in the Soviet Union. Beginning in 1971 and lasting through 1986, that is over a period of 15 years, its Salyut program consisted of a series of six crewed space stations positioned in earth orbit, four of them civilian, two military.[33]

Second, the 1970s were a surprisingly successful period for the robotic exploration of the solar system. Launched in May 1971, Mariner 9's orbital survey of Mars revealed entirely unexpected canyons, volcanoes and signs of massive floods in the planet's distant past. In March 1972 and April 1973, respectively, the Pioneer 10 and 11 probes were sent to Jupiter and Saturn, carrying the famous aluminum plaque picturing two naked human bodies against the backdrop of a map of pulsars. And in the fall of 1975, the Soviet Venera 10 probe survived all the way to Venus and returned a photograph during the 65 minutes of its operation on the surface. Together, these robotic missions gave rise to the 'Golden Age of planetary exploration,' as American science celebrity and media personality Carl Sagan (1934–96) had it, himself co-creator of the Pioneer plaque and one of the most influential 1970s space *personae*.[34]

However, the two most momentous of all these robotic undertakings were arguably the 1975 Viking missions to Mars and the two Voyager launches in 1977. While Voyager 1 was sent to Jupiter and Saturn and then continued on a trajectory beyond the solar system (see Figure 1.5 above), the photographs the two Viking probes sent back from barren Mars in 1976 found more immediate, if ambivalent, societal resonance. They contributed to the widespread recognition of humankind's cosmic isolation as much as they granted the sometimes ridiculed search for extraterrestrial intelligence new legitimacy.

Precisely because no immediate signs of life were detected anywhere on Mars, the Viking missions proved central for the emerging scholarly field of exobiology, soon to reinvent itself as astrobiology. Encouraged by opinion polls that nonetheless reported widespread belief in life on other planets, both futurist and bestselling *Future Shock* author Alvin Toffler (1928–2016) and noted anthropologist Sol Tax (1907–95) lent the contested discipline societal legitimacy by their sheer support.[35]

Third, fueled by the Club of Rome's gloomy predictions, triggered by a growing concern for imminent environmental disaster and inspired by the Skylab missions, grandiose space colonization scenarios witnessed a brief burst of popularity in the second half of the 1970s, in the United States more so than in Europe (see Figure 1.6). 'After you have landed 12 men at six locations on the moon to walk and jeep around scooping up rock samples, kept a space station manned for a total 171 days, and landed two robot spacecraft on a planet more than 200 million miles away from Earth, what do you do for an encore?,' one mid-1970s commentator could not help but wonder. To solve problems of overpopulation and counter the abiding energy crisis, for some the answer lay in bypassing the boundaries of a 'sharply limited planet' by transferring entire populations into space.[36] The leading advocate of such large-scale expansion scenarios soon became Princeton physicist Gerard K. O'Neill (1927–92). Having originally developed his concepts during the late 1960s, O'Neill first published a triad of articles in *Nature*, *Physics Today* and *Science* in 1974 before making headline news in mainstream media in 1976 and 1977. 'Is a planetary surface really the best place for an expanding technological civilization?,' O'Neill asked rhetorically in a *New York Times* article before going on to prophesy that 'thousands of people now alive may choose within the next two decades to live and work on a new frontier in space.'[37] Especially in his 1976 book *The High Frontier* O'Neill presented detailed concepts for a permanent human presence in outer space, envisioning large manned colonies at L5 – one of five points in space where the gravitational fields of the earth and the moon balance each other and where a space station could remain stable. Completed by the early 2000s, these human colonies would be constructed with unlimited raw materials from the moon and later the asteroid belt, spun to simulate gravity and employ light reflected from the sun for illumination, power and infinite energy. Proclaiming that 'water and food are no limits on the range of the human species in space,' O'Neill ultimately aimed to reinstate an idea of infinite boundlessness during the era of limits, what he addressed as the 'humanization of space.'[38]

Widespread as it was, the popularity of such colonization scenarios proved also short-lived, and by the end of the 1970s futuristic megastructures of this magnitude were largely transferred into virtual computer worlds. Despite popularization attempts by space advocates and architects such as Duncan Lunan (1945–) in Great Britain and Fritz Haller (1924–2012) in Switzerland, space colonies never seem to have found the same cultural resonance in Europe.[39] Nonetheless, the European Space Agency (ESA) responded dutifully when a high-school student asked for its particular stance on such

Figure 1.6 Interior of an imaginary post-Apollo space colony. The original caption read 'Main Street, Hometown, Cosmos finds colonists on the move, passing the stacked, modular habitations and shops of L-5. Fruit trees relieve the stark simplicity of a manufactured environment. The alumni of earth can order buildings, climate, and sunlight to suit. Yet L-5 is no playground in the void. Hardworking pioneers make it the latest outpost on a limitless frontier.'

Source: Isaac Asimov, 'The Next Frontier?,' *National Geographic* 150 (July–December 1976), 76–89, here 80. Painting by Pierre Mion.

expansion scenarios, seizing the opportunity to sell its own Spacelab project as a more feasible alternative:

> It is beyond doubt that space colonies are a real possibility for the future. Our Agency, however, has not been doing extensive work on such colonies but we are [...] developing [...] a space laboratory, the Spacelab, which will be a manned laboratory to be flown on the Space Shuttle. This Spacelab could very well evolve into a space station by linking different Spacelabs together in space.[40]

Fourth and finally, in the case of Western Europe, there was yet another ironic twist to the Post-Apollo paradox. One of NASA's main motives for seeking European involvement in its Post-Apollo Program was to counter a lack of public enthusiasm. Having more international partners would not only help to share some of the risks and expense, they reasoned, but also demonstrate that space exploration was the humanitarian task par excellence, only to be fulfilled on a truly global scale. European participation would be 'the most ambitious nonmilitary effort ever undertaken collectively by the West European nations,' the *New York Times* rejoiced somewhat prematurely.[41] For Western Europe, the invitation proved a double-edged sword as it allowed Europe to play a more active role on the international scene than ever before, but it also brought home the urgent need to completely reorganize the hitherto ill-starred European spaceflight program. After complex political struggle and organizational reform, ESA was officially established in 1975, replacing its predecessors, the European Launcher Development Organization (ELDO) and the European Space Research Organization (ESRO). While Europe's position, visibility and significance were indeed stronger than ever before – eventually emblematized by the successful launch of the first European-built Ariane 1 rocket on Christmas Eve 1979 from the spaceport in Kourou, after a decade of failures – the timing of its eventual lift-off in space was less than ideal. What space history veteran Walter McDougall has termed the 'rebirth of Space-Age Europe' took place when popular enthusiasm for space exploration and astroculture had long been in decline. Yet, from an institutional perspective the post-Apollo crisis largely meant a pre-Ariane *Aufbruchstimmung* or promise of departure, with Europe's much more limited participation in NASA's Post-Apollo Program itself providing a 'formative experience.'[42]

IV   Limiting outer space

If the 1970s were contemporaneously perceived as an age of boundaries impinging on man's project in outer space, at odds with formerly close connections between expansion fantasies and humankind's futurity, what effect did the general sense of crisis have on pre-existing imaginaries of outer space and extraterrestrial life? How were the new limits reflected, integrated and challenged by then-current visions of cosmic utopias and the disenchanted realities of spaceflight after Apollo? And were human boundaries effectively challenged, if not entirely transformed, in outer space? As a contribution to historical research on astroculture – a concept previously introduced and defined as the interplay of different social groups and heterogeneous cultural forms aiming to ascribe meaning to the infinite void that surrounds planet Earth – *Limiting Outer Space* focuses on what Arthur Clarke termed 'an interesting transition period' that allows one to 'compare the realities of space with earlier imaginings of artists.'[43] It is noteworthy that in private correspondence Clarke was far more blunt and less upbeat than this, and repeatedly lamented the 'present malaise' when referring to the situation post-Apollo. 'The human activity to which I have mainly devoted my life is in decline,' agreed his old friend Arthur 'Val' Cleaver (1917–77) from the British Interplanetary Society wholeheartedly.[44] Concentrating on this decade of crisis, disenchantment and reconfiguration, *Limiting Outer Space* explores a pivotal transition in imagining the cosmos and projecting utopian dreams into outer space. Inspired by and contributing to the ongoing historiographical reassessment of the 1970s, it argues that the post-Apollo period constituted a crucial, if hitherto underrated and understudied, era in the history of space, spaceflight and space thought that awaits closer scrutiny and smoother integration into mainstream historiography, just like space history itself.

While it would be unwise for a book that carries 'Apollo' in its subtitle to leave the most celebrated human spaceflight program aside, the geographical focus lies decidedly elsewhere, particularly in Western Europe, with all its complex transnational and intercontinental interdependencies.[45] It is, however, worth remembering that before the late 1970s, no human being from any nation other than the Soviet Union or the United States had left planet Earth. The first Eastern European in outer space was the Czech cosmonaut Vladimír Remek (1948–) onboard Russian Soyuz 28 spacecraft in March 1978, while French spationaut Jean-Loup Chrétien (1938–) followed four years later, in June 1982, onboard Soyuz T-6. Chrétien's seven-day mission to the Salyut 7 space station made him not only the first Western European but also the first Western non-American beyond the earth's atmosphere ever. Participating as payload specialist in STS-9, the ninth NASA Space Shuttle mission in November and December 1983, West German Ulf Merbold (1941–) was the first ESA astronaut proper in space. To date, no European has flown on a European-built spacecraft.[46]

This book's thirteen chapters – including this introduction and an epilogue – are grouped in three sections: 'Navigating the 1970s,' 'Reconfiguring Imaginaries' and 'Grounding Utopias.' The first part – 'Navigating the 1970s' – addresses the 1970s as the great division of the postwar years and aims to periodize the post-Apollo period accordingly. It includes a sweeping reconfiguration of some of the major conceptual issues associated with recent historiographical work on the 1970s as the 'in-between' period of twentieth-century change (Martin Collins); an essay on Great Britain's space program after the cancelation of its short-lived Black Arrow rocket program in July 1971, asking whether a particularly British variant of European astroculture can be identified (Doug Millard); and a chapter on the historical significance, societal impact and long shadow that the Apollo program cast over expectations for the future in the United States but also worldwide (Roger D. Launius).

The second section – 'Reconfiguring Imaginaries' – comprises four chapters dealing with manifestations and exemplars of 1970s European and global astroculture. It opens with a chapter tracing the history of Stanley Kubrick's 1968 feature film *2001: A Space Odyssey* from Arthur Clarke's 1951 science-fiction short story 'The Sentinel' through the novel *Journey Beyond the Stars* (1964) to the making of *2001* as a cult film, widely considered the most important space movie ever produced (Robert Poole); an analysis of the 1970s self-reflexive turn in English-language literature, as exemplified in the works of novelists Doris Lessing, A.S. Byatt and John Banville, all from non-spacefaring nations (Florian Kläger); a chapter on Legoland Space, the Danish toy company's hugely successful line created in 1978 as an example of the sacralization of modern consumer culture and intergenerational communication of values through material culture (Thore Bjørnvig); as well as a careful analysis of the international negotiations leading to the adoption of the United Nations' Outer Space Treaty in January 1967 and the Moon Agreement in December 1979, focusing on the competing normative and political rationales that informed the former's perceived success and the latter's failure (Luca Follis).[47]

Chronologically situated toward the end of the 'long' 1970s, that is the early 1980s, the third and final section – 'Grounding Utopias' – focuses on spacefarers, space stations and space colonies in science fiction and in science fact. This part features a chapter that examines three distinct collaborative moments – the Soviet Union's Interkosmos program created in 1970, the Apollo-Soyuz Test Project of 1975 and the establishment of the Association of Space Explorers in 1985 – as supranational attempts to promote international détente through spaceflight technology, propagating planetary consciousness as an alternative (Andrew Jenks); a comparative analysis of imaginary space architectures, be they located in ground-based laboratories, in outer space itself or as part of film sets, all probing the human-technology relationship during the post-Apollo period (Regina Peldszus); a chapter on the transnational media coverage of Spacelab, *the* European showcase project of the 1970s and early 1980s intended to signal Western Europe's active participation in, if not independent entry into, manned spaceflight (Tilmann Siebeneichner); and a contribution on the ways in which 1970s space-colony enthusiasts mingled with nuclear-weapon designers and military planners, effectively creating the foundations for Ronald Reagan's 1983 Strategic Defense Initiative (SDI), also known as Star Wars (Peter J. Westwick).[48] Finally, David A. Kirby's comprehensive epilogue reminds us how the grand expectations and celebrations of Space-Age accomplishments gave way to a growing awareness of the problems humankind faced on earth in the post-Apollo period. Dystopian, bleak and at times despairing science-fiction films set in space such as *Earth II* (1971), *Silent Running* (1972), *Solaris* (1972), *Soylent Green* (1973), *La Planète sauvage* (1973), *Dark Star* (1974), *The Man Who Fell to Earth* (1976), *Operation Ganymed* (1977) and *Alien* (1979) left no doubt that space exploration was no longer considered the key technology to solving terrestrial problems from without.[49]

Linking and interrelating the history of astroculture, space thought and spaceflight with recent scholarship on the social and political history of the 1970s, *Limiting Outer Space* aims to correct, complement and reorient the existing historiography on the post-Apollo period. Focusing on selected European countries – in particular Great Britain, France, West Germany and Denmark – its thirteen chapters examine the limiting of outer space and the grounding of utopia after the American moon landings. Rather than invoking oft-repeated narratives of a bipolar Cold War rivalry and an escalating Space Race between East and West, the book charts new historiographical ground by exploring a hitherto underappreciated decade in space history. With the rapid waning of what European observers termed *Apollo-Rausch* or Apollo frenzy, the classical Space Age gave way to an era of space fatigue and planetized limits: the post-Apollo period.[50]

Notes

1. J.G. Ballard and Lynn Barber, 'Sci-Fi Seer,' *Penthouse* [UK] 5.5 (May 1970), 26–30, here 27; Arthur C. Clarke, 'Foreword,' in Patrick Moore and David A. Hardy, eds, *The New Challenge of the Stars*, London: Mitchell Beazley, 1972, n.p. Thanks are due to Stephen Gross, Ruth Haake, Michael Neufeld, Robert Poole, Tom Reichard, Emily Rosenberg, Tilmann Siebeneichner and, as always, Anna Kathryn Kendrick.

2. Laurence Goldstein, 'Introduction,' in idem, ed., *The Moon Landing and its Aftermath*, Ann Arbor: University of Michigan, 1979 (= *Michigan Quarterly Review* 18.2), 153–4, here 153.

3. Tony Judt, *Postwar: A History of Europe Since 1945*, New York: Penguin, 2005, here 477; *Westdeutsche Allgemeine Zeitung* 161 (15 July 1978): 'allgemeine Weltraummüdigkeit.' Pertinent to the Apollo years yet self-limiting to the United States is Matthew D. Tribbe's recent *No Requiem for the Space Age: The Apollo Moon Landings and American Culture*, Oxford: Oxford University Press, 2014.

4. Eric Hobsbawm, *The Age of Extremes: A History of the World, 1914–1991*, New York: Pantheon, 1994, chapter 14: 'The Crisis Decades,' 403–32; Jean Fourastié, *Les trente glorieuses: Ou, la révolution invisible de 1946 à 1975*, Paris: Fayard, 1979.

5. Donella H. Meadows, Dennis L. Meadows, Jørgen Randers, William W. Behrens III and Club of Rome, *The Limits to Growth: A Report for the Club of Rome's Project on the Predicament of Mankind*, New York: Universe Books, 1972. The literature is vast but see, for example, Mauricio Schoijet, '*Limits to Growth* and the Rise of Catastrophism,' *Environmental History* 4.4 (October 1999), 515–30, and, above all, Helga Nowotny, 'Vergangene Zukunft:

Ein Blick zurück auf die "Grenzen des Wachstums,'" in Michael Globig, ed., *Impulse geben – Wissen stiften: 40 Jahre VolkswagenStiftung*, Göttingen: Vandenhoeck & Ruprecht, 2002, 655–94, here 663.

6. Tom Wolfe, 'The "Me" Decade and the Third Great Awakening,' *New York* (23 August 1976), 26–40, here 40; 'From the '60s to the '70s: Dissent and Discovery,' *Time* 94.25 (19 December 1969), 20–6, here 20, 22.

7. Richard Coopey and Nicholas Woodward, 'The British Economy in the 1970s: An Overview,' in eidem, eds, *Britain in the 1970s: The Troubled Economy*, London: UCL Press, 1996, 1–33, here 2; Patrick Kupper, 'Die "1970er Diagnose": Grundsätzliche Überlegungen zu einem Wendepunkt der Umweltgeschichte,' *Archiv für Sozialgeschichte* 43 (2003), 325–48, here 328. For review essays on 1970s historiography, see Rodney Lowe, 'Life Begins in the Seventies? Writing and Rewriting the History of Postwar Britain,' *Journal of Contemporary History* 42.1 (January 2007), 161–9; Martin H. Geyer, 'Auf der Suche nach der Gegenwart: Neue Arbeiten zur Geschichte der 1970er und 1980er Jahre,' *Archiv für Sozialgeschichte* 50 (2010), 643–69; Lawrence Black, 'An Enlightening Decade? New Histories of 1970s' Britain,' *International Labor and Working-Class History* 82 (Fall 2012), 174–86; and Barbara Keys, Jack Davies and Elliott Bannan, 'The Post-Traumatic Decade: New Histories of the 1970s,' *Australasian Journal of American Studies* 33.1 (July 2014), 1–17. Although US-centered and by no means as 'global' as its title promises, the best of the existing 1970s syntheses is arguably Thomas Borstelmann's *The 1970s: A New Global History from Civil Rights to Economic Inequality*, Princeton: Princeton University Press, 2012, with Andy Beckett's fabulous account of the UK, *When the Lights Went Out: Britain in the Seventies*, London: Faber & Faber, 2009, a close second.

8. For Europe, see in chronological order and without any claim to completeness: Gerhard A. Ritter, Margit Szöllösi-Janze and Helmuth Trischler, eds, *Antworten auf die amerikanische Herausforderung: Forschung in der Bundesrepublik und der DDR in den 'langen' siebziger Jahren*, Frankfurt am Main: Campus, 1999; Charles S. Maier, 'Two Sorts of Crisis? The "Long" 1970s in the West and the East,' in Hans Günter Hockerts, ed., *Koordinaten deutscher Geschichte in der Epoche des Ost-West-Konflikts*, Munich: Oldenbourg, 2004, 49–62; Hartmut Kaelble, 'Vers une histoire sociale et culturelle de l'Europe pendant les années de "l'après-prospérité,"' *Vingtième Siècle* 84.4 (2004), 169–79; idem, *The 1970s in Europe: A Period of Disillusionment or Promise?*, London: German Historical Institute, 2010; Anselm Doering-Manteuffel and Lutz Raphael, *Nach dem Boom: Perspektiven auf die Zeitgeschichte seit 1970*, Göttingen: Vandenhoeck & Ruprecht, 2008 (3rd edn 2012); Ingrid Gilcher-Holtey, Rainer Eckert, Etienne François, Christoph Kleßmann and Krzysztof Ruchniewicz, 'Die 1970er-Jahre in Geschichte und Gegenwart,' *Zeithistorische Forschungen* 3.3 (2006), 422–38; Konrad H. Jarausch, ed., *Die 1970er-Jahre: Inventur einer Umbruchzeit*, Göttingen: Vandenhoeck & Ruprecht, 2006 (= *Zeithistorische Forschungen* 3.3); idem, ed., *Das Ende der Zuversicht? Die siebziger Jahre als Geschichte*, Göttingen: Vandenhoeck & Ruprecht, 2008; Antonio Varsori, ed., *Alle origini delle presente: l'Europa occidentale nella crisi degli anni settanta*, Milan: Franco Angeli, 2007; Alwyn W. Turner, *Crisis? What Crisis? Britain in the 1970s*, London: Aurum, 2008; Philippe Chassaigne, *Les années 1970: Fin d'un monde et origine de notre modernité*, Paris: Armand Colin, 2008; Beckett, *When the Lights*

Went Out; Jeremy Black, *Europe Since the Seventies*, London: Reaktion Books, 2009; Laurel Forster and Sue Harper, eds, *British Culture and Society in the 1970s: The Lost Decade*, Newcastle-upon-Tyne: Cambridge Scholars, 2010; Marie-Janine Calic, Dietmar Neutatz and Julia Obertreis, eds, *The Crisis of Socialist Modernity: The Soviet Union and Yugoslavia in the 1970s*, Göttingen: Vandenhoeck & Ruprecht, 2011; Andreas Wirsching, Göran Therborn, Geoff Eley, Hartmut Kaelble and Philipp Chassaigne, 'The 1970s and 1980s as a Turning Point in European History?,' *Journal of Modern European History* 9.1 (2011), 8–26; Lawrence Black, Hugh Pemberton and Pat Thane, eds, *Reassessing 1970s Britain*, Manchester: Manchester University Press, 2013; and Martin H. Geyer, '"Gaps" and the (Re-) Invention of the Future: Social and Demographic Policy in Germany during the 1970s and 1980s,' *Social Science History* 39.1 (April 2015), 39–61.

9. For the United States, see only: Peter N. Carroll, *It Seemed Like Nothing Happened: The Tragedy and Promise of America in the 1970s*, New York: Holt, Rinehart and Winston, 1982; Elsebeth Hurup, ed., *The Lost Decade: America in the Seventies*, Aarhus: Aarhus University Press, 1996; Bruce J. Schulman, *The Seventies: The Great Shift in American Culture, Society and Politics*, New York: Free Press, 2001; Beth Bailey and David Farber, eds, *America in the Seventies*, Lawrence: University Press of Kansas, 2004; Bruce J. Schulman and Julian E. Zelizer, eds, *Rightward Bound: Making America Conservative in the 1970s*, Cambridge, MA: Harvard University Press, 2008; Jefferson Cowie, *Stayin' Alive: The 1970s and the Last Days of the Working Class*, New York: New Press, 2010; Niall Ferguson, Charles S. Maier, Erez Manela and Daniel J. Sargent, eds, *The Shock of the Global: The 1970s in Perspective*, Cambridge, MA: Harvard University Press, 2010; Daniel T. Rodgers, *Age of Fracture*, Cambridge, MA: Harvard University Press, 2011; Borstelmann, *1970s*; and Hallvard Notaker, Giles Scott-Smith and David J. Snyder, eds, *Reasserting America in the 1970s: U.S. Public Diplomacy and the Rebuilding of America's Image Abroad*, Manchester: Manchester University Press, 2016.

10. Carroll, *It Seemed Like Nothing Happened*; Jarausch, *Ende der Zuversicht?*; Rodgers, *Age of Fracture*; Doering-Manteuffel and Raphael, *Nach dem Boom*; Lutz Niethammer quoted after Calic, *Crisis of Socialist Modernity*, 8, 118; Ferguson et al., *Shock of the Global*; Borstelmann, *1970s*, 6. On 'decadology', see Joe Moran, 'Decoding the Decade,' *The Guardian* (13 November 2009); and Alexander C.T. Geppert and Till Kössler, 'Zeit-Geschichte als Aufgabe,' *Geschichte und Gesellschaft*. Sonderheft 25 (2015), 7–36, here 15–16.

11. Philippe Chassaigne, 'Why the 1970s Really Matter,' *Journal of Modern European History* 9.1 (2011), 21–3, here 23; idem, *Années 1970*. A comprehensive discussion of 'planetization' as a global consciousness concept, introduced and coined by French philosopher, geologist and Jesuit priest Pierre Teilhard de Chardin (1881–1955) in 1946, is beyond the scope of this chapter. Suffice to say, planetization denotes the historical process of transforming earth into one imagined entity (and community) and can be considered an alternative to 'globalization.' Teilhard de Chardin himself defined 'planetization' as the 'idea of the planetary totalisation of human consciousness.' See idem, 'Vie et planètes: Que se passe-t-il en ce moment sur la Terre?,' *Etudes: Revue de culture contemporaine* 248 (1946), 145–69; Eng. 'Life and the Planets: What is Happening at this Moment on Earth?,' in idem, *The Future of Man*, London: William Collins, 1964, 97–123, here 115.

12. Ballard, 'Sci-Fi Seer'; now more easily accessible in idem, *Extreme Metaphors: Selected Interviews, 1967–2008*, London: Fourth Estate, 2012, 22–35, here 25. For this inward turn, see also Judt, *Postwar*, 483, and, in particular, Florian Kläger's contribution, Chapter 6 in this volume.

13. 'Zukunft: Todlos glücklich,' *Der Spiegel* 20.53 (26 December 1966), 80–90, here 82–3; 'Zukunftsplanung: Ritt auf dem Tiger,' ibid. 24.1 (5 January 1970), 34–47, here 38; 'Skylab: Am Tag X eine Trümmerschleppe,' ibid. 33.27 (2 July 1979), 142–53. Skylab is not to be confused with Spacelab, the reusable laboratory flown on the US Space Shuttle during the 1980s and 1990s. On Skylab, see Regina Peldszus's contribution, Chapter 10 in this volume; on Spacelab, Tilmann Siebeneichner's article, Chapter 11.

14. J.G. Ballard and Christopher Evans, 'The Space Age is Over,' *Penthouse* [UK] 14.1 (January 1979), 39–42, 102, 106, here 40–1; also in *Extreme Metaphors*, 121–31, here 123; Svetlana Boym, *The Future of Nostalgia*, New York: Basic Books, 2001, xiv.

15. Among those few 'general' historians willing to consider, at least in passing, that the 'final frontier so recently opened up for exploration by the spacemen' might have had any effect at all on the 1970s and humankind's self-perception, are Thomas Borstelmann and Niall Ferguson. See Borstelmann, *1970s*, here 70–1, 138–9, 239–40; and Ferguson, 'Crisis, What Crisis? The 1970s and the Shock of the Global,' in idem et al., *Shock of the Global*, 1–21, here 2–3.

16. Thus, the periodization suggested here is more comprehensive than Matthew Tribbe's. Separating 1950s/1960s Space Age America from the 1968–72 moon landing years, Tribbe aims to introduce a specific 'Apollo era' for these four years only, even though the program itself began technically much earlier; see Tribbe, *No Requiem for the Space Age*, 14, 211. On conceptualizing the so-called Space Age as a historical period that lasted from 1942 through 1972 and was characterized by a specific temporal dimension, see Alexander C.T. Geppert, 'Die Zeit des Weltraumzeitalters, 1942–1972,' *Geschichte und Gesellschaft*. Sonderheft 25 (2015), 218–50.

17. John Noble Wilford, 'Last Apollo Wednesday: Scholars Assess Program,' *New York Times* (3 December 1972), 1, 68; idem, 'Meaning of Apollo: The Future Will Decide,' ibid. (21 December 1972), 21; Hoimar von Ditfurth, 'Kosmische Quarantäne,' and idem, 'Ein Schuß ins Leere,' both reprinted in *Zusammenhänge: Gedanken zu einem naturwissenschaftlichem Weltbild*, Reinbek: Rowohlt, 1977, 25–8, 76–9; 'Apollo: "Mann, ist der Berg groß,"' *Der Spiegel* 26.53 (25 December 1972), 76–8, quote Schlesinger on p. 78. See also Kurt Rudzinski, 'Die Monderoberung verweist auf die Erde zurück: Zum Abschluß des Apollo-Programms,' *Frankfurter Allgemeine Zeitung* (20 December 1972), 2. On worldwide reactions to the Apollo moon landings, see Roger Launius's contribution, Chapter 3 in this volume.

18. 'The Moon Age,' *Newsweek* (7 July 1969), 26–52; Wilford, 'Meaning of Apollo'; Norman Mailer, *Of a Fire on the Moon*, Boston: Little, Brown, 1970, 310.

19. Denis E. Cosgrove, 'Contested Global Visions: *One-World*, *Whole-Earth*, and the Apollo Space Photographs,' *Annals of the Association of American Geographers* 84.2 (June 1994), 270–94, here 271, 274; Robert Poole, *Earthrise: How Man First Saw the Earth*, New Haven: Yale University Press, 2008, 199; Benjamin Lazier, 'Earthrise; or, The Globalization of the World Picture,' *American*

Historical Review 116.3 (June 2011), 602–30; Horst Bredekamp, 'Blue Marble: Der Blaue Planet,' in Christoph Markschies et al., eds, *Atlas der Weltbilder*, Berlin: Akademie, 2011, 367–75, here 367, 372. Already vast, the literature on Earthrise and Blue Marble is ever-growing; see only Wolfgang Sachs, 'Satellitenblick: Die Ikone vom blauen Planeten und ihre Folgen für die Wissenschaft,' in Ingo Braun and Bernward Joerges, eds, *Technik ohne Grenzen*, Frankfurt am Main: Suhrkamp, 1994, 305–46; William Bryant, 'The Re-Vision of Planet Earth: Space Flight and Environmentalism in Postmodern America,' *American Studies* 36.2 (Fall 1995), 43–63; Sheila Jasanoff, 'Heaven and Earth: The Politics of Environmental Images,' in idem and Marybeth Long Martello, eds, *Earthly Politics: Local and Global in Environmental Governance*, Cambridge, MA: MIT Press, 2004, 31–52; Stefan Helmreich, 'From Spaceship Earth to Google Ocean: Planetary Icons, Indexes, and Infrastructures,' *Social Research* 78.4 (Winter 2011), 1211–42; Robin Kelsey, 'Reverse Shot: Earthrise and Blue Marble in the American Imagination,' *New Geographies* 4 (2011), 10–16; and most recently, even with a slightly different emphasis, Peter Sloterdijk, 'Starke Beobachtung: Für eine Philosophie der Raumstation,' in idem, *Was geschah im 20. Jahrhundert?*, Berlin: Suhrkamp, 2016, 177–84.

20. This is yet another argument that was put forward by contemporaneous observers rather than coined by latter-day historians. See, for instance, J.G. Ballard, 'Which Way to Inner Space?,' *New Worlds Science Fiction* 118 (May 1962), 2–3, 116–18 (reprinted in idem, *A User's Guide to the Millennium: Essays and Reviews*, London: HarperCollins, 1996, 195–8); and Bruce Mazlish, 'Historical Analogy: The Railroad and the Space Program and Their Impact on Society,' in idem, ed., *The Railroad and the Space Program: An Exploration in Historical Analogy*, Cambridge, MA: MIT Press, 1965, 1–52, here 41, 52. See also the contributions by Florian Kläger and Andrew Jenks, Chapters 6 and 9, respectively, in this volume.

21. Some, Ballard included, would argue that the three manned Skylab missions conducted between May 1973 and February 1974 should be categorized under the Apollo years. But both Skylab (1973–79) and the 1975 Apollo-Soyuz Test Project were different spaceflight programs. They were launched on a Saturn IB spacecraft, not the famous Saturn V rocket, and the Skylab workshop used a Saturn V with only two active stages

22. See United States Congress, Committee on Aeronautical and Space Sciences, *National Space Goals for the Post-Apollo Period: Hearings on Alternative Goals for the National Space Program Following the Manned Lunar Landing. Eighty-ninth Congress, First Session, August 23, 24, and 25, 1965*, Washington, DC: Government Printing Office, 1965; United States President's Science Advisory Committee, *The Space-Program in the Post-Apollo Period*, Washington, DC: Government Printing Office, 1967, here 45; United States Space Task Group, *The Post-Apollo Space Program: Directions for the Future*, Washington, DC: Government Printing Office, 1969; Richard Nixon, 'Statement Announcing Decision to Proceed with Development of the Space Shuttle,' 5 January 1972, *The American Presidency Project*, available at http://www.presidency.ucsb.edu/ws/?pid=3574 (accessed 1 October 2017). John M. Logsdon, *After Apollo? Richard Nixon and the American Space Program*, Basingstoke: Palgrave Macmillan, 2015, here 3, 103–15, 269, 275.

23. See, for example, the contributions by Thore Bjørnvig and David Kirby, Chapters 7 and 13, respectively, in this volume.

24. Jürgen Osterhammel and Niels P. Petersson, *Geschichte der Globalisierung: Dimensionen, Prozesse, Epochen*, Munich: C.H. Beck, 2003. 27; Martin Collins, 'One World... One Telephone: Iridium, One Look at the Making of a Global Age,' *History and Technology* 21.3 (September 2005), 301–24, here 304. See also Martin Collins's contribution to the present volume, Chapter 2.

25. United States Space Task Group, *Post-Apollo Space Program*, 6, 16.

26. '7600 Satelliten seit Sputnik I im Weltraum,' *Frankfurter Allgemeine Zeitung* (29 January 1975), 25. Despite their invisible omnipresence in everyday life, the history of artificial satellites is still dramatically under-researched. But see, in addition to Martin Collins's work, Pamela E. Mack, *Viewing the Earth: The Social Construction of the Landsat Satellite System*, Cambridge, MA: MIT Press, 1990; Edward R. Slack, 'A Brief History of Satellite Communications,' *Pacific Telecommunications Review* 22.3 (2001), 7–20; Hugh R. Slotten, 'Satellite Communications, Globalization, and the Cold War,' *Technology and Culture* 43.2 (April 2002), 315–50; Lisa Parks, *Cultures in Orbit: Satellites and the Televisual*, Durham: Duke University Press, 2005; and Laurence Nardon, 'Cold War Space Policy and Observation Satellites,' *Astropolitics* 5.1 (August 2007), 29–62.

27. See Martin Collins's contribution, Chapter 2 in this volume.

28. Konrad H. Jarausch, 'Verkannter Strukturwandel: Die siebziger Jahre als Vorgeschichte der Probleme der Gegenwart,' in idem, *Das Ende der Zuversicht?*, 9–26, here 22–3.

29. Sabine Höhler, *Spaceship Earth in the Environmental Age, 1960–1990*, London: Pickering & Chatto, 2015, 113; Asif A. Siddiqi, 'From Cosmic Enthusiasm to Nostalgia for the Future: A Tale of Soviet Space Culture,' in Eva Maurer et al., eds, *Soviet Space Culture: Cosmic Enthusiasm in Socialist Societies*, Basingstoke: Palgrave Macmillan, 2011, 283–306, here 284. The official record of the Apollo-Soyuz Test Project is Edward Clinton Ezell and Linda Neuman Ezell, *The Partnership: A History of the Apollo-Soyuz Test Project*, Washington, DC: NASA, 1978.

30. Nixon, 'Statement Announcing Decision'; George E. Mueller, 'In the Next Decade: A Lunar Base, Space Laboratories and a Shuttle Service,' *New York Times* (21 July 1969), 14; Logsdon, *After Apollo?*, 266–9.

31. Warren Smith, 'To Infinity and Beyond?,' *Sociological Review* 57.1 (May 2009), 204–12, here 205, 209.

32. John M. Logsdon, 'Evaluating Apollo,' *Space Policy* 5.3 (August 1989), 188–92, here 190.

33. Technically, there were even more than six: Salyut 2 aborted before it could be manned, and another station failed without taking the name of Salyut. On Skylab, see Courtney G. Brooks, Roland W. Newkirk and Ivan D. Ertel, *Skylab Chronology: The Story of the Planning, Development, and Implementation of America's First Manned Space Station*, Washington, DC: NASA, 1977; as well as William David Compton and Charles D. Benson, *Living and Working in Space: A History of Skylab*, Washington, DC: NASA, 1983. On Salyut, see Grujica S. Ivanovich, *Salyut – The First Space Station: Triumph and Tragedy*, New York: Springer, 2008; and, in particular, Cathleen Lewis, 'Space Spies in the Open: Military Space Stations and Heroic Cosmonauts in the Post-Apollo Period, 1971–77,' in Alexander C.T. Geppert, Daniel Brandau and Tilmann Siebeneichner, eds, *Militarizing Outer Space: Astroculture, Dystopia and the Cold War*, London: Palgrave Macmillan, forthcoming (= *European Astroculture*, vol. 3).

34. Carl Sagan quoted after Walter A. McDougall, 'A Melancholic Space Age Anniversary,' in Steven J. Dick, ed., *Remembering the Space Age: Proceedings of the Fiftieth Anniversary Conference*, Washington, DC: NASA, 2008, 389–95, here 394. See also Robert S. Kraemer, *Beyond the Moon: A Golden Age of Planetary Exploration, 1971–1978*, Washington, DC: Smithsonian Institution Press, 2000. Biographies of Carl Sagan include William Poundstone, *Carl Sagan: A Life in the Cosmos*, New York: Henry Holt, 1999; and Ray Spangenburg and Diane Moser, *Carl Sagan: A Biography*, Amherst: Prometheus, 2004. On the Sagan plaque, see William R. Macauley, 'Inscribing Scientific Knowledge: Interstellar Communication, NASA's Pioneer Plaque, and Contact with Cultures of the Imagination, 1971–1972,' in Alexander C.T. Geppert, ed., *Imagining Outer Space: European Astroculture in the Twentieth Century*, Basingstoke: Palgrave Macmillan, 2012, 285–303 (= *European Astroculture*, vol. 1).
35. 'World Poll Finds Wide Belief in Life on Other Planets,' *New York Times* (13 June 1971), 20; Magoroh Maruyama and Arthur Harkins, eds, *Cultures Beyond the Earth: The Role of Anthropology in Outer Space*, New York: Vintage Books, 1975. On the history of SETI, see George Basalla, *Civilized Life in the Universe: Scientists on Intelligent Extraterrestrials*, Oxford: Oxford University Press, 2006, here 120; and Steven J. Dick, 'Anthropology and the Search for Extraterrestrial Intelligence: An Historical View,' *Anthropology Today* 22.2 (April 2006), 3–7, here 6.
36. Terry White, 'What Do You Do for an Encore?,' in Goldstein, *Moon Landing and its Aftermath*, 275–7, here 277; Gerard K. O'Neill, *The High Frontier: Human Colonies in Space*, New York: William Morrow, 1976, 34.
37. Gerard K. O'Neill, 'A Lagrangian Community?,' *Nature* 250 (23 August 1974), 636; idem, 'The Colonization of Space,' *Physics Today* 27.9 (September 1974), 32–40; idem, 'Space Colonies and Energy Supply to the Earth,' *Science* 190 (5 December 1975), 943–7; and idem, 'Colonies in Orbit,' *New York Times Magazine* (18 January 1976), 11, 25–9, here 11, 25. The same 'letter from space' was also published in the introduction of his *High Frontier*, 13–17. Declaring that 'space exploration is inevitable and those who prepare now will ride a wave of the future,' a contemporaneous bibliography listed more than 60 articles and essays on Gerard O'Neill and his space colonization scenarios; see Michael E. Marotta, *Space Colonization: An Annotated Bibliography*, Mason: Loompanics, 1979, here 4. See also Michael A.G. Michaud, *Reaching for the High Frontier: The American Pro-Space Movement, 1972–84*, New York: Praeger, 1986, 60–3; Peder Anker, 'The Ecological Colonization of Space,' *Environmental History* 10.2 (April 2005), 239–68; and Douglas Murphy, *Last Futures: Nature, Technology and the End of Architecture*, London: Verso, 2016, 77–9, here 77, 135, in addition to the contributions by Tilmann Siebeneichner and Peter Westwick, Chapters 11 and 12, respectively, in this volume.
38. O'Neill, *High Frontier*, 36, 46.
39. Duncan Lunan, *Man and the Stars: Contact and Communication with Other Intelligence*, London: Souvenir Press, 1974, 104–17; Fritz Haller, *Umweltgestaltung einer prototypischen Raumkolonie*, Karlsruhe: Universität Karlsruhe, 1980.
40. Walter M. Thiebaut to Pam Dattilo, 12 April 1979, Historical Archives of the European Union/European Space Agency (hereafter HAEU/ESA), 9970.

41. 'Europe Assured on Role in Space,' *New York Times* (23 September 1970), 9.
42. For the 'rebirth,' see Walter A. McDougall, 'Space-Age Europe: Gaullism, Euro-Gaullism, and the American Dilemma,' *Technology and Culture* 26.2 (April 1985), 179–203, here 195; Guy Collins, *Europe in Space*, Basingstoke: Macmillan, 1990, here 22, 30–2; Helmuth Trischler, *The 'Triple Helix' of Space: German Space Activities in a European Perspective*, Noordwijk: ESA, 2002, 17–18; and idem, 'Contesting Europe in Space,' in Martin Kohlrausch and idem, *Building Europe on Expertise: Innovators, Organizers, Networkers*, Basingstoke: Palgrave Macmillan, 2014, 242–75, here 269. See also Burl Valentine, 'Obstacles to Space Cooperation: Europe and the Post-Apollo Experience,' *Research Policy* 1.2 (April 1972), 104–21, and John Krige, Angelina Long Callahan and Ashok Maharaj, *NASA in the World: Fifty Years of International Collaboration in Space*, Basingstoke: Palgrave Macmillan, 2013, 65–124. For the United Kingdom's space policy in particular, see Doug Millard's contribution, Chapter 4 in this volume.
43. Clarke, 'Foreword.' As this concept has been introduced, defined and discussed in detail in the first volume, there is no need for repetition here; see Alexander C.T. Geppert, 'European Astrofuturism, Cosmic Provincialism: Historicizing the Space Age,' in idem, *Imagining Outer Space*, 3–24, here 6–9; and idem, 'Rethinking the Space Age: Astroculture and Technoscience,' *History and Technology* 28.3 (September 2012), 219–23. See also De Witt Douglas Kilgore, 'Exploring Astroculture,' *Science Fiction Studies* 41.2 (July 2014), 447–50; and the epilogue to the forthcoming third volume, Geppert et al., *Militarizing Outer Space*.
44. Arthur C. Clarke to Julian Scheer, 10 April 1971; Val Cleaver to Arthur C. Clarke, 9 January 1971, both in Smithsonian National Air and Space Museum Archives, Arthur C. Clarke Collection (hereafter NASMA/ACCC), 007/06 and 007/07, respectively.
45. On the United States, see the contributions by Roger Launius and Peter Westwick, Chapters 3 and 12, respectively, in this volume.
46. For further observations on the simultaneity of comprehensive space enthusiasm and the decades-long abstinence from independent human spaceflight in Space-Age Europe, see Geppert, 'European Astrofuturism, Cosmic Provincialism,' 9–13.
47. Space Treaty is short for 'Treaty on Principles Governing the Activities of States in the Exploration and Use of Outer Space, including the Moon and Other Celestial Bodies'; Moon Agreement stands for 'Agreement Governing the Activities of States on the Moon and Other Celestial Bodies.' As of 1 January 2016, 104 states have signed the former, but only 16 the latter; see http://www.unoosa.org/documents/pdf/spacelaw/treatystatus/AC105_C2_2016_CRP03E.pdf (accessed 1 October 2017).
48. See also the contributions to the forthcoming third volume in this European Astroculture trilogy: Geppert et al., *Militarizing Outer Space*.
49. See also H. Bruce, Franklin 'Don't Look Where We're Going: Visions of the Future in Science-Fiction Films, 1970–82,' *Science Fiction Studies* 10.1 (March 1983), 70–80.
50. Dieter Vogt, 'Die gute alte Zeit der Mondrakete: Amerikas Weltraumbahnhof nach dem Apollo-Rausch,' *Frankfurter Allgemeine Zeitung* (31 January 1976), BuZ 1.

PART I

Navigating the 1970s

CHAPTER 2

The 1970s: Spaceflight and Historically Interpreting the In-Between Decade

Martin Collins

The 1970s has become a peculiar adventure. However defined chronologically, whether in precise decadal fashion or as slopping over its numerical boundaries, the 1970s has risen in scholarly evaluation as a time of refashioning and restructuring through which a vast scale of change may be assigned or read. A not insignificant swath of researchers now argue that the 1970s loosened, unraveled, recast or merely brought into clearer view the world of the immediate postwar decades – at least as seen through the lens of US-Western European experience. Decline of Fordist modes of political economy; diminished coherence of national economic markets; heightened interdependence in international relations; amplified and geographically more extensive consumer cultures; a fading of respect for expertise; changed valuations of key background social-cultural tropes such as trust and risk; and, not least, a perceived enhancement of emphasis on the individual and individuality in cultural life – in various combinations and collectively all these pointed to the 1970s as a period of dramatic reorientation, restitching micro and macro phenomena on a transnational scale.[1] Such analysis gained added heft, as these vectors of change only seemed to intensify and become the organizing motifs of the post-1980 period. The effort to characterize the history of the last several decades thus provided critical motivation to revisit these earlier years. Among historians, Charles Maier's study *Among Empires: American Ascendancy and Its Predecessors* deftly summarizes this 1970s

Martin Collins (✉)
Smithsonian National Air and Space Museum, Washington, DC, USA
e-mail: collinsm@si.edu

© The Author(s) 2018
Alexander C.T. Geppert (ed.), *Limiting Outer Space*
European Astroculture, vol. 2
https://doi.org/10.1057/978-1-137-36916-1_2

transformation, capturing both the emphasis of change and its scale, with its strong US slant, as one from an empire of production to an empire of consumption.[2]

Connecting most of these dots was the analytic judgment that capitalism had undergone a substantive shift, of which the expansions of the transnational and mass consumption were critical elements. As corollary, the role of the nation state shifted too, to become facilitator of market-driven phenomena, thus diminishing its own authority in what had been a prior area of state power. As is well-known, such changes, and their import, have been given conceptual energy through capacious terms such as 'postmodernity' and 'globalization.'[3] Through such conceptual terms, and in this listing of changes, one can discern a key intellectual problem: Were these changes, à la a quasi-Marxian perspective, bound together, in which structures of political economy were joined to culture and individual subjectivity? And, if so, how and with what stakes? Or, put more generally, how do we characterize the causal map presented by such literature, issuing from across the disciplinary spectrum of the humanities? In making this summary, it is to note that much of this writing, though attentive to critical events such as the 1973 oil crisis or the diminished standing of liberal political thought and the commensurate rise in conservative thought (especially in the United States, but also in the United Kingdom), is seeking an explanatory frame that is related to but not dependent on these specific developments.[4] Rather, it is to see the primary engines of change encapsulated within the 'logics' of capital, characterized by the attributes already listed, but then increasingly informed and justified by neoliberal economic ideology, on the rise through the 1970s and after.[5]

This chapter takes these questions as points of departure for sorting through the 1970s as historical problem, and the place of spaceflight – broadly construed – therein. It is to argue, more specifically, that these interpretive angles – drawn from different literatures – intersect substantively with the history of spaceflight, in its relations to period transformations of political economy and of cultural imaginaries that were reconfiguring relations between self and society on local and global scales. The period categories of the transnational and the global were fundamentally entwined with the practices and perspectives attached to spaceflight.[6] This stance too provides purchase on key (and interlocking) tropes of the period and of this volume: utopia and limits. Especially for utopia (or its opposite, dystopia), such conceptions were someone's or some group's shorthand assessments of the 1970s condition, of the connections among political economy, culture and individuality, of their beneficial possibilities or deep failings. That is, such terms were bound to specific historical positions, contexts and purposes. Utopia and limits, then, perhaps become terms for tracing the response of different historical actors to the conditions of the 1970s – more epiphenomena than descriptive of separate, distinct realities. Interestingly, the two terms speak to the intellectual tension already sketched: utopia/dystopia relates largely to the matrix of culture and subjectivity; limits to claims grounded in political economy. Thus, though, these terms provide vital windows on the 1970s, they are each and together (as viewed through these literatures) bound up with the larger, more complex historical terrain of period change. Such analysis again highlights the causal connection between space-flight and the large-scale change under consideration. Spaceflight becomes not merely a cultural screen on which to project notions of utopia and limits, but a ground for understanding period political economy as well as a resource for various historical actors as they created new notions of self with new frames of reference and meaning.

Such claims point to an issue in the literatures reviewed and in the framing of this chapter: Do the various investigations sketched above all relate, fit into a common explanatory framework? When assessing the state of cultural studies in the early 1990s, historian Carolyn Steedman has raised the question of whether practitioners had developed a tendency to assert that 'any one aspect of society is related to any other' and 'everything connects to everything else,' to assume too readily 'a principle of arbitrary connectedness.'[7] But her very question, a minority concern at that moment in the field, only makes sense if the reigning perspective already posited that in the historical condition of the 1970s (and after) individual subjectivities, social collectives and political economy were inti-mately integrated, bound together causally (though not wielding equal power). Steedman's cautionary critique sought to query the validity of such an 'out on a limb' theoretical assumption but then to wonder if such relatedness, indeed, was historically manifest in the very time period under consideration.

In the following, my aim is to suggest contours of historical and theoretical writing on the 1970s, to focus on organizing ideas, meanings and interpretations presented therein rather than to provide a critique of such work. This effort thus is not concerned with the accuracy of claims, whether empirically or as intellectual motifs. Nor, within this limitation, is this effort intended as a full-blown historiographical essay, covering the mul-tiple by-ways of the various relevant literatures. Rather it is to tease out some assumptions in the writing about the 1970s – as expressed at the time and subsequently – and place spaceflight in conversation with this broader scope of writing about the decade. As such, this chapter seeks to engage a long-standing challenge of integrating the history of spaceflight with the broader field of research in the humanities. More specifically, though, it aims to make explicit what is only partially articulated in the literatures under consideration: that spaceflight as a period category and field of action became fundamental to the making of – and, thus, to our understanding of – the 1970s and after.

I   Theorizing the 1970s

To make the historical stakes attached to the 1970s more tangible, consider this image: the photo that was on the cover of Victoria de Grazia's *Irresistible Empire: America's Advance through Twentieth-Century Europe* (Figure 2.1).[8] If one had to pick a single image to bring forward the seemingly abstract relations among political economy, US power, the US-European geopolitical nexus,

Figure 2.1   The original description of this image was 'Model sitting atop the engine of Pan Am's 707 Clipper wearing pink printed sharkskin jacket and matching shorts by Randol Juniors and Emme hat.' Ca. 1959.
Source: Courtesy of Condé Nast Archive/Corbis.

the emerging postcolonial world, mass consumption and its attendant individual imaginaries, this might be it. Though this image and de Grazia's thesis might be deconstructed in any number of ways, together they help foreground capitalism, consumer culture and individuality as historical agents and, thus, as problems for our understanding of the 1960s, 1970s and 1980s.

Despite the rambunctious title, the book is a careful calibrating of local and varied European responses to the American 'advance' (countering its own argument with regard to the United States as 'irresistible empire') and thus is consonant with this volume's interest in exploring the historical question of a European-inflected twentieth-century astroculture.

The image also serves as a relevant temporal marker. Taken in 1959, it highlights continuities (capitalism, the state, consumption) from the immediate postwar years to the 1970s, but helps set the context for change in the literatures presented here. More specifically, for example, the presence of a jet – a Pan Am jet – and a posed, exuberant human (female) body, offered as an exemplar of self-fashioning, point to a key theme that emerges from the 1970s: a perceived intensification of the collapse of space and time as vectors of individual and cultural experience, and the enhanced ease of circulation of things and images on a global scale, a condition materially bound to the development of space applications, especially communications. Such change was augmented by new tropes of globality; especially those images of earth from space that energized a variety of historical actors in the United States and Europe to rethink relations between political economy and self, between the local and literally global, between limits and utopia. Such shifts speak to the question of individual subjectivity, but also something related and deeper: a claim, advanced by theorists such as Jean Baudrillard (1929–2007) and Fredric Jameson (1934–) in different forms from the late 1960s through the early 1980s, that the very order of perceptual experience – of the ontology of everyday life in the United States and Europe – had changed via this amplified circulation of market and media-created stuff and images, had become in Jameson's words 'the omnipresent raw material of our cultural ecosystem.'[9]

Such a claim today perhaps seems less jarring than before, now that worldwide web-ness and information-ness seem our natural epigenome. But the contrast then, still, largely, was with the local. The standard exemplar has been Walter Benjamin's reflections on an urban citizen's experience in a shopping arcade in Paris of the nineteenth century and the resulting reorientation of perceptual space engendered by new practices of consumption and for creating cultural meaning. In these more recent theoretical reflections, such reoriented space now extended to every corner of life experience and to the totality of the globe.[10] Marshall McLuhan (1911–80), in his frantic literary way, captured this shift, only beginning to happen, in his notion of the 'global village' and through aphorisms such as 'in the electric age, we wear all mankind as our skin.'[11] For the historian – and even for the theorist – such a claim, specific to the post-1960s, about a changed condition of experience is disorienting, in part, because it lessens the agency of traditional big movers, such as the state, the corporation or the institutions of religion, as discrete actors, and reforegrounds capitalism (or more precisely, in its incarnation as 'late-capitalism'), especially in this period, as complex totality and logic, engine and mediator of this condition, as the vital frame in which things happen.

De Grazia's image and thesis thus provide an opening to explore the issue of spaceflight's relation to the phenomena she engages, of finding the appropriate causal frame, in particular of how American-style capitalism and consumer culture connect with the historiography of state-centered accounts, including those that build on the Cold War narrative of technopolitics. For it is in this late 1960s, early 1970s moment that spaceflight, primarily but not only under US aegis, achieved its multi-faceted mature form: as significant genre of postwar technology; as site and instrumentality of knowledge creation; as state activity, business undertaking, military venture and global utility; as a cultural zone for contesting or repurposing dominant values and beliefs; and as national and international trope extraordinaire, one that resonated with universalistic values in first, second and third worlds. Such occurrence was coeval with the 1970s reorientation already sketched.

To get at this complex of issues, consider the earlier characterization of the 1970s as peculiar adventure. Why peculiar? In part, because the scholarly insights derive from a historical conjunction: 1970s as a change-decade coincides with a reorientation of the academy in which the disciplines of sociology, anthropology and comparative literature, and the regard for 'theory' and interdisciplinarity rose as sites of intellectual activity and influence. In 1980, in an essay titled 'Blurred Genres: The Refiguration of Social Thought,' anthropologist Clifford Geertz (1926–2006) sought to survey this recent change in the academy, its excitements and its ensuing confusions, particularly regarding approaches in explanation of social, cultural or historical change. In his concluding thoughts, Geertz suggested that 'the refiguration of social theory represents, or will if it continues, a sea change in our notion not so much of what knowledge is, but of what it is we want to know' and poses 'the question of the relationship of such [knowledge] systems to what goes on in the world.'[12] He was, of course, reflecting, in part, on the march of French theory into the US academy, dissatisfactions in these disciplines with scientistic models and the rise of new areas of critique such as feminist studies. In ensuing years, the refiguration Geertz outlined moved from conditional 'if' to robust phenomenon, perhaps best represented by the ascendance of cultural studies by the end of the 1980s. Geertz, of course, was not alone in offering broad assessments of period change; as a sampling, Daniel Bell, Raymond Williams, Jean-François Lyotard, Fredric Jameson and Ulrich Beck also entered this disciplinary reorientation, each in different ways and, in the discussion to follow, will stand in for the larger period notion of theory.[13]

Though he did not express it as such, Geertz was marking the diminished authority in humanities thinking of Marxist explanatory categories of base, superstructure and class; or perhaps more accurately their substantial complication as to modes of agency, their interrelation and relative role in shaping historical orders and change. Here was a seeming irony, given the emerging concern in the 1970s with late-capitalism, but not really. For the main tension in reassessing 'what goes on in the world' was how to characterize the relation between dominant modes of political economy and individual subjectivity, how persons are constituted, self-fashioned and theorized as having (or not having) agency – a problem that traditional Marxist critique was not well equipped to engage. If one had to identify *the* major problem undergirding the Geertzian-described refiguration and its vast theoretical elaboration during the 1980s, 1990s and after, it would be this deep question.[14] Giving the political economy/subjectivity problem added saliency was, as noted above, the marked postwar expansion of mass consumption, the cultural values that phenomenon embodied, and the perception that life in the market was *a*, if not *the*, significant means through which individuals fashioned their identities. The market thus stood as prevailing ontology and had as its organizing principle, at least rhetorically, the individual as fundamental causal agent. Part of the glue that linked the market and individual subjectivity was the notion of imaginary – a space in which cultural images and meanings overlapped with and were mutually constituted by individual self-fashioning.[15] In this historical condition, what indeed then was the balance between classic structures of dominance and individual sovereignty and agency? And then for the thesis of this chapter: In what ways did spaceflight enter into either side of this equation?

As this brief discussion of Geertz suggests, the 1970s decade was more actively theorized before it was historicized; a situation that predominated through most of the 1990s. Indeed, such work, it might be argued, set the intellectual predispositions through which the 1970s would eventually be assessed; in particular, the foundational question of how to understand the relation between political economic structure and individuality. Which leads to another peculiarity of 1970s scholarship: writers like Geertz and the other theorists mentioned made strong historical claims about the 1970s and offered analyses of broad explanatory scope or implication, but were – for the most part – little interested in historical or empirical research per se. And for that most elusive component of their large-scale accounts – individual subjectivity in the context of late-capitalism and its causal meaning – the methodological guidance offered was more ambiguous compared to their reflections on political economy. An illustrative example is Michel Foucault (1926–84). In the 1970s his own methodology underwent a reinvention: from a focus on the individual as a subject that was acted *on* – think of his concept of disciplinarity – to one in which the individual was if not a self-governing agent then at least resistive to fields of social force, a shift embodied in Foucault's notions of biopolitics and governmentality. It was not an altogether coincidental alignment with the market-oriented, neoliberal times.[16] But Foucault, despite this shift, still rejected the idea of the individual as Enlightenment self-fashioning actor. These various tensions as to the status of individual agency (moving along a spectrum of negligible agency, delimited by social forces, to that fully empowered rational agent envisioned in neoliberal and liberal theory) would play out in cultural studies as it developed through the 1980s and 1990s. As is well known, investigating subjectivity became a major preoccupation in this domain, but with a focus that typically has tended toward micro rather than macro analysis, blurring the expansive theoretical claims as to a reoriented condition of subjectivity. In short, whether along the axes of the political-economic or subjectivity, theory offered broad explanatory templates, but research practice favored deep readings, delimited in scope.[17]

II   The relation of history to theory

Historians, not surprisingly, first approached the 1970s in cautious, fragmentary fashion. In 1982 one of the first US historical assessments was entitled *It Seemed Like Nothing Happened: America in the 1970s*; in 2006 the arc of subsequent evaluation was captured by the title *Something Happened: A Political Cultural Overview of the 1970s*.[18] Each took up culture as central subject, but as this playful citing of titles suggests each highlights the absence of a strong, synoptic interpretation.[19] In the 2000s, historians began to give concerted attention to the 1970s, but in many cases, still, without particular intersection with theoretical literature.[20] As an example, *Shock of the Global: The 1970s in Historical Perspective*, edited by scholars Niall Ferguson and Charles Maier, and probably the best single historical treatment on the decade, only makes glancing connection with these literatures. Rich in detail on political and financial changes in the international arena, it pays scant attention to the interpretive significance of Maier's argument of a shift to an empire of consumption and its meaning for identity-making as a historical force. Yet, Ferguson makes a passing reference to Tom Wolfe's characterization of the 1970s as the 'Me Decade,' a period Wolfe argued was marked by a crisis of the self and the 'spirit of the age was introspective' – one marker, even if critical, of the preference for individual agency as causal axis rather than the notion of Foucauldian subject. Then Ferguson notes, but that was 'the view from within [...] viewed from without [...] from the final frontier so recently opened up by the spacemen – the crisis of the seventies was as much global as it was personal.' And then leaves it there, never addressing the potential relationship between the two, between a global, space-derived perspective and the question of subjectivity. By omitting the historical relevance of theory to its analysis – of the relation between political economy and subjectivity – *Shock of the Global* diminishes its own argument as to what is at stake in understanding the 1970s.[21]

By the mid-2000s there was a striking consilience among these various literatures, historical and theoretical, claiming that the late 1970s to early 1980s inaugurated a significant political/economic and cultural shift, roughly defined as described previously.[22] The main differences have concerned the weight given the issue of subjectivity, and the relative ambition of explanatory frameworks – on all of which history has been more modest than theory. One might say that history came belatedly to accept the claims of pre-existing theory. On this broad terrain of scholarship I can only suggest an outline of interconnections among political economy, subjectivity and spaceflight.

There is one point, though, in reinterpreting the literature through a lens that includes spaceflight, on which one can be emphatic: a perceived, intimate relation between space applications, especially but not only communications, and a reoriented framework of 1970s-era political economy.[23]

There is perhaps no better illustration of the perceived difference between the worlds of the 1950s and 1960s and that of the 1970s than the work of liberal-cum-conservative sociologist Daniel Bell (1919–2011). In 1960, in the *End of Ideology: The Exhaustion of Political Ideas*, Bell argued that this 'end of' condition was the consequence of ideological accommodation among the state, capital and labor post-Second World War; it was, to use my·earlier analytic characterization, the triumph of Fordism as the dominant mode of political economy. Thirteen years later, in 1973, Bell argued for a fundamental reordering of that accommodation in the *Coming of Post-Industrial Society*.[24] It is important to emphasize that the oft-used 'post-industrial' was not for Bell the decline per se of Ford-style industrialization, but the ascent of a new structure for knowledge and the social relations that supported it: 'the novel and central feature of post-industrial society is the codification of theoretical knowledge and the new relation of science to technology.' Discounting Bell's naïve appraisal of the science–technology relationship, the critical underlying issue was the social structure of post-industrial society and the shift in power it implied; such society 'is made up of estates (professional groups – technological, scientific, administrative and cultural) and situses [*sic*], which are the institutional locations (corporate, government, military, the universities and research complexes) where the estates are located.'[25] It was, in short, the story of how the Cold War state had, through its policies, inadvertently diminished the role of a labor-capital economic social structure in national life, relocating it to that vast cadre of military and NASA employees, contractors and university researchers who supported a wide array of research and development interests. It was for Bell, explicitly, a critique of the relevance of Marxist categories of analysis, and it was, contra to period thinking about limits, utopian. The post-industrial was a story of technocratic optimism.

Significantly, this story was not merely US-centric. At nearly the same time French sociologist Alain Touraine (1925–) also invoked the term and idea of 'post-industrialism' to describe developments in Europe, particularly in France.[26] Space, as one element in the assemblage of US–Europe Cold War research and development activity, was a lesser, but emblematically important, note in this larger transformation of professions and their enhanced economic standing. But the *Coming of Post-Industrial Society* also was about the importance of new technologies, particularly those relating to information; not just in and of themselves but for their role in stitching together these knowledge communities and amplifying their political economic influence. Introduced here but taken up with vigor through the next decade, Bell became one of the great expounders of the transformative importance of the information revolution; a creature with many parts, but one of whose muscular elements was the expanding domain of satellite communications. Such communications, transnational from their inception, gave a vastly increased sense of spatial scale to the possibilities of computer-related innovation.

Though Bell was one of the first and most influential to describe the 1970s condition, he was ambivalent on how to situate capitalism's role in this analysis. The *Coming of Post-Industrial Society* was mostly silent on this point, reflecting his assessment that recent change derived from a nexus of state–industry relations. Three years later, in 1976, Bell voiced this ambivalence in the *Cultural Contradictions of Capitalism*, a work of interest, in part, because Bell did not yet emphasize the relative shift from regulated to deregulated markets under the banner of neoliberalism, which already had succeeded in elevating ideologically and via policy the status of markets, nationally and transnationally.[27] What he did see, though, is a core problem of that change – that of subjectivity, or the relation of individuals and the market. But unlike the post-1970s period in which this relation was taken as fundamental, as a social fact to be described and given a positive valence, Bell regarded it as problematic. After all how can capitalism be an effective means of economic advancement if sober, bourgeois values are in decline, if cultural values of discipline and thrift are supplanted by individual projects of self-fashioning, if, for example, that fashionista sitting on a jet engine becomes not just an adornment but the face and soul of capitalism? It could not – and hence his argument about cultural contradiction.

It is important to note the different regard in which each of these two works of Bell has been held. *Post-Industrial Society* had and still has deep influence as a statement about the political-economic condition attached to the 1970s; *Cultural Contradiction* is remarkable for the way in which his capitalism was jettisoned in favor of the notion of late capitalism, with its presumption of deep and necessary relations among markets, consumption and subjectivity, of a field of experience that all inhabit and in which all participate, masses and elites, consumers and producers, a basic reorganization of the perceptual and social order. This latter assumption, embedded in the late 1980s and early 1990s in classic works by geographer David Harvey and literary theorist Fredric Jameson, took as a given spaceflight's fundamental integration into the political-economic structure and as a signature cultural motif.

One can see early formations of this view in 1970s social theorists such as Raymond Williams (1921–88). In the 1960s, Williams was further pursuing his interest in the problem of culture, shifting his attention from the nineteenth and early twentieth centuries to the then present. One result that conjoined these interests was the publication of *Keywords: A Vocabulary of Culture and Society* in 1976.[28] The substantial legacy of this work, not least for historians of technology, made culture a broadly ecumenical concept, embracing ways of doing and meaning-making at all levels of the social. It pushed into the background Williams's more agonistic conception of culture as terrain of contestation between what he called minority cultures and entrenched cultures of the state and capitalism – his attempt to adapt a Marxist critique and a view of culture as causal agent. It was that latter framework that would inform his shift in the early 1970s in analyzing television and communications developments broadly, with satellites occupying a critical techno-political position:

> There is a clear intention, in the strongest centers, to use this technology to override – literally to fly over, existing national cultural and commercial boundaries. The satellite is seen as the perfect modern way of penetrating cultural and commercial areas hitherto controlled or regulated by 'local' national authorities: that is to say societies with their own arrangements and governments.[29]

Like Bell, Williams was trying to reconcile assumptions that were in tension – in his case between the political value of the local and larger structures of production and cultural dissemination, whether conceived in social or geographic terms. Williams was seemingly torn by the notion of culture as a tool for drawing boundaries or as a social solvent, the local and global dimensions of which were being further complicated by and repositioned through the planetary frames of reference enabled by communications satellites and spaceflight more broadly.

In his *From Counterculture to Cyberculture*, historian Fred Turner threads together Bell's two arguments and touches on Williams's perspective, and neatly for this purpose, makes spaceflight integral to the story. Turner's essential contribution is to elucidate how historical actors – such as the *Whole Earth Catalog*'s Stewart Brand – saw a vital relation between post-industrial capitalism, rooted in Cold War innovation ideology and practices, and a private-market *mentalité*. For Brand and others, this conjunction was the means to transport counterculture values into the larger society and to invigorate Enlightenment beliefs in the individual as the measure of all things, to make the self-fashioned individual acting within the market the kernel of social change. Thus, in the 1970s, one sees how emerging constructs of the global and capitalism provided a home for seemingly contrary belief systems, for the 1960s multinational corporation *and* for its fundamental transformation through counterculture notions of business and self – thus lining up with a neoliberal ideology just gaining ascendance.[30] An interesting question, following Turner and Brand, is to what degree and in what way did space as one site of post-industrial reorientation and as cultural resource for self-fashioning and universalistic values participate in such conflation. On one level it was to give a new cast to Cold War systems thinking via the concept of ecosystems (especially at the planetary scale) and the latter's radically different moral presuppositions.[31] On another, and related level, it was to see the significance of space not primarily for its role as a political economic instrumentality, but rather as a key imaginary for relating subjectivity, values and capitalism into a *literal* world view meant to bind together and make sense of individual, local and global registers of experience and action.

Such linkage substantiated the idea that space-based infrastructures were integral to the changed political economy of the 1970s and after – as material modality and as trope for the compression of space and time. The importance of spaceflight, along these several vectors, became a staple of theory, captured in the complementary analyses of Harvey, who tracked the spatial configurations of the new capitalism, its reliance on flexible production and the rapid ubiquitous flow of information, and of Jameson, who also promoted the theme of global flows and circulation, but emphasized the role of images, particularly those attached to consumption and the imaginaries of consumers, collectively and individually. Space-based infrastructure too came to serve as a background fact for historians seeking to characterize the change of the 1970s and after, using it as quick shorthand, occasionally overstating its importance, when it still was a developing domain of activity in the 1960s and 1970s.[32]

Oddly, perhaps, the connection between theory and space-inflected political economy makes an appearance in Jean-François Lyotard's 1978 *Postmodern Condition*. In a work better known for its declaration of the death of metanarratives and its critique of modernity as an intellectual framework, Lyotard, à la Williams, considered rising business interest in space-based communications, a consequence of period deregulation policies, as one of the few concrete examples in his argument. He wonders:

> Suppose for example that a firm such as IBM is authorized to occupy a belt in the earth's orbital field and launch communications satellites or satellites housing data banks. Who will have access to them? Who will determine which channels or data are forbidden? The state? Or will the state be one user among others? New legal issues will be raised, and with them the question: Who will know?[33]

Much as Bell's *Cultural Contradictions*, Lyotard's *Postmodern Condition* was grappling with a reconfigured capitalism in which existing authorities of the state were moving to the market. Such movement was one element, in his view, of how the foundations of knowledge production and their integration into cultural life were changing, with space as one of multiple sites that highlighted that change. The intellectual dimensions of this change thus were not merely derived from critique of prior theory or of modernity's guiding assumptions, but were embedded in the more market-centered modes of knowledge production and control emerging in the 1970s.

In the 1970s and 1980s, the fact and theme of space-based infrastructure and perspectives took on different valences, causally and ethically, as they entered into the calculus of accounting for and expressing relations among theory, political economy, individual subjectivity and the seemingly enhanced problematic of the local and global. The depth of this grappling is highlighted in particular by Ulrich Beck's idea of risk society. He was deeply influenced, as was Lyotard, by Bell's notion of the post-industrial and the turn toward markets – risk society, in his argument, emerges with post-industrial society. Though space capabilities are not mentioned explicitly in his work, they are there as fundamental background, part of the aggregate presence of infrastructure; those technologies of broad geographical or global distribution that enable post-industrial life but whose pervasive daily influence only becomes visible when they fail. Importantly, as converse, such space technologies have been part of the means by which we become aware of and produce knowledge of risks, especially of the environmental variety. The very pervasiveness of risk – variously encountered, potentially and actually – results, in Beck's view, in a new human condition, one requiring a new mode of sociology. His theory thus makes strong claims about the structural character of the post-industrial and about subjectivity, but the self-fashioning invoked is tinged with dread and defensive reaction. This condition is the flip-side of Brand's 'Whole Earth'-ism in which related-ness and system-ness on a global scale are utopian, liberating concepts for individual action and redefinition of the social, rather than a condition of existence in which individuals are embedded and the status of individual agency ambiguous.

III   Subjectivity as historiographic challenge

Connecting period themes and theorizing about political economy to space-flight is relatively straightforward – the latter readily maps onto structural arguments of the role of states and markets in historical change. But the subjectivity issue, especially in its relation to political economy, is more elusive, indicated by the varying theoretical approaches to framing issues of self and identity – and thus of how spaceflight may enter into the problematic of understanding the balance between claims of rampant, powerful global forces and individual agency in the 1970s and after. This can be approached via the market and consumer culture and consumption, as already highlighted. It can be assayed through the study of communities or small groups. Or, it can be engaged by looking at cultural presuppositions that seem to have wide currency, such as the aforementioned strategy of 'imaginary' as conceptual tool. Each of these angles bears on the broader problem outlined: how the turn to individuality, associated with political-economic transformation assigned to the 1970s and after, can be investigated, how the small scale effectively might be related to the large; that is, after all, the core historical claim of globalization and postmodernity, as well as period invocations of utopia/dystopia. One can see such an effort in Fredric Jameson's more recent work in which he seeks to translate his theoretical claims to readings of science-fiction literature and film, sites which he sees as particularly evocative of individual and cultural responses to and refashioning of contemporary capitalism and the status of the utopian therein.[34] As he offers:

> Communicational and information technologies [are] scientific machineries of reproduction rather than production. [...] They foreground and dramatize this transformation. [...] But they themselves serve as allegories of something else, of the whole unimaginable decentered global network itself. [...] All thinking today, whatever else it is, is an attempt to think the world system as such.[35]

Subjectivity, for Jameson, thus is about how the vast geopolitical – and its emphasis on the circulation of images and things, detached from circumstances of production – enters into the unconscious. But such analysis walks an awkward line. On one level, it is to say that the individual unconscious or modes of interiority are mirrored and presented in narrative conventions, especially of film, leaving unclear how to characterize the boundary between the self and the condition of experience in recent decades. On another level, it presents film as a cultural site in which historical actors explore these relations, thus serving more as an empirical record about the producers of such cultural artifacts than as a theoretical exemplification of a collective experience of how and in what ways selves are fashioned in postmodernity.

A finer-grained view of the relations among space, the global and subjectivity, in the US context, emerges in Sam Binkley's community-oriented study *Getting Loose: Lifestyle Consumption in the 1970s.*[36] Amid the dislocations of the 1970s, Binkley argues that a broad swath of middle-class Americans felt the need to 'become loose' from the social strictures of prior decades and in this process embarked on projects of creating a more 'durable sense of self.' His analysis focuses on understanding 'everyday patterns of choice, inscribing them as expressive vehicles of a developing subjectivity.'[37] For his historical actors, the means was the production and consumption of a literature of lifestyle manuals and related media, often non-mainstream, that stimulated a sense of individual agency and refashioning as a daily ongoing project. One critical vector in this lifestyle movement was a reoriented relationship to nature, as represented in the imaginaries of Brand's 'Whole Earth' and Buckminster Fuller's 'Spaceship Earth,' which for the subjects in Binkley's study, resulted in a 'new ecological world view [...] in which man and nature were seen in a holistic, integrated totality – a shift in consciousness lived out in a set of ecologically informed styles of life.'[38] These spaceflight tropes, in broad cultural circulation, thus became vital instruments in individual projects of self-definition.

The meaning and import of such images and imaginaries, widely reproduced and consumed, are seen not only as bound up with the problem of subjectivity at the micro level but in its intimate and multiple relations to period capitalism. This connection is particularly explicit in geographer Denis E. Cosgrove's 2004 essay 'Apollo's Eye: A Cultural Geography of the Globe.' He makes the claim that Apollo 8's 'Earthrise' image, taken by American astronauts, highlighted how fundamentally the globe (and the space vantage that produces it) is an imagined object – rather than directly experienced, except for those few who have traveled sufficiently far from earth. Thus:

> If the globe is experienced solely through graphic representations we should pay special attention to the construction of global images and the meanings attached to them. [...] The whole earth especially became a poster for two somewhat divergent visions of globalism. It was appropriated by the environmental movement of the 1970s and 1980s, which drew upon an interpretation of the globe as a vulnerable home of life

– human and non-human – that was in need of protection from the ravages of unrestrained technology, economic development, human greed and exploitation of nature. It was similarly popular with some of the most urgent promoters of technology and development: airlines, communications firms and international finance which drew upon the idea of a global population drawn together in a world without borders and advancing towards a utopian future.[39]

The globe thus becomes an imaginary that both reorders experience and serves as an instrument for advancing particular social or political economic ends – but does so at the very historical moment when these particular images and this imaginary connect with deep questioning of practices of representation and its relation to a globally robust capitalism. These are phenomena, if you will, that are born and given meaning together, a historically specific interpenetration of spaceflight, political economy and subjectivity.

Cosgrove's claim loops together the two strands of this chapter: the import of 1970s and early 1980s theoretical framings on subsequent historical assessments of the 'in-between' decade, and the importance of spaceflight as a motif in both. The relations between these strands are evident in an article by Benjamin Lazier, 'Earthrise; or, The Globalization of the World Picture,' an effort seeking to identify cultural presuppositions that take hold after the Apollo 8 'Earthrise' image in 1968.[40] Lazier's argument is not that the iconic image is merely a defining, constituent part of the post-1970 reality à la Jameson or Cosgrove. The claim is stronger: that *two* images, Earthrise and Blue Marble, in particular, defined and constituted an 'Earthrise era': 'Within the span of a decade, something had changed – evident in philosophical discussion and in Western culture writ large. The Earthrise era had begun.' These images have acquired 'an iconic power that helps organize a myriad of political, moral, scientific and commercial imaginations.'[41] From a different angle, it is an argument that Jameson might make regarding a geopolitical unconscious. But there is a deeper point here: I suspect the *American Historical Review* would not have given credence to the core premise of this argument had it not been for this 1970s-originated scholarship; for the multi-faceted theorizing about the connections between political economy and subjectivity built up over several decades, literature that Lazier barely references but is instrumental to his claim that spaceflight stands at the center of period change.

IV   Theory, history and spaceflight

The 1970s theoretical works considered here (and the many left out) form a genetically related set of projects, all grappling with how to describe and theorize a world many scholars, from a range of disciplines, saw emerging in the 1970s, a world substantively different than that in prior decades. This reconnaissance aimed at presenting two interrelated points. One was to bring forward the main lines of scholarly thought as to what is at stake historically in understanding the 1970s and subsequent decades; on framing core problems of political economy and subjectivity, and, especially, how to conceptualize their interrelation as the panoply of changes wrought by globalization and postmodernity became manifest. The other was to highlight the ways in which spaceflight, as multidimensional period enterprise, became deeply implicated in such explanatory efforts, even if not explicitly invoked.

This overview analysis also has sought to frame a historiographic agenda, in two related parts. For the broader field of historical inquiry on the post-1970 world there has been an emphasis on globality, circulations, flows, the refashioning of spatial and temporal orders and their meaning at varying scales of experience, seeking to relate the individual, the local, regional and planetary. Yet, the obvious relevance of spaceflight to the perceived large-scale transformations of recent decades has been only episodically foregrounded in mainstream historiography.[42] A tighter integration of spaceflight into theorizing on political economy and subjectivity in recent historical experience remains to be done. For specialists in space history, their opportunity is the flipside of this scholarly condition: to assess more systematically the problematic posed by the 1970s and, as critical corollary, take on the not insignificant historical stakes attached to integrating spaceflight into the broader explanatory concerns of history.[43] Such effort, though, on either side is intimately bound up with intellectual developments in the academy through the very same decades under consideration. Change and our characterization and attempts at explanation of such change have been intertwined. Thus, integrating history and spaceflight is not just envisioning the terrain of possible interconnections but understanding the history of the tools and methods through which scholars have sought to make sense of this new order of experience.

Notes

1. For a sampling of scholarship on these various points, see Charles S. Maier, *Among Empires: American Ascendancy and Its Predecessors*, Cambridge, MA: Harvard University Press, 2006; Niall Ferguson, Charles S. Maier, Erez Manela and Daniel J. Sargent, eds, *The Shock of the Global: The 1970s in Perspective*, Cambridge, MA: Harvard University Press, 2010; Reinhold Wagnleitner and Elaine Tyler May, eds, *Here, There, and Everywhere: The Foreign Politics of American Popular Culture*, Hanover: University Press of New England, 2000; Rob Kroes, *If You've Seen One, You've Seen the Mall: Europeans and American Mass Culture*, Urbana: University of Illinois Press, 1996; Ulrich Beck, *Risikogesellschaft: Auf dem Weg in eine andere Moderne*, Frankfurt am Main: Suhrkamp, 1986 (Eng. *Risk Society: Towards a New Modernity*, London: Sage, 1992); and Fredric Jameson, *Postmodernism, or, the Cultural Logic of Late Capitalism*, Durham: Duke University Press, 1991.

2. Maier, *Among Empire*. It should be noted that Maier is using the term 'empire' provisionally as a shorthand for comparing the scale and impact of US power in the twentieth century (particularly post-Second World War) with historical examples of other hegemonic regimes or nation states.

3. The literature on both these organizing motifs now is overflowing. On postmodernity, one of the iconic texts is David Harvey, *The Condition of Postmodernity: An Enquiry into the Origins of Cultural Change*, Oxford: Blackwell, 1989. For a broad assessment of globalization in its post-Second World War incarnation, see George Ritzer, ed., *The Blackwell Companion to Globalization*, Oxford: Blackwell, 2007.

4. It is important to emphasize that this literature sees a strong consonance between change in the United States and Europe; that at a broad structural level each participated, with national variation, in the same political economic and cultural ecosystem. This is the implied argument of *Shock of the Global*. This view is also represented in works of scholars of Europe; see, for example, Hartmut Kaelble, *The 1970s in Europe: A Period of Disillusionment or Promise?*, London: German Historical Institute, 2010; Andreas Wirsching et al., 'The 1970s and 1980s as a Turning Point in European History?,' *Journal of Modern European History* 9.1 (April 2011), 8–26; and Charles S. Maier, 'Two Sorts of Crisis? The "Long" 1970s in the West and the East,' in Hans Günter Hockerts, ed., *Koordinaten deutscher Geschichte in der Epoche des Ost-West-Konfliktes*, Munich: Oldenbourg, 2004, 49–62. See also Alexander Geppert's introduction to this volume.

5. On the neoliberal, see David Harvey, *A Brief History of Neoliberalism*, New York: Oxford University Press, 2005, as well as Philip Mirowski and Dieter Plehwe, eds, *The Road from Mont Pèlerin: The Making of the Neoliberal Thought Collective*, Cambridge, MA: Harvard University Press, 2009. On the surge of interest in history of capitalism, directly derived from the transformation outlined here, see Jeffrey Sklansky, 'The Elusive Sovereign: New Intellectual and Social Histories of Capitalism,' *Modern Intellectual History* 9.1 (April 2012), 233–48. On developments in the academy since 1970, especially those relating social and cultural history to capitalism, see Michael Denning, *Culture in the Age of Three Worlds*, London: Verso, 2004; and William Hamilton Sewell, *Logics of History: Social Theory and Social Transformation*, Chicago: University of Chicago Press, 2005; as well as idem, 'AHR Forum: Crooked Lines,' *American Historical Review* 113.2 (April 2008), 393–405. The latter is an engagement with Geoff Eley, *A Crooked Line: From Cultural History to the History of Society*, Ann Arbor: University of Michigan Press, 2005. Together, these works provide an important tracing of changes from the 1970s forward in academic humanities thought and practice as they relate to the post-1960s condition sketched herein. For a European-centered perspective, also following the chronology outlined here, see Luc Boltanski and Eve Chiapello, *The New Spirit of Capitalism*, London: Verso, 2005. For broader views of capitalism in the twentieth century, see Giovanni Arrighi, *The Long Twentieth Century: Money, Power, and the Origins of Our Times*, London: Verso, 1994; and Jeffry A. Frieden, *Global Capitalism: Its Fall and Rise in the Twentieth Century*, New York: Norton, 2006.

6. The author has taken up this theme in several essays; see, for instance, Martin Collins, 'Community and Explanation in Space History (?),' in Stephen J. Dick and Roger D. Launius, eds, *Critical Issues in the History of Spaceflight*, Washington, DC: NASA, 2006, 603–13.

7. Carolyn Steedman, 'Culture, Cultural Studies, and the Historians,' in Lawrence Grossberg, Cary Nelson and Paula A. Treichler, eds, *Cultural Studies*, New York: Routledge, 1992, 613–22, here 616. In making these points Steedman was drawing on the work of Carolyn Porter and Dominic LaCapra. Steedman's

critique was, in context, particularly apt: *Cultural Studies* was deemed, by practitioners, as a foundational text of the emerging field.

8. Victoria de Grazia, *Irresistible Empire: America's Advance through Twentieth-Century Europe*, Cambridge, MA: Harvard University Press, 2005.

9. Quote from Fredric Jameson, 'The End of Temporality,' *Critical Inquiry* 29.4 (June 2003), 695–718, here 697. Baudrillard, from the 1960s on, moved from a primarily Marxian framework of analysis to one in which culture and communications were his main axis of theoretical engagement; see Jean Baudrillard, *Selected Writings*, Stanford: Stanford University Press, 1988. This point is the consistent theme of Jameson's body of work; as one example, see Fredric Jameson and Masao Miyoshi, eds, *The Cultures of Globalization*, Durham: Duke University Press, 1998.

10. For a useful, synoptic assessment of Benjamin, see Vanessa R. Schwartz, 'Walter Benjamin for Historians,' *American Historical Review* 106.5 (December 2001), 1721–43.

11. Marshall McLuhan, *Understanding Media: The Extensions of Man*, New York: McGraw-Hill, 1964, 47.

12. Clifford Geertz, 'Blurred Genres: The Refiguration of Social Thought,' *American Scholar* 49.2 (Spring 1980), 165–79, here 178–9.

13. The justification for this sampling is these authors' concern with large-scale issues of political economy and its relation to culture and individual identity. A more comprehensive analysis would also bring in theories relating to race and gender. The latter, in contrast to period emphasis on limits, had a strong orientation toward the utopian; see, for example, Tatiana Teslenko, *Feminist Utopian Novels of the 1970s: Joanna Russ and Dorothy Bryant*, New York: Routledge, 2003.

14. As one 1970s period example of this consider Christopher Lasch, *The Culture of Narcissism: American Life in an Age of Diminishing Expectations*, New York: Norton, 1979. Lasch sought explicitly to connect post-industrial political economy with reoriented cultural and personal valuations of life choices, but regarded enhancement of individual agency in less than a positive light. Importantly, his argument, though focused on the United States, saw the same condition in Europe. Though Lasch did not use the word 'postmodern,' the book is an extended lament on the transition from a world organized around modern values to one in which postmodern values were ascendant. Narcissism, for Lasch, stood as an overarching symbol and symptom of the degree to which individual subjectivity had risen as a significant historical phenomenon as part of this condition.

15. This point is central to the works of Jameson already cited. For a periodization of the connection between consumption and cultural and individual imaginaries, see Emily S. Rosenberg, 'Consumer Capitalism and the End of the Cold War,' in Melvyn P. Leffler and Odd Arne Westad, eds, *The Cambridge History of the Cold War*, vol. 3: *Endings*, Cambridge: Cambridge University Press, 2010, 489–512. Tracking the argument here on the 1970s as change-decade and of capitalism's transnational ascendance, Rosenberg offers: 'By the 1980s many variants of consumerist imaginaries circled the globe. [...] Consumerism had become so globalized and diversified that it no longer automatically stirred visions of "Americanization." In many localities the idea of consumer-led growth became incorporated into nationalist programs, and material abundance seemed a test of national success and pride. "Multilocal" consumer revolutions, powered by diverse forms

of consumer nationalism, seemed consistent not only with US-style capitalism but with systems emphasizing varying models of social democracy and even with China's "market socialism."' Ibid., 490, 511. See also Peter N. Stearns, *Consumerism in World History: The Global Transformation of Desire*, New York: Routledge, 2001. On the concept of imaginaries more broadly, see Charles Taylor, *Modern Social Imaginaries*, Durham: Duke University Press, 2004.

16. On this point, see especially the work of Michael C. Behrent, 'Liberalism Without Humanism: Michel Foucault and the Free-Market Creed, 1976–1979,' *Modern Intellectual History* 6.3 (November 2009), 539–68; and idem, 'Foucault and Technology,' *History and Technology* 29.1 (May 2013), 54–104.

17. This observation is not true for some fields of history, particularly intellectual history. With a focus on the United States, see, for example, James Livingston, *Pragmatism and the Political Economy of Cultural Revolution, 1850–1940*, Chapel Hill: University of North Carolina Press, 1994; and Wilfred M. McClay, *The Masterless: Self and Society in Modern America*, Chapel Hill: University of North Carolina Press, 1994. An account that integrates the rise of theory in the 1970s, the arts, and conceptions of self is J. David Hoeveler, *The Postmodernist Turn: American Thought and Culture in the 1970s*, New York: Twayne, 1996. A work including the European context is Jerrold E. Seigel, *The Idea of the Self: Thought and Experience in Western Europe Since the Seventeenth Century*, Cambridge: Cambridge University Press, 2005.

18. Peter N. Carroll, *It Seemed Like Nothing Happened: The Tragedy and Promise of America in the 1970s*, New York: Holt, Rinehart and Winston, 1982; Edward D. Berkowitz, *Something Happened: A Political and Cultural Overview of the Seventies*, New York: Columbia University Press, 2006.

19. This seems to be the case for the British literature on the 1970s; see the review essay by Lawrence Black, 'An Enlightening Decade? New Histories of 1970s' Britain,' *International Labor and Working-Class History* 82 (Fall 2012), 174–86.

20. This is not to say that compelling, quality histories have not been written or that there are not counter examples to this claim. On the former, see, for example, Judith Stein, *Pivotal Decade: How the United States Traded Factories for Finance in the Seventies*, New Haven: Yale University Press, 2010. On a work that falls in-between in the handling of the political economic/subjectivity question, see Jefferson Cowie, *Stayin' Alive: The 1970s and the Last Days of the Working Class*, New York: New Press, 2010. Cowie does indeed seek to link the political economic (particularly regarding labor issues) to the cultural but largely treats the latter as a mirror of the former. As a counter example to my statement, see especially Sam Binkley, *Getting Loose: Lifestyle Consumption in the 1970s*, Durham: Duke University Press, 2007. Binkley's look at lifestyle as a 1970s category takes as his central problem the relation between subjectivity and the political economic. On the 1970s in relation to the 1980s and 1990s, see James Livingston, *The World Turned Inside Out: American Thought and Culture at the End of the Twentieth Century*, Lanham: Rowman & Littlefield, 2010.

21. Quotes from Ferguson et al., *Shock of the Global*, 2. The title of the book was meant to be evocative of an older phrasing, 'shock of the new,' a coinage that referred to the combined impact of industrialization in the 1910s and 1920s, *and* the response to those developments by period artists; that is, it was a shorthand for the relation between political economy and subjectivity in another historical

moment, that of the shock of modernity. See also Alexander Geppert's introduction to the present volume.

22. Another characterization of this transformation related to but covering different intellectual terrain is Paul Forman, 'The Primacy of Science in Modernity, of Technology in Postmodernity, and of Ideology in the History of Technology,' *History and Technology* 23.1 (March/June 2007), 1–152. The shift to a higher cultural regard for technology over science can be ascribed, within the framework I have outlined here to, first, an emphasis of the Cold War state on weapons research and then the rise of a neoliberal, market-oriented value system. More explicit on this connection is Philip Mirowski, *Science-Mart: Privatizing American Science*, Cambridge, MA: Harvard University Press, 2011.

23. Bruce Mazlish, *Conceptualizing Global History*, Boulder: Westview Press, 1993. The relationship between modalities of spaceflight and globalization became a prominent motif of the globalization literature. Not atypical was Mazlish's view that: 'The starting point for global history lies in the following basic facts of our time (although others could be added): our thrust into space, imposing on us an increasing sense of being in one world – "Spaceship Earth" as seen from outside the earth's atmosphere; satellites in outer space that link the peoples of earth in unprecedented fashion'; ibid., 2–3.

24. Both in the United States and Europe the idea of the post-industrial predated the early 1970s, including Bell's own trajectory of interest in the subject. See Howard Brick, 'Optimism of the Mind: Imagining Postindustrial Society in the 1960s and 1970s,' *American Quarterly* 44.3 (September 1992), 348–80.

25. Daniel Bell, *The Coming of Post-Industrial Society: A Venture in Social Forecasting*, New York: Basic Books, 1973.

26. Alain Touraine, *The Post-Industrial Society: Tomorrow's Social History. Classes, Conflicts and Culture in the Programmed Society*, New York: Random House, 1971.

27. See, for example, Mirowski and Plehwe, *Road from Mont Pèlerin*.

28. Raymond Williams, *Keywords: A Vocabulary of Culture and Society*, New York: Oxford University Press, 1976.

29. Raymond Williams and Tony Pinkney, *The Politics of Modernism: Against the New Conformists*, London: Verso, 1996, 121.

30. Fred Turner, *From Counterculture to Cyberculture: Stewart Brand, the Whole Earth Network, and the Rise of Digital Utopianism*, Chicago: University of Chicago Press, 2006.

31. Sabine Höhler, 'The Environment as a Life Support System: The Case of Biosphere 2,' *History and Technology* 26.1 (March 2010), 39–58; idem, *Spaceship Earth in the Environmental Age, 1960–1990*, London: Pickering & Chatto, 2015.

32. Williams's reflections on space infrastructure, already described, are one instance of this tendency. More common is giving salience to spaceflight as postwar historical vector but then not problematizing its integration into period accounts. As a significant example of this, see Akira Iriye, ed., *Global Interdependence: The World after 1945*, Cambridge, MA: Harvard University Press, 2014, esp. the chapter 'The Transnationalism of Humanity,' 681–848.

33. Jean-François Lyotard, *The Postmodern Condition: A Report on Knowledge*, Minneapolis: University of Minnesota Press, 1984, 8.

34. See Fredric Jameson, *The Geopolitical Aesthetic: Cinema and Space in the World System*, Bloomington: Indiana University Press, 1992; and idem, *Archaeologies of the Future: The Desire Called Utopia and Other Science Fictions*, New York: Verso,

2005. It should be noted that many of these readings focus on examples of non-Western cinema and thus are concerned with indigenous responses to the global.

35. Jameson, *Geopolitical Aesthetic*, 4, 13.

36. Sam Binkley, *Getting Loose: Lifestyle Consumption in the 1970s*, Durham: Duke University Press, 2007. Such analysis is mirrored in the British literature as well; see, for example, Laurel Forster and Sue Harper, eds, *British Culture and Society in the 1970s: The Lost Decade*, Newcastle-upon-Tyne: Cambridge Scholars, 2010, especially the essay by Gwilym Thear, 'The Self-Sufficiency Movement and the Apocalyptic Image in 1970s British Culture,' 35–50.

37. Binkley, *Getting Loose*, 129, for all quotes.

38. Ibid., 132.

39. Denis E. Cosgrove, 'Apollo's Eye: A Cultural Geography of the Globe,' Hettner Lecture 2005, available at www.sscnet.ucla.edu/geog/downloads/418/45.pdf, 15 (accessed 1 October 2017). Cosgrove's argument is more fully developed in idem, *Apollo's Eye: A Cartographic Genealogy of the Earth in the Western Imagination*, Baltimore: Johns Hopkins University Press, 2001.

40. Benjamin Lazier, 'Earthrise; or, The Globalization of the World Picture,' *American Historical Review* 116.3 (June 2011), 602–30.

41. Ibid., 605–6.

42. One indicator of this, in the works cited here, is the contrast between Mazlish's *Conceptualizing Global History* and Iriye's *Global Interdependence*; the former gives spaceflight a foundational place in his explanatory model, the latter considers it barely relevant to the global as a historiographic problem.

43. Given this chapter's focus on the broader terrain of thought on the 1970s, it has not sought to provide a review of the space history literature, in which there are many examples of scholarship speaking to the larger concerns of historical inquiry. The point is that, as a field, space history has not given concerted attention to its intellectual agenda vis-à-vis history or the humanities writ large.

CHAPTER 3

Responding to Apollo: America's Divergent Reactions to the Moon Landings

Roger D. Launius

The Apollo program of the 1960s and early 1970s cast a long shadow over expectations for the American exploration of space, a shadow that has yet to be overcome. This chapter focuses on five major responses to Apollo that gained credence in the United States in the 1970s. They remain central to interpreting NASA's human spaceflight efforts since that time. First, the memory of Apollo evoked especially powerful connotations for the majority of Americans as a key moment in the triumphalist, exceptionalist history of the nation overcoming adversity in the face of the Soviet Union's rivalry. American Exceptionalism as a concept emphasized the qualitative difference of the United States to other nations, created through its revolutionary origin, its focus on liberty, egalitarianism, individualism and laissez-faire capitalism. The successful Apollo program became immediately an exemplar of the exceptionalist trope in American history. Indeed, this has become the primary interpretation of the Space Race. It revolves around an initial shock to the system as a challenge from a powerful Soviet Union seemed to overcome American capability and then a whirlwind of activity to recapture the initiative in the realm of space. Ultimately, America is both justified and purified through the challenge of the successful moon landings.[1]

Second, at the beginning of the 1960s, Americans expressed a strong consensus that science and technology, coupled with proper leadership and the inspiration of a great cause, could solve almost any problem of society. It

Roger D. Launius (✉)
Launius Historical Services, Auburn, AL, USA
e-mail: launiusr@gmail.com

Alexander C.T. Geppert (ed.), *Limiting Outer Space*
European Astroculture, vol. 2
https://doi.org/10.1057/978-1-137-36916-1_3

was that faith, as well as the Cold War necessity of undertaking something spectacular to overshadow the Soviet Union, that sparked the 1961 Kennedy decision to go to the moon and to empower experts, in this case aerospace engineers, with the decision-making responsibility and wherewithal to execute the Apollo program.[2] By 1970 this commitment to scientific and technological answers had waned and a resurgence of an almost populist belief in the right, indeed the duty, of ordinary citizens to control all affairs, including those of a scientific and technological nature, was gaining pre-eminence. This came about for many reasons; these ranged from the loss of a technological war in southeast Asia to a 'small is beautiful' dominant ideal to an emerging sense of resource limitations. All of this had a profound impact on the course of human space exploration in the 1970s as NASA curtailed its expansive human space exploration agenda.

Third, the true believers of human space exploration and colonization – having been energized by the success of Apollo and the promise inherent in it of becoming a multi-planetary species – just as roundly had their hopes dashed by the decision of American political leaders to retreat from the aggressive space exploration agenda of the 1960s into something much less adventuresome and closer to home. These orphans of Apollo responded by creating pressure groups to lobby elected officials and NASA leadership; the result was the organization of such interest groups as the Committee for the Future, the L5 Society and the National Space Institute. They failed to achieve their ends through the public policy arena, and over time some abandoned support for the government's spaceflight program and charted a course toward private-sector space efforts. A libertarian strain of spaceflight advocacy emerged from some of these early supporters in the 1970s, and accelerated in the 1980s and 1990s, that questioned the official NASA programs while pursuing more commercial efforts to reach space. These efforts emerged full-blown in the last part of the twentieth century as entrepreneurial firms sought to recapture the initiative lost at the end of the Apollo program.[3]

Fourth, almost from the point of the first Apollo missions, some have denied that it had taken place at all. In some instances this became official state policy, as in Cuba, in which Castro refused to acknowledge the American success and denied that the United States had ever landed on the moon. This idea expanded in the America of the 1970s, as NASA withdrew from its deep space human exploration efforts in favor of earth orbital operations and methods of communication enhanced the ease of spreading such ideas. This perspective has risen in importance as the events of Apollo recede into history. Aided by a youth movement that does not remember what went down in the Apollo era and for whom distrust of government runs high, it is among that cadre of Americans where those who are skeptical have proliferated.

Finally, there arose in the 1970s – and there are vestiges of this dynamic in every other aspect of the themes discussed in this chapter – an Apollo nostalgia as Americans look back toward an increasingly distant time. This nostalgia glorifies a sense of the past in which society, culture, economics, politics and other attributes of the public sphere seemed to work better. Apollo nostalgia manifests itself in several ways. It revolves around the issue of mythical recreations of a bygone era before the world went off its course. The expression of public comfort with a white male establishment is palpable throughout these recountings of the story of Apollo. The quintessential company man worked for NASA during Apollo. The engineering 'geeks' of Mission Control, with their short-sleeved white shirts, narrow black ties, slide rules hung on their belts like side arms, and their pocket protectors complete with compass and ruler and myriad pens and mechanical pencils all personified a conservative America that many looked back on with fondness and nostalgia. The astronauts reflected the cultural norms of the time and served as virtuous knights battling the forces of the communist empire. Collectively, NASA's moon program is nostalgically remembered as a time in which the nation could accomplish any task it set for itself, and that nostalgia has grown with time.

I   Apollo as an expression of American exceptionalism

Apollo represented, for many observers since the first landing, and perhaps consistently for the majority of the American public to the present, an epochal event that signaled the opening of a new frontier in which a grand visionary future for Americans might be realized. It represented, most Americans have consistently believed, what set the United States apart from the rest of the nations of the world. American exceptionalism reigned in this context, and Apollo is often depicted as a great event in American history, one that must be revered because it shows how successful Americans can be when they try. Militarism and imperialism are the fundamental building blocks of the exceptionalist idea in American history, and Apollo fulfilled both of them in ways that were unique and inviting. The Space Race, of which Apollo was the critical component, was war by another means and the besting of the Soviet Union in reaching the moon served as a statement of military capability. At the same time, the moon landings represented a form of imperialism – such as the planting of the American flag on the moon – but bald-faced imperialism was muted by such statements as that included on the plaque affixed to the Lunar Module that read, 'We Came in Peace for All Mankind.'[4] At a basic level, therefore, Apollo offered an exceptionalist perspective that has dominated its public characterizations from the beginning to the present, regardless of the form of those characterizations.[5]

The moon landings have persistently represented a feel-good triumph for the nation and its people. Certainly Apollo represented this in the imagery that became iconic in the public consciousness – an astronaut on the moon saluting the American flag served well as a patriotic symbol of what the nation had accomplished (Figure 3.1). This self-image of the United States as a successful nation gained affirmation in the Apollo program.[6] At sum,

Figure 3.1 Astronaut John W. Young (1930–2018), commander of the Apollo 16 lunar landing mission, jumps up from the lunar surface as he salutes the US flag at the Descartes landing site during the first Apollo 16 extravehicular activity (EVA-1) on 21 April 1972. The planting of the flag and the astronaut salute were two of the most significant ceremonial events of every landing and have remained a powerful reminder of American exceptionalism and nostalgia for the past. The flag, seemingly blowing in the breeze, has been questioned by moon landing deniers as evidence that the scene was filmed in a sound stage and not on the moon. In actuality, with the 1/6 gravity and the lack of an atmosphere on the moon, the flag continues to wave for some time after it was planted on the surface for this photograph.
Source: Courtesy of NASA.

Americans largely viewed Apollo as a result of a grand visionary concept for human exploration that could be directly traced to the European voyages of discovery beginning in the fifteenth century.[7] Celebrants of Apollo have long argued that returns on investment in this age of exploration changed Americans' lives.[8] As President Lyndon B. Johnson remarked at the time of the third Gemini flight in August 1965, 'Somehow the problems which yesterday seemed large and ominous and insoluble today appear much less foreboding.' Why should Americans fear problems on earth, he believed, when they had accomplished so much in space?[9] In this triumphalist narrative, the moon landings demonstrated that anything we set our minds to we could accomplish. 'If we can put a man on the moon, why can't we [...]?' entered the public consciousness as a statement of unlimited potential.[10]

Apollo has remained a powerful trope of American exceptionalism to the present. It seems obvious that the fierceness of debates over the nature of national character has arisen from a desire to secure a national identity of one-nation, one-people coupled with a concern that the bulwarks of appropriate conceptions may be crumbling.[11] The interpretation of Apollo as an expression of American exceptionalism fits beautifully into a larger narrative of American awesomeness. It is one of an initial shock to the system, surprise and ultimately recovery with success after success following across a broad spectrum of activities. It offers general comfort to the American public as a whole and an exceptionalistic, nationalistic and triumphant model for understanding the nation's past. Small wonder that this interpretation of Apollo emerged to dominate the 1970s and since. It offered a subtle, usable past for the nation as a whole.

Two anecdotes drawn from television situation comedies airing thirty years apart emphasize the American exceptionalism present in the moon landings. Actor Carroll O'Connor perhaps said it best in an episode of *All in the Family* in 1971. Portraying the character of Archie Bunker, the bigoted working-class American whose perspectives were more common in American society than many observers were comfortable with, O'Connor summarized well how most Americans responded to Apollo. He observed that he had 'a genuine facsimile of the Apollo 14 insignia. That's the thing that sets the US of A apart from [...] all them other losers.'[12] In very specific terms, Archie Bunker encapsulated for everyone what set the United States apart from every other nation in the world: success in spaceflight. At a basic level Apollo provided the impetus for the perception of spaceflight as a great positive for the nation.

The second anecdote, made thirty years later, suggests that not much has changed. In the critically acclaimed television situation comedy about a team that produces a nightly cable sports broadcast, *Sports Night*, one episode included simply as a sidebar a discussion of space exploration. The fictional sports show's executive producer, Isaac Jaffee, played by renowned actor Robert Guillaume, was recovering from a stroke and disengaged from the daily hubbub of putting together the nightly show. His producer, Dana Whitaker, played by Felicity Huffman, kept interrupting him in this episode as he was reading a magazine about space exploration. The exchange is noteworthy. Isaac tells her, 'They're talking about bio-engineering animals and terraforming Mars. When I started reporting Gemini missions, just watching a Titan rocket liftoff was a sight to see. Now they're going to colonize the solar system.' Dana suggests that perhaps Isaac is obsessing about this and he agrees. So Dana asks why he is so obsessed. Quietly, Isaac responds, 'Because I won't live to see it.' It is a touching conversation about hope and aspirations and mortal limitations. But more than that, Isaac Jaffee affirms his fundamental faith in the importance of space exploration and the America's capabilities. 'You put an X anyplace in the Solar System,' he says, 'and the engineers at NASA can land a spacecraft on it.'[13] Nothing more effectively states the sense of American exceptionalism present in the Apollo experience.

That exceptionalist narrative found expression not just in the United States, but also around the world. One of the objectives of the Apollo program was to demonstrate American technological superiority as a means of bringing allies to the Western camp during the Cold War. It was successful in achieving this objective. The flight of Apollo 11 met with an ecstatic reaction around the globe, as everyone shared in the success of the astronauts. The front pages of newspapers everywhere suggested how strong worldwide enthusiasm was. NASA estimated that because of nearly worldwide radio and television coverage, more than half the population of the planet was aware of the events of Apollo 11. Although the Soviet Union tried to jam *Voice of America* radio broadcasts, most living there and in other countries learned about the adventure and followed it carefully. Police reports noted that streets in many cities were eerily quiet during the moon walk as residents watched television coverage in homes, bars and other public places.[14] The Apollo missions to the moon caused that type of experience for millions of others in other nations. The world paused in July 1969 when Neil Armstrong first set foot on the moon. One seven-year-old boy from San Juan, Puerto Rico, said of the first moon landing: 'I kept racing between the TV and the balcony and looking at the Moon to see if I could see them on the Moon.'[15] His experiences proved typical.

Official congratulations poured in to the US president from other heads of state, even as informal ones went to NASA and the astronauts. All nations having regular diplomatic relations with the United States sent their best wishes in recognition of the success of the mission. By demonstrating technological success in the moon landings NASA served a valuable foreign policy objective as it helped keep other nations in the United State's corner during the Cold War. The experience of the People's Republic of China is instructive. Those without diplomatic relations with the United States, such as China, made no formal statement on the Apollo 11 flight to the United States, and the mission was reported only sporadically by its news media because Mao Zedong refused to publicize successes by Cold War rivals. It was not until February 1972, when Nixon flew to China and met with Mao Zedong, that the United States established formal diplomatic relations with the nation. China now seeks to go to the moon, fully recognizing the success of the Apollo program.[16]

II   Apollo and the waning of technocratic faith

The 'New Frontier' of the Kennedy Administration and the 'Great Society' of the early Johnson years represented what might be characterized as the high tide of a technological hubris that had been building for much of the twentieth century, and its manifestation may be found in the Apollo program that took Americans to the moon in the 1960s and early 1970s (Figure 3.2). Nurtured in the political climate of progressivism at the turn of the century, with its emphasis on professionalism and expertise, Americans had long been taught that scientific and technological knowledge could solve almost any problem. Social reformer Edward Scribner Ames (1870–1958), for instance, reflected just after the First World War that the best way to prevent such massive destruction again was to pursue science and technology. He believed scientists had 'saved the western world' from conquest with their weapons. Moreover, science and technology gave American society all types of new conveniences, medical capabilities and improvements to life. Indeed, no more significant evidence of the value of science and technology to resolve national ills could be found than their service in the Second World War.[17]

The immediate postwar era found the application of wartime mobilization models for science also being applied to peacetime problems. In 1952 Edward Everett Hazlett (1892–1958) wrote to presidential candidate Dwight D. Eisenhower about declaring 'War on Untimely Death.' He suggested that a widespread government effort to 'smash the atoms' of disease 'seems no more likely to fail than did that on the atom. It has, in addition, the spiritual advantage of being a campaign to save life and not to take it.'[18] Such faith in science and technology motivated all manner of activities in the twenty years after the Second World War, and government officials yielded to the authority of experts with something akin, according to James B. Conant, to 'the old religious phenomenon of conversion.'[19]

These perspectives were also present in the Kennedy Administration. David Halberstam has shrewdly observed that 'if there was anything that bound the men [of the Kennedy Administration], their followers, and their subordinates together, it was the belief that sheer intelligence and rationality could answer and solve anything.'[20] This translated into an ever-increasing commitment to science and technology to resolve problems and point the direction for the future. They took that approach with international relations, and the space program and the techno-war in Vietnam were two direct results.[21] The NASA Administrator in the 1960s, James E. Webb (1906–92), became the high priest of technological efforts to resolve national problems. He argued for a scientific management approach that could be used to reduce all problems to a technological common denominator and then to overcome them. Webb wrote as late as 1969 that 'our Society has reached a point where its progress and even its survival increasingly depend upon our ability to organize the complex and to do the unusual.' Proper expertise, well-organized and led, and with sufficient resources could resolve the 'many great economic, social, and political problems' that pressed the nation.[22]

Figure 3.2   The handsome German émigré Wernher von Braun (1912–77) became one of the most recognized individuals in the 1950s and 1960s as he championed an aggressive spaceflight program that found realization in Project Apollo. Here he stands in front of one rocket that his team built, the Saturn IB, at Kennedy Space Center launch complex in January 1968. After building the V-2 for Nazi Germany during the Second World War, von Braun led a team of German rocket engineers to the United States where he became a leader in the US space program. As director of NASA's Marshall Space Flight Center in Huntsville, Alabama, he oversaw development of the Mercury-Redstone, which put the first American into space; and later the Saturn rockets, Saturn I, Saturn IB, and Saturn V. The Saturn V launch vehicle put the first human on the surface of the moon, and a modified Saturn V vehicle placed Skylab, the first US space station, into earth orbit. An eloquent spokesperson for space exploration, von Braun quipped that the hardest thing about it may have been the paperwork. *Source*: Courtesy of NASA.

By the time of the completion of Apollo, however, perspectives had shifted during a complex debate over the vision of American as it had been articulated for more than a generation. This sustained criticism of national character and meaning plunged the United States into fundamental changes, political turmoil and activism of all stripes, and a counter culture that rejected middle-class perceptions and social construction during the 1960s. Especially significant was the violence that erupted both at home and abroad for all to see. Indicative of this discord, in March 1968, President Lyndon B. Johnson – who had been both a hardheaded politician and committed to US primacy overseas and social reform at home – formally abandoned any hope of reelection citing a lack of support. Fritjof Capra's representative definition of a social paradigm is appropriate when considering the changes that took place in the decade: a radical alteration in the 'constellation of concepts, values, perceptions and practices shared by a community, which forms a particular vision of reality that is the basis of the way the community organizes itself.'[23]

NASA Administrator James C. Fletcher (1919–91) was the first to state publicly that the disarray present in American culture during the 1960s held specific negative aspects for NASA. In September 1971 he identified what he considered the most pressing problem facing his agency in the 1970s, 'the fact that this country seems to be on an antitechnology kick.' He commented: 'I admit we might have spent too much on technology during the peak period of the Apollo program and in connection with Vietnam. Now we are overreacting.'[24] Fletcher had hit upon an important debate that was taking place in the late 1960s and early 1970s over the proper role of electoral authority versus technological expertise in the democratic process. At least by the time of the Tet Offensive in Vietnam in 1968, however, it was clear to many Americans that science and technology did not hold all the answers that had been promised. All the bomb tonnage, all the modern military equipment and all the supposed expertise had not defeated the simply clad and armed North Vietnamese. Neither had it been capable of eradicating disease, ending world hunger, resolving racial strife, stamping out poverty, fostering human equality, enhancing the level of education and settling a host of energy and ecological issues (Figure 3.3).

In too many instances, science and technology were viewed as fundamental parts of the problems and not as the solutions that they had once seemed. Scientist/philosopher Jacob Bronowski (1908–74) voiced concerns about the promise of science and technology in a democratic environment. 'A world run by the specialists for the ignorant is, and will be,' Bronowski insisted, 'a slave world.'[25] Economist John Kenneth Galbraith (1908–2006), certainly not a radical bent on completely altering American society, aired similar concerns. 'If we continue to believe that the goals of the industrial system – the expansion of output, the companion increase in consumption, technological advance, the public images that sustain it – are coordinate with life, then all of our lives will be in the service of these goals,' he wrote in 1975. 'What is

Figure 3.3 This juxtaposition of a mule pulling a plow at NASA's Mississippi Test Facility where rockets were fired suggests the deep ironies of a futuristic program in the 1960s at the same time that the Civil Rights Crusade sought to overcome centuries of prejudice and racism. This image was taken at the Mississippi Test Facility in March 1970.
Source: Courtesy of NASA.

consistent with these ends we shall have or be allowed; all else will be off limits.'[26] Representatives' more radical perspectives were even more condemnatory of technical and scientific expertise and its application in modern society. This partly was the result of a belief that the expansion of the federal government's role in the lives of individual Americans represented bad government and that the Jeffersonian tradition of local politics and democratic responsiveness to the needs of individuals was being seriously threatened. The most radical critics called for, among other things, a return to a more simple existence – the most drastic application of which were communes of all types – and a retaking of the political system from the experts.[27]

When Richard M. Nixon was elected to the American presidency in the fall of 1968, it seemed that anti-technology bias had gained ascendancy in the White House. He was widely viewed as a reactionary 'cold warrior' who was both 'intellectually and emotionally at odds with what he viewed as a liberal east-coast-based media, academic and bureaucratic establishment.'[28] Part of that mistrust was aimed at scientific and technological experts for not being responsive to the needs of the nation. They were unable to meet social and economic needs, and he posited a return to direct political action to wrest control from the technocrats. Nixon stated that a major part of his program, known collectively as 'The New Federalism,' 'is to help regain control of our national destiny by releasing a greater share of control to state and local governments and to the people.'[29] Nixon's abolition of the Office of the Science Advisor in 1973 signaled for many, said one newspaper specializing in these issues in Washington, 'a reversal of the emphasis on science and technology generated 15 years ago.'[30] At the same time, closer to home for NASA and Space Shuttle fortunes, he abolished the National Aeronautics and Space Council, the body that had overseen all space activities of the federal government.[31]

James Fletcher, Nixon's appointee as NASA Administrator, entered the position, at least in part, because he was believed to be politically malleable and would accept Nixon's priorities for a lesser role for the space agency. In a memo to one of the President's staff, White House staffer Clay Whitehead commented in 1971 that 'NASA is – or should be – making a transition from rapid razzle-dazzle growth and glamour to organizational maturity and more stable operations for the long term.' In seeking a new head for NASA, Whitehead wrote:

> We need a new Administrator who will turn down NASA's empire-building fervor and turn his attention to (1) sensible straightening away of internal management and (2) working with OMB and White House to show us what broad but concrete alternatives the President has that meet all his various objectives. In short, we need someone who will work with us rather than against us, and will seek progress toward the President's stated goals, and will shape the program to reflect credit on the President rather than embarrassment.[32]

In such a setting, there was no prospect for NASA undertaking a broad space exploration agenda that followed on the success of the Apollo program. As a result, NASA thrashed about for the development of a new human spaceflight program for the post-Apollo era.

The result was a much less ambitious Space Shuttle effort that flew for thirty years (1981–2011) but kept humanity locked in low-earth orbit. The revulsion toward technological perspicacity fueled this decision. Moreover, the Space Shuttle decision has been condemned by many in the pro-space community as a 'policy failure' that forced NASA into a strange orbital purgatory for more than a generation. A major statement of that position came in the fall of 2005 when the NASA Administrator of some six months, but a long-time member of the space community, publicly called the Space Shuttle a mistake. Michael D. Griffin (1949–) commented that NASA had pursued the wrong path with the Shuttle when conceived in the 1960s and developed in the 1970s, and persisted with it long after its flaws had been discovered. That poor decision now had to be corrected, he noted, albeit more than thirty years after the fact. 'It is now commonly accepted that was not the right path,' Griffin told USA Today in an interview that appeared as a page-one story on 28 September 2005: 'We are now trying to change the path while doing as little damage as we can.' When asked pointedly if the Shuttle had been a mistake the NASA Administrator responded: 'My opinion is that it was […] a design which was extremely aggressive and just barely possible.'[33] Griffin's assertion that the Shuttle had been the 'wrong path,' a mistake persisted in for more than a generation, set off a firestorm of debate within the spaceflight community to the extent that NASA issued a point paper explaining what Griffin had meant.[34] The technological faith that had taken humanity to the moon faced a declension throughout the 1970s and thereafter, in the view of those such as Michael Griffin.

III   Losing the space boosters

As Project Apollo, the remarkable national commitment to land an American on the moon by the end of the 1960s, ended with the Apollo 17 mission in December 1972, space enthusiasts believed they were on the verge of a new golden age in which anything could be accomplished. Apollo raised the hopes of those dreaming of a great human migration into space. Its transcendental qualities were not lost on those who believed that the human race could eventually attain this end. Movement into space, first with exploring expeditions and later with colonies, offered an opportunity for humanity to rise above the issues that divided it, to overcome the centuries of baggage that confounded it, and to defeat the inequities of the political and economic systems that restricted it. Society could move outward and start anew on a pristine planet. Apollo had shown it was possible. It suggested that America had both the capability and the wherewithal to accomplish truly astounding goals. All humans needed was political will.

That political will did not exist, for reasons both great and small. The major program of NASA for the decade became the development of a reusable Space Shuttle that was intended to travel back and forth between earth and space more routinely and economically than ever before. The rationale for this vehicle was decidedly utilitarian.[35] At the same time, the NASA budget continued its decline and the prospects for human exploration of the solar system seemingly went to nil. The Shuttle became the largest, most expensive and highly visible project undertaken by NASA after its first decade and it continued to be a central component in the US space program until July 2011. The Space Shuttle accomplished much, but none of it excited either the general public or the space advocate community in the way that the moon landings had. There are arguably four core accomplishments of the

Space Shuttle era. First, the Space Shuttle provided more than two decades of significant human spaceflight capability and stretched the nature of what could be accomplished in earth orbit much beyond what it had previously been. Second, the Space Shuttle served as a relatively flexible platform for scientific activities. Third, and perhaps most significantly since the American human spaceflight program has always been focused on national prestige, the Space Shuttle served well as a symbol of American technological excellence.[36]

By far the most important accomplishment, however, was the one that Richard Nixon promised in approving it in 1972. It made trips to and from low-earth orbit routine. It succeeded in incorporating into the normal realm of human operations the area from 100 to 500 kilometers about the earth. In an irony of the first magnitude for those who insist that humans are not truly engaged in space exploration because they have been confined to low-earth orbit for the last 40 years since the end of Apollo, the Space Shuttle turned orbital space into a place that was no longer a frontier. The first astronauts and cosmonauts, of course, truly were explorers in the traditional sense of the term, pushing back the frontiers of knowledge about this unusually strange, harsh and hostile environment. They learned how to operate there and to do useful things in the region. The astronauts and cosmonauts of the Shuttle era made possible the mundane activities of spaceflight. In essence, NASA's Space Shuttle program succeeded in making the most exciting activity imaginable seem boring.[37]

Space boosters sought to overcome what they viewed as a poor state of space exploration in the early 1970s by organizing political action groups, known collectively as the 'pro-space movement,' to lobby for increased federal funding for spaceflight. Most of these elements joined with the traditional space lobby in Washington, made up of major interest groups that had long been arguing for increased governmental funding. Scientists, military officials, aerospace industrialists and members of the federal space policy-making community welcomed these activists into their ranks at first, but soon they recoiled from the more radical notions expressed in their ideology. Many observers made fun of the wildest elements in the pro-space movement for their 'weirdness,' criticizing their affection for 'Star Trek' uniforms and belief in UFOs.[38]

The formal beginnings of the modern 'pro-space movement' may be best traced to the June 1970 formation of the Committee for the Future (CFF), a small group of space activists, dreamers and misfits. Meeting in the home of Barbara Marx Hubbard (1929–), daughter of the toy king, and her husband, artist-philosopher Earl Hubbard, in Lakeville, Connecticut, they proposed creating a lunar colony. They unabashedly offered this as a great experiment in which humanity, free from the constraints of everyday society, could create a perfect community. The CFF's charter voiced these ideals clearly:

Earth-bound history has ended. Universal history has begun. Mankind has been born into an environment of immeasurable possibilities. We, the Committee for the Future, believe that the long-range goal for Mankind should be to seek and settle new worlds. To survive and realize the common aspiration of all people for a future of unlimited opportunity, this generation must begin now to find the means of converting the planets into life support systems for the race of Men.

They concluded, 'A Challenge of this magnitude can emancipate the genius of Man.' They also offered shares in the lunar colony to millions of investors, immediately creating a constituency that could lobby Congress for funding for bold space ventures.[39]

They convinced Representative Olin E. Teague (1910–81), a lion within the Democratic Party and a long-time supporter of Apollo, to sponsor a resolution calling for a study of the feasibility of this lunar effort. When NASA, the aerospace industry and the science community opposed the resolution, fearing that it might jeopardize other plans, it died a prompt death in Congress. CFF then rewrote the bill to propose a 'citizens in space' mission in low-earth orbit, called 'Mankind One,' but NASA opposed that as well and it met a similar fate. Many within NASA apparently agreed with the ideology of the CFF, although they eschewed its political strategies. Barbara Hubbard wrote how upon first meeting Christopher C. Kraft, director of the Manned Spacecraft Center (renamed the Johnson Space Center in 1973), he told her, 'This step into the universe is a religion and I'm a member of it.' Hubbard was deeply troubled, however, by the reaction of NASA to the CFF's proposals. She wrote, 'The corporate decision of NASA as a government agency was less responsive than the decision of any of its individual members.'[40] One may trace to this incident the beginnings of a deep-seated wariness of government as a whole and NASA in particular to 'do the right thing' in opening the space frontier that is presently held by many in the pro-space movement.

Organizing symposia, called 'synergistic convergences' or SYNCONs, and publishing literature about a hopeful future in space, members of the CFF converted a sizable group of mostly young people to a hopeful utopian future in space. Space groupies came from everywhere to participate in the SYNCONs, some wearing 'Star Trek' uniforms, energizing a loyal base of activists who firmly believed that only through space settlement would the human destiny of a perfect society be realized.[41] While the Committee for the Future ceased to exist as a separate organization in the mid-1970s, Barbara Hubbard has continued her commitment to a utopian future in space to the present.[42] This comment from 1995 explained her position on the promise of space and her belief that governments have held back its attainment:

Perhaps the last great function of existing centralized power, as in the United States and the Soviet Union, is the establishment of the first productive foothold in the Universe. After the step, the higher frontier will open to cooperative, free enterprise and self-selected groups of pioneers. The opportunity for new life styles, new wealth, new knowledge gained by highly motivated people beyond the planet will reduce the control of Earth-bound governments everywhere. [...] Imagine a world in which individual men, women and children are liberated from the past phase of creature human functions or maximum reproduction and survival tasks. Our daily survival needs for food and shelter are provided with minimum human effort. The productive capacity of a universal species, utilizing intelligent machines and renewable resources in an Earth-space environment, is astronomical.[43]

Most assuredly the space professionals at NASA and in industry considered Barbara Hubbard and the CFF, both then and now, 'wacky' and without substance. Some considered her a cult leader.

The CFF represented a strain of spaceflight enthusiasm that could not be ignored, one that emphasized individual activism apart from NASA. It gained a greater respectability when espoused by Princeton academic Gerard K. O'Neill (1927–92). Infected with the enthusiasm of Apollo, O'Neill undertook a set of studies aimed at answering the question, 'Is a planetary surface the right place for an expanding technological civilization?' He found that possibilities for human colonies in free space seemed limitless, as he calculated the technical issues of energy, land area, size and shape, atmosphere, gravitation and sunlight necessary to sustain a colony in an artificial living space. Rather than live on the outside of a planet, settlers could live on the inside of gigantic cylinders or spheres of roughly one-half to a few miles in each dimension. These would hold a breathable atmosphere, all the ingredients necessary for sustaining crops and life, and include rotating habitats to provide artificial gravity. While the human race might eventually build millions of these space colonies, each settlement would of necessity be an independent biosphere with trees and lakes and blue skies spotted with clouds along each colony's inner rim where all oxygen, water, waste and other materials could be recycled endlessly. Animals and plants endangered on earth would thrive on these cosmic arks; insect pests would be left behind. Solar power, directed into each colony by huge mirrors, would provide a constant source of non-polluting energy.[44]

This bold vision catapulted O'Neill into the spotlight of the space community and prompted a collective swoon from those attracted by the CFF but repelled by its zaniness. They formed the L5 Society in 1975, and adopted the slogan, 'L-5 in 1995.' A particularly attractive group of space activists, one of their members wittily opined that they intended to 'disband the Society in a mass meeting at L-5.'[45] The space settlement mission also received a major boost from numerous science-fiction and science-fact writers, among them Arthur C. Clarke (1917–2008), who popularized O'Neill's concept for colonies in space.[46] The strongly utopian impulse present in the O'Neill movement found voice in the words of aerospace writer Thomas A. Heppenheimer (1947–2015). 'On Earth it is difficult for [...] people to form new nations or region[s] for themselves. But in space it will become easy for ethnic or religious groups, and for many others as well to set up their own colonies,' Heppenheimer wrote in 1977. 'Those who wish to found experimental communities, to try new social forms and practices, will have the opportunity to strike out into the wilderness and establish their ideals in cities in space.'[47]

O'Neill's vision of a practical and profitable colony in space found an audience in many quarters of NASA even as it did so in the larger pro-space movement. He received funding from NASA's Advanced Programs Office – but only $25,000 – to develop his ideas more fully. Senior NASA officials such as Administrator James Fletcher and Ames Research Center Director Hans Mark (1929–) encouraged his efforts.[48] In the summers of 1975 and 1976, NASA officials took O'Neill's ideas seriously enough to convene a study group of scientists, engineers, economists and sociologists at the Ames Research Center, near San Francisco, to review the idea of space colonization, and followed it up with a study the next summer. Surprisingly, they found enough in the scheme to recommend it. Although budget estimates of $100 billion accompanied any colonization project, the authors of this study concluded, 'in contrast to Apollo, it appears that space colonization may be a paying proposition.' For them, it offered 'a way out from the sense of closure and of limits which is now oppressive to many people on Earth.' The study recommended an international project led by the United States that would result in the establishment of a space colony at L5.[49]

Following closely on the heels of these two space colonization summer studies, in 1977 O'Neill founded the Space Studies Institute at Princeton University with the intention of organizing small groups into teams to develop the 'tools of space exploration independently of governments and to prove that private groups could get things done enormously cheaper and quicker than government bureaucracies.' He never strayed from his belief that the private sector would be the only organization to open the space frontier. Indeed, he was appointed to the 1985 National Commission on Space in no small part because of his expansive vision of space colonization and because of his increasingly dogged commitment to space colonization and entrepreneurship.[50]

O'Neill's ideas of space colonization were born in the first half of the twentieth century and made seemingly achievable with the Apollo moon landings; the declension NASA accepted in the 1970s prompted a pro-space movement that argued for a bountiful future on a new and pristine planet, with or without government involvement. These individuals are the orphans of Apollo. They found their way into myriad economic, political and social camps. They evinced distrust of authority, especially government authority, and celebrated the entrepreneurial spirit of such individuals as Elon Musk and Sir Richard Branson whom they believe will finally open the space frontier. They may support NASA's efforts when they converge with their own, but they are deeply critical of much of what large government bureaucracies attempt.

IV   Denying the moon landings

One of the truly most remarkable outcomes of the 1970s was the emergence of a small cadre who denied the reality of the Apollo moon landings. It began slowly but as methods of communication have proliferated, and standards of journalism have atrophied, the conspiracy ideologues and believers have proliferated. Of course almost from the point of the first Apollo missions a small group of Americans have denied that it had taken place at all. Apollo 8 astronaut William 'Bill' Anders (1933–) thought live television would help convince skeptics since watching 'three men floating inside a spaceship was as close to proof as they might get.'[51] He could not have been more wrong. Fueled by conspiracy theorists of all stripes, this number has grown over time. In a 2004 poll, while overall numbers remained about the same, among Americans between 18 and 24 years old, '27% expressed doubts that NASA went to the Moon,' according to pollster Mary Lynne Dittmar. Doubt is different from denial, but it was a trend that seemed to be growing over time among those who did not witness the events.[52]

How widespread were the skeptics about the moon landings in the 1960s? That is almost impossible to say. For example, *New York Times* science reporter John Noble Wilford remarked in December 1969 that 'a few stool-warmers in Chicago bars are on record as suggesting that the Apollo 11 moon walk last July was actually staged by Hollywood on a Nevada desert.'[53] More important, the *Atlanta Constitution* led a story on 15 June 1970, with: 'Many skeptics feel moon explorer Neil Armstrong took his "giant step for mankind" somewhere in Arizona.' It based its conclusion that an unspecified 'many' questioned the Apollo 11 and 12 landings, and presumably the April 1970 accident aboard Apollo 13, on an admittedly unscientific poll conducted by the *Knight Newspapers* of 1,721 US citizens in 'Miami, Philadelphia, Akron, Ohio, Detroit, Washington, Macon, Ga., and several rural communities in North and South Carolina.' Those polled were asked, 'Do you really, completely believe that the United States has actually landed men on the moon and returned them to earth again?' While numbers questioning the moon landing in Detroit, Miami and Akron averaged less than five percent, among African Americans in such places as Washington, DC, a whopping 54 percent 'doubted the moon voyage had taken place.' That perhaps said more about the disconnectedness of minority communities from the Apollo effort and the nation's overarching racism than anything else. As the story reported, 'A woman in Macon said she knows she couldn't watch a telecast from the moon because her set wouldn't even pick up New York stations.'[54]

The first conspiracy theorist to make a sustained case for denying that the United States had landed on the moon was Bill Kaysing (1922–2005), a journalist who had been employed for a few years in the public relations office at Rocketdyne, Inc., a NASA contractor, in the early 1960s. His 1974 pamphlet, *We Never Went to the Moon*, laid out many of the major arguments that have been followed by other conspiracy theorists since then.[55] Kaysing's rationale for questioning the Apollo moon landings offered poorly developed logic, sloppily analyzed data and sophomorically argued assertions. Kaysing believed that the failure to land on the moon all sprang from the fact that NASA lacked the technical expertise to accomplish the task, requiring the creation of a massive cover-up to hide that fact. He cited as evidence optical anomalies in some imagery from the Apollo program, questioned the physical features of certain objects in the photographs (such as a lack of a star field in the background of lunar surface imagery and a presumed waving of the US flag in an airless environment), and challenged the possibility of NASA astronauts surviving a trip to the moon because of radiation exposure.[56]

Others have followed in Kaysing's footsteps, arguing one conspiracy or another, none with compelling evidence and often with nothing that might be considered anything more than assertions. Skepticism has proliferated within a youth movement that does not remember what went down in the Apollo era and for whom distrust of government runs high. Jaded by so many other government scandals and such real conspiracies as those that took the United States into war in Iraq and fostered an economic crisis, these younger members of society (whose recollection of Apollo is distant to begin with and receding into the background quickly as time progresses) find it easy to believe the seemingly rational questioning they are exposed to from myriad moon hoax advocates. Lack of understanding of science and failure to employ critical analytical skills make them more susceptible to this type of hucksterism.

While it is impossible for most Americans to take this denial seriously and opinion surveys show consistently that few do so, for those raised in the postmodern world of the present where the nature of truth is so thoroughly questioned, it is more likely to gain a footing. Indeed, postmodernism suggests that reality is more a suggestion of meaning rather than an absolute. It blurs the line between fact and fiction, between realism and poetry, between the unrecoverable past and our memory of it.[57] This raising of the inexact character of historical 'truth,' as well as its relationship to myth and memory and the reality of the dim and unrecoverable past, have foreshadowed deep fissures in the landscape of identity and what it means to be American. Truth, it seems, has differed from time to time and place to place with reckless abandon and enormous variety. Choice between them is present everywhere both in the past and the present; my truth dissolves into your myth and your truth into my myth almost as soon as it is articulated. We see this reinforced everywhere about us today, and mostly we shake our heads and misunderstand the versions of truth espoused by various groups about themselves and about those excluded from their fellowship. They have given and continue to give meaning and value to individual human lives and to create a focal point for explaining the sufferings and triumphs of the group.

At some level there is no absolute; instead everything is constructed. If so, what might be the case of the moon landings? This has happened in history repeatedly, as versions of the past have replaced earlier versions that once seemed so true. The denials of the moon landings excite the response of crank and crackpot from most who hear them. But so too do many other conspiracy theories that are now major elements of the memory of the nation. For example, how many Americans believe that John F. Kennedy was assassinated by means of a massive conspiracy that involved the national security establishment? More than fifty years of a persistent churning over the data, near data, and wishful thinking has forced massive fissures in the conclusions of the Warren Commission. Might this happen in the future in relation to the moon landings?

V   Apollo nostalgia

In the end, the cultural debate continues in the first part of the twenty-first century over the meaning of the Apollo program. Much of the recollection of Apollo's legacy revolves around ideas of 'progress' for the American nation.[58] At the same time, Apollo signals nostalgia for the past in which society, culture, economics, politics and other attributes of the public sphere seemed to work better. Apollo nostalgia manifests itself in several ways. It may be found in numerous popular conceptions of Apollo, especially in film, literature, music, theater and advertising. In each of these arenas, three great themes played out in the nostalgic past of Apollo. Apollo nostalgia hearkens back to an era of the 1960s in which order ruled and all seemed in its place. Central to this, in the pre-Great Society and pre-social reformation era, white men oversaw America in a 'Leave it to Beaver' type of existence where women were docile helpmates, ethnic and race relations favored American-born whites, and all understood their place in the system.

Most important for reinforcement of this issue, the system worked and in memory enjoyed efficiencies lost in a postmodern, multicultural setting. Even Norman Mailer (1923–2007), as much an embodiment of the 1960s counter culture as anyone, ranted about this aspect of Apollo while covering the moon landings in 1969. Mailer expressed fascination and not a little perplexity with the time warp that he witnessed at the Manned Spacecraft Center in Houston. He railed against an overwhelmingly white male NASA steeped in middle-class values and reverence for the American flag and mainstream culture. Mailer grudgingly admitted, however, that NASA's approach to task accomplishment – which he viewed as the embodiment of the Protestant Work Ethic – and its technological and scientific capability got results with Apollo. Even so, he hated NASA's closed and austere society, one where he believed outsiders were distrusted and held at arm's length with a bland and faceless courtesy that betrayed nothing. For all of his skepticism, for all of his esotericism, Mailer captured much of interest concerning rocket technology and the people who produced it in Project Apollo.[59]

Mailer's critique foreshadows by twenty-five years a powerful nostalgia that has grown up around Apollo as a program that was done right, in no small part because it took place within the cultural confines of an era before the social revolution of the 1960s. Nothing captures this nostalgia more effectively than the feature film *Apollo 13*, a 1995 docudrama directed by Ron Howard. Set in 1970 when an explosion crippled a lunar landing mission and NASA nearly lost astronauts Jim Lovell, Fred Haise and Jack Swigert, it has been recast as one of NASA's finest hours, a successful failure. At 56 hours into the flight, an oxygen tank in the Apollo service module ruptured and damaged several of the power, electrical and life-support systems. People throughout the world watched and waited and hoped as NASA personnel on the ground and the crew worked to find a safe way home. It was a close-run thing, but the crew returned safely on 17 April 1970. The near disaster served several important purposes for the civil space program, especially prompting reconsideration of the propriety of the whole effort while also solidifying in the popular mind NASA's collective genius.[60] While one must give the NASA flight team high marks for perseverance, dedication and an unshakable belief that they could safely bring the crew home, it is quite strange that no one seems to realize that the mission had already failed, and failed catastrophically, by the time of the accident. The fact that Apollo 13 is now viewed as one of NASA's shining moments says much about the ability of humanity to recast historical events into meaningful morality plays.[61]

In this instance, the *Apollo 13* film became a vehicle for criticism of the social order that emerged from the 1960s and a celebration of an earlier age. When the film appeared in 1995, reviewer John Powers, writing for the *Washington Post*, commented on its incessant nostalgia for 'the paradisiacal America invoked by Ronald Reagan and Pat Buchanan – an America where men were men, women were subservient, and people of color kept out of the way.' In addition, Powers wrote, 'Its story line could be a Republican parable about 1995 America: A marvelous vessel loses its power and speeds toward extinction, until it's saved by a team of heroic white men.'[62] If anything, Powers underemphasized the white America evoked in *Apollo 13*. The only women with speaking parts of substance were Marilyn Lovell (Kathleen Quinlin), wife of the Apollo 13 commander, whose role is distinctly one of offering proud support while privately fearing the worst, and their daughter whose role seems to be as spokesperson for the social revolution underway while consistently reflecting its least important elements. For example, she complains in a shrieky voice that the Beatles had just broken up and her world has accordingly collapsed.

The heroes of *Apollo 13* are the geeks of Mission Control, with the astronauts aboard the spacecraft as spirited but essentially and metaphorically emasculated characters to be saved. Lovell, Haise and Swigert must wait to be rescued in a manner not unlike Rapunzel, as an active helper but unable to accomplish the task alone. As historian Tom D. Crouch wrote of this film's depiction of the 'studs' in Mission Control:

> The real heroes of this film are either bald or sporting brush cuts; wear thick glasses; are partial to rumpled short sleeve shirts; and chain-smoke an endless string of cigarettes, cigars, and pipes. For all of that, these slide rule-wielding technonerds solve all of the difficult problems required to bring the crew home. They are, in the words of one of the astronauts portrayed in the film, 'steely-eyed missile men.'[63]

Apollo 13 the film, accordingly, venerates a long past era in American history. Indeed, it may have been an era already gone by the time of the actual mission in 1970. It is a hallowing of masculinity in a nostalgic context.

VI   Meanings

At a fundamental level, Apollo served as a means of defining and transmitting values both to Americans and to other peoples. Psychologist David C. McClelland argues in *The Achieving Society* that the celebration of success and uniqueness and exceptionalism are fundamental to perceptions of well-being. Apollo offers a cooperative example of McClelland's achieving society. It represented a time and place when leaders set moderately difficult but potentially achievable goals that were then translated into individual achievable tasks with deliberate but manageable risks. In essence, Apollo symbolized a moderate degree of risk that highly skilled and thoughtful people overcame through their diligent efforts and abilities. The Apollo team, characteristic of achievement-motivated people in McClelland's study, was more concerned with achievement than with the ordinary rewards of wealth and fame. They did not reject rewards, but those rewards were not as valued as the accomplishment itself. Object lessons in this are everywhere apparent in the Apollo program. These are some of the ingredients that Americans want to recapture with their remembering of Apollo.[64]

A core legacy, Americans hearkened back to that brief moment in time when the United States went to the moon and want to recreate Apollo as the twenty-first century dawns. But the difficulty with emphasizing this legacy of accomplishment is that it recalls a time that no longer exists. Apollo was born out of Cold War rivalries long gone, and indeed they did not exist much beyond the mid-1960s. The global setting that emerged in the 1970s presented a shock to order that existed at the time.[65] Demonstration of American technological capability was no longer required and nothing else compelled the level of resource expenditure that had been required earlier. The idea that Apollo would lead to an age of space exploration was a dream to some who grew disenchanted with the results and struck out on their own toward libertarian visions of the future. All of the seeds of the present space effort for the United States were sown during the Apollo era, nurtured in the far different climate of the 1970s, and reaped in the more recent past for both positive and negative results.

The moon landings worldwide represented a watershed in human history. Capping the decade of the 1960s with the moon landings offered a seemingly collective catharsis for humanity, representing for many the first tentative steps toward becoming a cosmic species. Certainly Apollo was a Cold War initiative; it was a surrogate for war – the primary goal of the program when first envisioned in 1961. At the same time spaceflight conjured the best in human spirit and served, in the words of journalist Greg Easterbrook, as 'a metaphor of national inspiration: majestic, technologically advanced, produced at dear cost and entrusted with precious cargo, rising above the constraints of the earth.' It 'carries our secret hope that there is something better out there – a world where we may someday go and leave the sorrows of the past behind.'[66] All of this made it difficult to accept that this program was ending.

Notes

1. On American exceptionalism, see Charles Lockhart, *The Roots of American Exceptionalism: Institutions, Culture and Policies*, Basingstoke: Palgrave Macmillan, 2003; David W. Noble, *Death of a Nation: American Culture and the End of Exceptionalism*, Minneapolis: University of Minnesota Press, 2002; and Godfrey Hodgson, *The Myth of American Exceptionalism*, New Haven: Yale University Press, 2010.

2. This deference to the authority of expertise, or the lack thereof, was also seen in many other technical arenas. See Thomas L. Haskell, ed., *The Authority of Experts: Studies in History and Theory*, Bloomington: Indiana University Press, 1984; Douglas Walton, *Appeal To Expert Opinion: Arguments From Authority*, State College: Penn State University Press, 1997; Steven Yearley, 'Making Systematic Sense of Public Discontents with Expert Knowledge: Two Analytical Approaches and a Case Study,' *Public Understanding of Science* 9 (April 2000), 105–22; Timothy L. O'Brien, 'Scientific Authority in Policy Contexts: Public Attitudes about Environmental Scientists, Medical Researchers, and Economists,' *Public Understanding of Science* 22 (October 2013), 799–816.

3. See Michael A.G. Michaud, *Reaching for the High Frontier: The American Pro-Space Movement, 1972–84*, New York: Praeger, 1986; Roger D. Launius, '"And Now for Something Completely Different": Creating Twenty-First Century Space Access,' *Space Times* 51 (March/April 2012), 4–11; Roger D. Launius

and Dennis R. Jenkins, 'Is It Finally Time for Space Tourism?,' *Astropolitics* 4.3 (Winter 2006), 253–80.

4. Thomas O. Paine to Associate and Assistant Administrators and Center Directors, 25 February 1969, NASA Historical Reference Collection, NASA Headquarters, Washington, DC; Anne M. Platoff, *Where No Flag Has Gone Before: Political and Technical Aspects of Placing a Flag on the Moon*, Houston: Lyndon B. Johnson Space Center, 1993.

5. This may be seen in many books. A recent example is Claude A. Piantadosi, *Mankind Beyond Earth: The History, Science, and Future of Human Space Exploration*, New York: Columbia University Press, 2012.

6. I made this argument in relation to Apollo in Roger D. Launius, 'Perceptions of Apollo: Myth, Nostalgia, Memory or all of the Above?,' *Space Policy* 21 (May 2005), 129–39.

7. The best example of this is Stephen J. Pyne, 'Space: A Third Great Age of Discovery,' *Space Policy* 4.3 (August 1988), 187–99.

8. Stephen J. Pyne, *The Ice: A Journey to Antarctica*, Iowa City: University of Iowa Press, 1986; Nathan Reingold, ed., *The Sciences in the American Context: New Perspectives*, Washington, DC: Smithsonian Institution Press, 1979; Norman Cousins et al., *Why Man Explores*, Washington, DC: NASA, 1976; Sarah L. Gall and Joseph T. Pramberger, *NASA Spinoffs: 30 Year Commemorative Edition*, Washington, DC: NASA, 1992.

9. Lyndon B. Johnson, 'President's News Conference at the LBJ Ranch,' 29 August 1965, in Lyndon B. Johnson, *Public Papers of the Presidents*, Washington, DC: Government Printing Office, 1966, 945; available at http://quod.lib.umich.edu/p/ppotpus/4730960.1965.002/417?page=root;rgn=full+text;size=100;view=image;q1=august+1965 (accessed 1 October 2017).

10. To determine how widespread this question is, in 2001 I undertook a search of the DowJones database, which includes full text of more than 6,000 newspapers, magazines, newswires and transcripts in search of 'If we can put a man on the moon [...]?' Some of the publications go back to the 1980s but most have data only from the 1990s. Except for perhaps Lexis-Nexis, DowJones is the largest full-text database available. There are more than 6,901 articles using this phrase, or a variation of it, in the database. Among them was a statement by former White House Chief of Staff, Mack McClarty concerning Mexico on National Public Radio's 'All Things Considered,' entitled, 'Analysis: President Bush to visit Mexico and its President.' Maria Elena Salinas, co-anchor at Miami-based, Spanish-language, cable network Univision, used this phrase when discussing her decision to list the Apollo moon landings as first in the top 100 news events of the twentieth century; see A. Levinson, 'Atomic bombing of Hiroshima tops journalists' list of century's news,' Associated Press, 24 February 1999.

11. Roger D. Launius, 'American Memory, Culture Wars, and the Challenge of Presenting Science and Technology in a National Museum,' *The Public Historian* 29 (Winter 2007), 13–30; and idem, 'Public History Wars, the "One Nation/One People" Consensus, and the Continuing Search for a Usable Past,' *OAH Magazine of History* 27.1 (January 2013), 31–6.

12. 'Carroll O'Connor Obituary,' *Morning Edition*, National Public Radio, 22 June 2001. This report by Andy Bowers is available at http://www.npr.org/templates/story/story.php?storyId=1124754 (accessed 1 October 2017).

13. 'The Sweet Smell of Air,' *Sports Night*, first aired 25 January 2000; available at http://www.imdb.com/title/tt0707457/ (accessed 1 October 2017).

14. See Harlen Makemson, *Media, NASA, and America's Quest for the Moon*, New York: Peter Lang, 2009; Michael Allen, *Live from the Moon: Film, Television, and the Space Race*, London: I.B. Tauris, 2009; and David Meerman Scott and Richard Jurek, *Marketing the Moon: The Selling of the Apollo Lunar Program*, Cambridge, MA: MIT Press, 2014.

15. 'One Small Step for Man,' *On This Day*, BBC.co.uk; available at http://news.bbc.co.uk/onthisday/hi/witness/july/21/newsid_3058000/3058833.stm (accessed 1 October 2017).

16. John Krige, Angelina Long Callahan and Askok Maharaj, *NASA in the World: Fifty Years of Collaboration in Space*, Basingstoke: Palgrave Macmillan, 2013, 3–5; Joan Johnson-Freese, 'China's Manned Space Program: Sun Tzu or Apollo Redux?,' *Naval War College Review* 56.3 (Summer 2003), 51–71; James R. Hansen, 'The Great Leap Upward: China's Human Spaceflight Program and Chinese National Identity,' in Steven J. Dick, ed., *Remembering the Space Age: Proceedings of the Fiftieth Anniversary Conference*, Washington, DC: NASA, 2008, 109–20.

17. See Susan Curtis, *A Consuming Faith: The Social Gospel and Modern American Culture*, Baltimore: Johns Hopkins University Press, 1991, 270–85; Wayne K. Hobson, 'Professionals, Progressives and Bureaucratization: A Reassessment,' *The Historian* 39 (August 1977), 639–58.

18. Edward Everett Hazlett to Dwight D. Eisenhower, 24 September 1952, Eisenhower Personal Papers (1916–1953), Eisenhower Presidential Library, Abilene, KS, as quoted in Brian Balogh, 'Reorganizing the Organizational Synthesis: Federal-Professional Relations in Modern America,' *Studies in American Political Development* 5 (Spring 1991), 119–72, here 164.

19. James B. Conant, 'The Problems of Evaluation of Scientific Research and Development for Military Planning,' speech to the National War College, 1 February 1952, quoted in James G. Hershberg, '"Over My Dead Body": James B. Conant and the Hydrogen Bomb,' paper presented to the Conference on Science, Military and Technology, Harvard/MIT, June 1987, 50.

20. David Halberstam, *The Best and Brightest*, New York: Viking, 1973, here 57, 153.

21. Charles S. Maier, 'Two Sorts of Crisis? The "Long" 1970s in the West and the East,' in Hans Günter Hockerts, ed., *Koordinaten deutscher Geschichte in der Epoche des Ost-West-Konflikts*, Munich: Oldenbourg, 2004, 49–62.

22. James E. Webb, *Space Age Management: The Large Scale Approach*, New York: McGraw-Hill, 1969, 15. See also Roger D. Launius, 'Managing the Unmanageable: Apollo, Space Age Management and American Social Problems,' *Space Policy* 24.3 (August 2008), 158–65.

23. Fritjof Capra, 'Paradigms and Paradigm Shifts,' *ReVision* 9.1 (Summer/Fall 1986), 11. Capra's definition was closely related to Thomas S. Kuhn, *The Structure of Scientific Revolutions*, Chicago: University of Chicago Press, 1970, esp. 175–6.

24. 'Antitechnology Bias,' *Air Force Magazine*, September 1971, 53. This was confirmed in a James C. Fletcher interview by the author, 19 September 1991.

25. Quoted in Ralph E. Lapp, *The New Priesthood: The Scientific Elite and the Uses of Power*, New York: Harper & Row, 1965, 228.

26. John Kenneth Galbraith, *Economics and the Public Purpose*, New York: New American Library, 1975, 405.

27. See Haskell, *Authority of Experts*; Langdon Winner, *Autonomous Technology: Technics-Out-of-Control as a Theme in Political Thought*, Cambridge, MA: MIT Press, 1977.

28. David McKay, *Domestic Policy and Ideology: Presidents and the American State, 1964–1987*, Cambridge: Cambridge University Press, 1989, 65.

29. *The Public Papers of the President, 1969*, Washington, DC: Government Printing Office, 1970, 696.

30. *Space Daily* (16 February 1973), 71.

31. *Presidential Documents: Richard Nixon, 1973*, vol. 9, Washington, DC: Government Printing Office, 1974, 75–6.

32. Clay T. Whitehead, White House Staff Assistant to Peter M. Flanigan, Assistant to the President, 8 February 1971, Record Group 51, Series 69.1, Box 51-78-32, National Archives and Records Administration, Washington, DC; Fletcher interview by author, 19 September 1991.

33. Traci Watson, 'NASA Administrator Says Space Shuttle was a Mistake,' *USA Today* (28 September 2005), 1A.

34. NASA Press Release, 'NASA Memo: Griffin Point Paper on USA Today Article, 9/28/05,' 29 September 2005, NASA Historical Reference Collection, NASA History Division, NASA Headquarters, Washington, DC.

35. See Thomas A. Heppenheimer, *The Space Shuttle Decision: NASA's Search for a Reusable Space Vehicle*, Washington, DC: NASA 1999; Roger D. Launius, 'NASA and the Decision to Build the Space Shuttle, 1969–72,' *The Historian* 57 (Fall 1994), 17–34; John M. Logsdon, 'The Decision to Develop the Space Shuttle,' *Space Policy* 2 (May 1986), 103–19; idem, 'The Space Shuttle Program: A Policy Failure,' *Science* 232 (30 May 1986), 1099–105.

36. Roger D. Launius, 'Assessing the Legacy of the Space Shuttle,' *Space Policy* 22 (2006) 226–34; John M. Logsdon, *After Apollo? Richard Nixon and the American Space Program*, Basingstoke: Palgrave Macmillan, 2015, chapter 4, 'Space and National Priorities,' 83–102.

37. Roger D. Launius, 'Lessons from Terrestrial Exploration for Earth Orbit,' *Space News* 24.41 (21 October 2013), 19, 21.

38. Two really interesting cogitations on the weirdness of some elements in the pro-space movement are Jodi Dean, *Aliens in America: Conspiracy Cultures from Outerspace to Cyberspace*, Ithaca: Cornell University Press, 1998; and Constance Penley, *NASA/Trek: Popular Science and Sex in America*, New York: Verso, 1997.

39. Edward S. Cornish, 'A Quest for the Meaning of Life (a review of *The Hunger of Eve*),' *The Futurist* (December 1976), 336–40, here 340.

40. Barbara Marx Hubbard, *The Hunger of Eve*, Harrisburg: Stackpole Books, 1976, here 144, 150.

41. Hubbard even produced what is known as a 'SYNCON' wheel; see William Sims Bainbridge, *The Spaceflight Revolution*, New York: John Wiley, 1976, 173.

42. Michaud, *Reaching for the High Frontier*, 43–5.

43. Barbara Marx Hubbard, 'The Future: Previews of Coming Attractions,' *First Foundation News* 1 (August 1995), 1–4.

44. Gerard K. O'Neill, 'The Colonization of Space,' *Physics Today* 27.9 (September 1974), 32–40; idem, *The High Frontier: Human Colonies in Space*, New York: William Morrow, 1976; Peter E. Glaser, 'Energy from the Sun: Its Future,' *Science* 162 (1968), 857–60; idem, 'Solar Power via Satellite,' *Astronautics and Aeronautics* (August 1973), 60–8; and idem, 'An Orbiting Solar Power Station,' *Sky and Telescope* (April 1975), 224–8. See also the contributions by Alexander Geppert, Thore Bjørnvig, Tilmann Siebeneichner and Peter Westwick, Chapters 1, 7, 11 and 12 in this volume.

45. Michaud, *Reaching for the High Frontier*, 57–102.

46. Arthur C. Clarke, *Rendezvous with Rama*, New York: Bantam Books, 1973. See also Robert Poole's contribution, Chapter 5 in this volume.

47. Thomas A. Heppenheimer, *Colonies in Space*, Harrisburg: Stackpole Books, 1977, 279–80.

48. This would be completely consistent with their ideology. See Roger D. Launius, 'A Western Mormon in Washington, DC: James C. Fletcher, NASA, and the Final Frontier,' *Pacific Historical Review* 64.2 (May 1995), 217–41; 'Colonies in Space,' *Newsweek* (27 November 1978), 95–101.

49. See Richard D. Johnson and Charles Holbrow, eds, *Space Settlements: A Design Study in Colonization*, Washington, DC: NASA, 1977, a study sponsored by NASA Ames, ASEE, and Stanford University in the summer of 1975 to look at all aspects of sustained life in outer space. See also John Billingham, William Gilbreath, Gerard K. O'Neill and Brian O'Leary, eds, *Space Resources and Space Settlements*, Washington, DC: NASA, 1979.

50. Freeman Dyson, 'Obituary, Gerard Kitchen O'Neill,' *Physics Today* 46 (February 1993), 97–8.

51. Andrew Chaikin, *A Man in the Moon: The Voyages of the Apollo Astronauts*, New York: Viking, 1994, 100.

52. Mary Lynne Dittmar, 'Building Constituencies For Project Constellation: Updates To The Market Study Of The Space Exploration Program,' Presentation at Building and Maintaining the Constituency for Long-Term Space Exploration Workshop, George Mason University, Fairfax, VA, 31 July–3 August 2006; The Gallop Poll, 'Did Men Really Land On The Moon?,' 15 February 2001; available at http://www.gallup.com/poll/1993/did-men-really-land-moon.aspx (accessed 1 October 2017).

53. John Noble Wilford, 'A Moon Landing? What Moon Landing?,' *New York Times* (18 December 1969), 30.

54. 'Many Doubt Man's Landing on Moon,' *Atlanta Constitution* (15 June 1970), 1.

55. Bill Kaysing and Randy Reid, *We Never Went to the Moon: America's Thirty Billion Dollar Swindle*, Pomeroy: Health Research Books, 1976.

56. Janet Barth, *The Radiation Environment*, NASA Goddard Space Flight Center; available at http://radhome.gsfc.nasa.gov/radhome/papers/apl_922.pdf (accessed 1 October 2017); Jim McDade, 'Glowing in the Dark: The Glaring Ignorance of Apollo Hoax Hucksters,' 2002, copy in author's personal collection.

57. See the discussion of myth and history in Hayden White, *Metahistory: The Historical Imagination in Nineteenth-Century Europe*, Baltimore: Johns Hopkins University Press, 1973; and Roland Barthes, 'The Discourse of History,' *Comparative*

Criticism: A Yearbook 3 (1981), 3–20; Dominick LaCapra, *Rethinking Intellectual History*, Ithaca: Cornell University Press, 1983; Brook Thomas, *The New Historicism: And Other Old-Fashioned Topics*, Princeton: Princeton University Press, 1991.

58. See Svetlana Boym, *The Future of Nostalgia*, New York: Basic Books, 2002, 33–41.

59. Norman Mailer, *Of a Fire on the Moon*, Boston: Little, Brown, 1970.

60. NASA Office of Public Affairs, *Apollo 13: 'Houston, We've Got a Problem,'* Washington, DC: NASA, 1970.

61. Gene Kranz, *Failure is Not an Option: Mission Control from Mercury to Apollo 13 and Beyond*, New York: Simon & Schuster, 2000.

62. Jonathan Powers, 'The Wrong Stuff,' *Washington Post* (9 July 1995), C1.

63. Tom D. Crouch, 'Apollo 13,' *Journal of American History* 84.3 (December 1997), 1180–2.

64. David C. McLelland, *The Achieving Society*, New York: Free Press, 1985.

65. Niall Ferguson, 'Introduction: Crisis, What Crisis? The 1970s and the Shock of the Global,' in idem, Charles S. Maier, Erez Manela and Daniel J. Sargent, eds, *The Shock of the Global: The 1970s in Perspective*, Cambridge, MA: Harvard University Press, 2010, 1–24; see also Alexander Geppert's introduction, Chapter 1 in this volume.

66. Gregg Easterbrook, 'The Space Shuttle Must Be Stopped,' *Time* (10 February 2003); available at http://content.time.com/time/magazine/article/0,9171,1004201,00.html (accessed 1 October 2017).

CHAPTER 4

A Grounding in Space: Were the 1970s a Period of Transition in Britain's Exploration of Outer Space?

Doug Millard

On the cool, fresh South Australian morning of 28 October 1971, the United Kingdom's Black Arrow R3 rocket rose from its pad and climbed steadily into a clear blue sky.[1] Some 40 minutes later the receiving station at Fairbanks, Alaska, received a signal from the rocket's payload, the X3 satellite, indicating it had reached orbit. X3 would now be known as 'Prospero' and made the United Kingdom the sixth nation to orbit its own spacecraft. Ironically, the Black Arrow rocket program had already been canceled and so, in the minds of some, its space program had been too.[2] Did this event signal a fundamental break in the British space program? Was it representative of a broader transition during the 1970s in the nation's exploration of outer space?

For over 15 years the United Kingdom had been involved in the development of space launch vehicles and payloads, including satellites. Shortly before the creation of NASA in 1958, US State Department officials had hoped that the United Kingdom would be the next nation to launch a satellite after the two superpowers, so demonstrating the West's inherent strength in space.[3] The United Kingdom's already functioning Skylark rocket and its developmental Blue Streak and Black Knight missile programs suggested a technological capability for doing so.[4] Yet, such fundamental technical competency was not accompanied by any strong political willingness to direct these rockets toward an indigenous British satellite-launching program. Indeed,

Doug Millard (✉)
Science Museum, London, UK
e-mail: doug.millard@nmsi.ac.uk

Alexander C.T. Geppert (ed.), *Limiting Outer Space*
European Astroculture, vol. 2
https://doi.org/10.1057/978-1-137-36916-1_4

successive governments' attitudes to space during the following decade seemed unclear, if the number of reports and enquiries into space policy during that period are deemed a measure.[5] How then can we gauge the significance of the 1970s in regard to the United Kingdom's exploration of outer space?

Historian Alexander Geppert points to the need for broad cultural histories of outer space; ones that transcend the hitherto largely political and technological studies of the subject that, furthermore, are rooted largely in contemplations of the Cold War and its two chief protagonists (the United States and the former Soviet Union) to the oft near exclusion of other perspectives.[6] Yet, he drills down further, seeking to 'disrupt and overcome worn dichotomies such as science/fiction, real/imaginary and human/environment,' by suggesting also the concepts of 'astroculture' and 'technoscience' as alternative and more informative analytical tools with which to build such histories.[7] This chapter employs these concepts of astroculture, looking specifically at the British Interplanetary Society, comic strips, radio and television, and technoscience in an attempt to understand better the United Kingdom's explorations of outer space and any changes that might be observed to have occurred by and during the post-Apollo years of the 1970s. In so doing it also interrogates Geppert's proposition that outer space, 'was key to the self-image of public, governmental and technical elites, and to modernist narratives of progress' in the twentieth century.

I   Towards the technoscientific: UK space activities, 1957–75

In September 1970 Britain's third Black Arrow vehicle to be launched, and the second with a satellite payload, failed to reach orbit. Soon after, Sir William Penney (1909–91), Rector of Imperial College, London, and the former leading luminary of Britain's nuclear program, accepted an invitation to chair a report into the future viability of the Black Arrow satellite launcher. In his report, Penney remarked of the Black Arrow program that, paradoxically, there had been too few rocket launches for it to mature as a viable system yet too many vehicles produced (one a year) for a planned satellite production rate of one every three years.[8] In other words, if the market for Black Arrows was experimental satellites then there were too few forthcoming to render the rocket's production cost-effective. Penney recommended cancelation of the Black Arrow program after the next orbital launch attempt, and purchasing future launches on United States rockets instead. The Edward Heath Government accepted the recommendations on 6 July 1971, and in the House of Commons later that month the Minister for Aerospace, Frederick Corfield (1915–2005), gave a written answer to a question on the future of Black Arrow, announcing the termination of the program, with the next launch to be the final one in the series. Following the successful orbiting of the Prospero satellite by the R3 vehicle, its successor Black Arrow R4 was

Figure 4.1 Larch rocket engine chamber, ca. 1970. The Larch (Large Rocket Chamber) was an up-rated Gamma engine intended for Britain's Black Arrow mark 2 satellite launch vehicle. It burned kerosene fuel in hydrogen peroxide at a greatly increased combustion chamber pressure. Development stopped when the entire Black Arrow program was canceled in 1971. *Source*: Courtesy of Science Museum/Science and Society Picture Library, London.

duly dispatched to the Science Museum in London where it can be seen on display to this day (Figure 4.1).[9] Miranda, the next satellite in the series, was launched by a US Scout rocket on 9 March 1974 from Vandenberg Air Force Base, California.

Black Arrow had been developed by the United Kingdom's Royal Aircraft Establishment and Saunders Roe Company from the Black Knight rocket test vehicle, used to test re-entry bodies for the Blue Streak ballistic missile in the late 1950s.[10] The Black Knight trials swiftly produced the necessary data and subsequent vehicles were redirected toward investigations of re-entry phenomena, these experiments being conducted with both offensive and anti-ballistic missile defense roles in mind. Much of the impetus was provided by the United States' close interest and involvement in the trials which, unlike any of the US equivalent programs, could deliver data swiftly and efficiently over land rather than at sea.[11]

But following Sputnik's launch in 1957, UK design studies were carried out by the Guided Weapons Department of the Royal Aircraft Establishment and in industry for combining Blue Streak and Black Knight as part of a British satellite launch vehicle program.[12] Indeed, the US State Department had hoped the United Kingdom would be the third nation to launch a satellite, following those launched by the two superpowers, thus demonstrating the depth of Western capability in space.[13] Further, senior scientists in the United Kingdom with an interest in developing a satellite research program argued for an autonomous national launch capability.[14] In reality, the United Kingdom lacked the resources to both develop Blue Streak as a weapon system and as the basis for a satellite launch vehicle simultaneously. At the second meeting of the International Council for Science's Committee for Space Research (COSPAR), in March 1959, NASA offered its rockets for launching science experiments from other countries.[15] The UK government accepted with alacrity, the decision welcomed also by the science research community which had already been designing upper atmosphere and space experiments for launching on the Skylark sounding rockets.[16] Thus was the Ariel program of science satellites born; the first launched on 26 April 1962 as the world's first international orbital collaboration, and the last (Ariel 6) in 1979.

The United Kingdom embarked upon a steep learning curve of satellite design and production with increasing levels of responsibility being passed from government research establishments (and from the United States) to British industry. Ariel 3 (1967) was the first satellite to be designed and built entirely in the United Kingdom. With this rapid accumulation of engineering and technical experience – supplementing that already amassed by the government research establishments and university departments – the United Kingdom was ideally placed to play a leading role in the nascent European space endeavor. The United Kingdom was a founding member of both the European Space Research Organisation (ESRO, 1964) and the European Launcher Development Organisation (ELDO, 1962), whose Europa satellite launch vehicle was based around the Blue Streak rocket.[17] Both endeavors suited the United States, which was anxious to encourage European collaboration and avoid unilateral proliferation.[18] Elsewhere, the United Kingdom's global defense communication needs were met by highly active collaboration with the United States on what became Skynet, the first geostationary military satellite program.

Yet, progress in ELDO proved slow and increasingly costly. In 1966 the UK government announced its intention to cease contributing to the funding of ELDO.[19] There was developing also in government a broader concern about the worth and organization of space activities; the very successful Skylark sounding rocket program, for example, survived a review in part because of its value to foreign policy rather than any intrinsic belief in its scientific research value.[20] Calls for the creation of an agency or council to coordinate space affairs and render them more cost-efficient, especially with European programs, were rejected with Prime Minister Wilson opting in 1968 to appoint instead the Lord President as a ministerial 'space co-ordinator,' a return to the arrangement adopted by the Conservative Government of 1959–64. In September 1972 the United Kingdom announced finally that it would be leaving ELDO on 1 January 1973.[21]

And yet the Minister for Aerospace in the Heath Government, Michael Heseltine, then pushed hard for a new agency combining ELDO and ESRO, in which member states could choose which projects to participate in.[22] In effect, he wanted to 'invert the logic and have the national authorities decide first how much money they wanted to spend on space in the light of their overall R&D budgets and then choose the programs they wanted to join and at what financial level.'[23] This indeed is what transpired, with member states signing the European Space Agency Convention in May 1975, the new Agency based on Heseltine's 'program à la carte' concept. Heseltine tells a more entertaining story in which he secured a major role for the United Kingdom in building the prospective European communications satellites by using money to buy into the new (French) launch vehicle program and the (German) NASA Post-Apollo Program, thereby securing those nations' support for the United Kingdom's own preference for investing heavily in the satellite program.[24] The United Kingdom had in effect now moved far closer to a coherent space policy (in Europe) while maintaining its skeptical stance on space as a distinct policy area, a standing it was to maintain for the next forty years.[25]

II   The British Interplanetary Society

There had been Britons who in the 1950s and before had believed passionately in the need for spaceflight and who had espoused and explained the means by which it could be achieved. Young British males (and one or two females) either side of the Second World War, many strongly influenced by transatlantic popular literature, had set about envisioning, theorizing and disseminating information about the means by which humans would soon reach and explore space. Many became members of the British Interplanetary Society (BIS), established in 1933. This organization, with its extensive program of meetings, communications, publications and demonstrations, perhaps presented as full a complement of Geppert's requisite characteristics of an astroculture as any other in Great Britain.[26] It was the BIS, more than any other entity, which epitomized British 'outer space' in the decades either side of the Second World War.[27] The Society had paused its activities during the war but reconvened shortly after its end. The effects and aftermath of the conflict catalyzed further the members' cosmic aspirations, through considerations of the new technologies developed in the course of hostilities and their potential for allowing humans to reach and travel through space – rocket, radar, nuclear in particular. Also strengthened was their belief that space exploration could play an instrumental part in bringing humanity together so as to avoid further, probably terminal, conflict.

It was the V-2 rocket, the world's first long-range missile, used to attack occupied Europe and southeast England from the fall of 1944, that, more than any other wartime technology, demonstrated a latent capability for reaching space. It stood 14 meters high, weighed twelve and a half tons, and had a range of over 300 kilometers, touching space as it climbed to a height of 88 kilometers before dropping in a ballistic path on to its target. It had been developed by the army at the Peenemünde research establishment on the Baltic coastline of Germany. Designated originally as the *Aggregat 4* or A-4, it was the latest in a series of new rockets designed by the army as an alternative to and improvement upon long-range artillery. *Reichspropagandaminister* Joseph Goebbels (1897–1945) later renamed it *Vergeltungswaffe 2* (Vengeance Weapon 2), then widely abbreviated to V-2.[28] The V-2s demonstrated the likely validity of the spaceflight predictions of Konstantin Tsiolkovsky in Tsarist and Soviet Russia, Hermann Oberth in Germany and Robert Esnault-Pelterie in France, whose works the BIS members had immersed themselves in during the 1930s.

The BIS melded imagination and enthusiasm as it set about reporting the new technologies that had emerged from the war and that were now evolving at breakneck speed during the years that followed. Some of the Society's increasing membership was working in the United Kingdom's defense industries and research establishments. Its extensive program of meetings and publications fed the public's interest in an apparently impending era of space travel. Scans of the UK press, radio and television of the time repeatedly throw up the names of BIS people as they wrote and spoke about the coming 'Space Age,' in particular engineers Archibald Low (1888–1956) and Arthur 'Val' Cleaver (1917–77), physicist Les Shepherd (1918–2012) and, above all, science-fiction author Arthur C. Clarke (1917–2008). The concepts and designs described and illustrated in BIS publications were then routinely carried in the national press.[29]

The Society was a founding member of the International Astronautical Federation (IAF), a postwar assembly of similar groups around the world dedicated to the development of space travel for peaceful means. The IAF, its founders intended, would be a holding organization until the United Nations, as it was assumed and hoped, would take charge. The BIS membership itself was drawn from around the world and reflected the internationalist philosophy prominent in the Society's thinking. In 1951 the Federation's second congress was held in London and themed on the artificial satellite.[30] Members and delegates outlined a range of thoughts and concepts on the desirability and likely nature of earth-orbiting satellites and satellite vehicles (space stations), the latter judged an essential prerequisite for human interplanetary flight – such missions being easier to effect from earth orbit than from the ground. Extracts from the papers delivered were reproduced in the December 1951 edition of the Society's *Journal* and a special booklet containing all of the contributions was published in 1952 as *The Artificial Satellite*.[31] This thinking on artificial satellites influenced greatly the American scientific attaché to the United Kingdom, S. Fred Singer (1924–). He drew up and published his own plans for a Minimum Orbital Unmanned Satellite. This design, known also as MOUSE, was intended to show the simplest type of artificial satellite that could be built at the time. It was drum-shaped and about the size of a domestic waste paper basket. It would contain some simple instruments including a Geiger counter, photoelectric cells and telemetry circuits for communication. Singer's design coincided with US military studies into similar minimum vehicles including the Army-Navy project Orbiter.[32] The Society's membership increased by a factor of five during the 1950s. In 1959 it organized a British Commonwealth Spaceflight Symposium in London. The program included discussions on the potential technical and geographical resources from across the Commonwealth that could be combined with the United Kingdom's developing rocket capability, this to forge a new age of cosmic exploration in which the nations of the old British Empire would gather and join in under the beneficent leadership of the United Kingdom.

For many in the BIS, their visions of a future in space were overtly utopian. Spaceflight was no mere option, it was a necessity if society were to progress. Historian of religion Thore Bjørnvig has described 'the language of aspiration' of such pro-space movements, but notes also, with particular reference to Arthur C. Clarke, the apocalyptic resonances of their strivings toward a better world.[33] Clarke, so prominent a member of the postwar BIS, allowed his imagination to leap across the aeons, grappling with the limitations of gravity in his stories and writing also of mankind's evolution beyond the bodily form toward a true cosmic intelligence, a theme introduced in his novella *The Sentinel*, and then exploited fully for the cinema in his collaboration with film director Stanley Kubrick in *2001: A Space Odyssey*.[34] Harry Ross (1904–78), another BIS luminary, thought deeply about the philosophical necessities of expansion into the cosmos, while the Society's artists, such as R.A. Smith (1905–59), gave visual form to such future worlds.

Toward the end of the 1950s, the British Interplanetary Society's vision appeared nearer to being realized. In 1950 scientists had met in the United States to discuss ideas for a new year of international scientific collaboration equivalent to those of the International Polar Years of 1882–83 and 1932–33. This new year would coincide with the next period of heightened solar activity (1957–58) and should, the scientists concluded, include investigations of increased solar radiation on the earth and its atmosphere, these experiments conducted by instruments carried by balloons and the newly developed sounding rockets.[35] This International Geophysical Year (IGY) might also see the first satellites launched to assist in these investigations, and such a resolution was duly adopted by the organizing committee of the IGY in October 1954. This public decision supplemented the US military studies on satellites while in the Soviet Union missile program chief designer Sergey Korolyov (1907–66) began his and his colleagues' push to persuade the authorities to back a Soviet satellite program.[36]

The astroculture espoused and practiced by the BIS – those 'images and artifacts, media and practices'– was starting to become informed by a putative technoscience of 'high-tech projects funded and directed by state agencies, or even by state-backed academic consortia.'[37] Alongside the erudite articles of the *Journal of the British Interplanetary Society* and *Spaceflight* there appeared an increasing number of 'factual' publications. When Sputnik had been launched, the BIS had started, inevitably perhaps, to focus in increasing detail on the 'reality' of what had happened, was happening and what was being planned to happen as the superpowers set about their explorations of outer space.[38] The field where imaginaries had previously enjoyed free rein was being slowly infiltrated by the overtures to and preparations for the first actual space launches, and then invaded by reportage of the first, headline-grabbing events – Sputnik, Explorer, Vostok, Mercury and so on. Clearly defined limits were now constraining the utopias of the previously untrammeled imaginaries. Yet, this was not necessarily apparent at the time. With the arrival of the Space Age it was all too easy to assume that the events unfolding were a vanguard for the visions and dreams and imaginings of outer space that had gone before. The details and characteristics of the Soviet and NASA space shots were delivering a paradigm of entering and moving into outer space with imagination its only precedent. Who, indeed, was in a position to doubt that a utopia in space was now beginning? For the duration of the 1960s the Society continued in its delivery of articles and events, management and membership assimilating, digesting and repurposing the vast quantities of information produced in the international, but mostly American, space sector.[39] Further, Society members were probing the far less accessible Soviet program and offering a succession of scoops on what was going on behind the Iron Curtain.[40]

This new reality of the Space Age was fascinating yet demanding and time-consuming; there was less space available and perhaps desired for stoking imaginations. By the 1970s, Ken Gatland (1924–97), the then editor of *Spaceflight* magazine, suggested that with the disappointment of the intended post-Apollo activities for the United Kingdom and the rest of Europe it was indeed time to look forward again and consider the 'more imaginative aspects of space travel,' in order to rekindle the informed, yet speculative articles that appeared routinely in the pre-Space Age editions of the journal.[41] Gatland was acknowledging, perhaps, a destination reached and the need to chart a new journey; to anticipate the Society's soon to be adopted motto, 'reality' had seemingly arrived but with it a concomitant need for a renewed 'imagination.'[42]

Thus did the articles on interstellar travel and communication start to appear, first in *Spaceflight* and then in dedicated issues of the Society's *Journal*. From this new avenue of thinking Project Daedalus took shape – a minimal mission to a nearby extra solar star (Barnard's) using known technologies.[43] Similar, futurist articles continued to be carried in the journal, but the paradigm of reality was hard to shake off and 'imagination' represented a relatively small component of the BIS total output. Only in recent years has there been a resurgence in the study of similarly exotic space concepts.[44] UK astroculture, as mediated by the BIS, was changing even as Sputnik was launched. If the nation's exploration of outer space is viewed through lenses of the astrocultural and the technoscientific, then the proportions of each component were shifting. The imaginaries of the former were giving way to the actualities of the latter, and – indeed – some time before the 1970s.

III   Astroculture in comic strips and on the radio: *Dan Dare* and *Journey into Space*

Perhaps it was inevitable that as the United Kingdom's world role, and its cultural representation, changed as dramatically as it did between the 1950s and 1970s, so too would its cultural depictions of outer space. *Dan Dare* was a strip cartoon that ran from 1950 through to 1967 in the *Eagle*, a children's comic periodical that included also adventure stories and factual articles. In recent decades it has achieved seminal status in the minds of those movers and shakers in British scientific, technological, academic, political and cultural life who read the *Dan Dare* adventures as young (in the main) boys. When in 2008 the Science Museum used *Dan Dare* as a cultural hook upon which to hang an exhibition of postwar industrial and domestic technology, it invited political historian Peter Hennessy to open the display. At the ceremony he confirmed his own passion for the comic book hero while asking, 'So what went wrong? Surely we were meant to be all whizzing around in space ships by now?'[45]

Dan Dare was a colonel in the 'Interplanet Space Fleet' of the 1990s who piloted his loyal crew on adventures across the solar system, often engaging with the extraterrestrials of the planets. The quality of the strip's artwork was exceptionally high – skilled, quality drawing and the use of vivid colors made it stand out from the rest on the newsstand. Tellingly, perhaps, the *Dan Dare* strips sat alongside an array of formats used by *Eagle*: text, technical cutaways, drawings and illustrations; a mix of the imaginary and the actual. For many, however, *Dan Dare* became the epitome of a future in outer space, yet all too clearly fashioned from the United Kingdom's recent past.

With his chiseled features, uniform and jaunty cap, Dare was an extension of the all-British wartime hero, ever the explorer of Empire, who flew dangerous missions across space to parley and work with friendly or threatened aliens (Figure 4.2). His arch-enemy, the Mekon, was the green, bulbous-headed leader of the Treen race, inhabitants of Venus's northern hemisphere and separated from the Theron race of southern latitudes by the 'flame belt.' In the 9 March 1951 edition of *Eagle* the Treens switch off their ray barrier over the flame belt and threaten invasion in an apparent echo of the Korean War. Similarly, the character's creator and artist Frank Hampson likened the evil Treens to the Nazis.[46] The *Eagle* creator, Marcus Morris, had been inspired to create the comic in order to counter the increasing volume of American popular-fiction magazines that had been crossing the Atlantic since the 1930s. He was anxious to provide a popular comic of more robust ethical and indeed Christian values for young readers than that provided by the sometimes gratuitous American pulp-fiction titles.[47]

Dan Dare's heyday was the 1950s. Such was his influence that competing comics like *Lion* ran *Dan Dare*-like stories: Captain Condor 'whizzed' through space in the fourth millennium rather than Dare's 1990s, yet he looked like Dare, sported the same type of peaked cap, and flew a spaceship that, like Dare's *Anastasia*, replicated the Flying Fortress gun turret in its forward design. Hampson left *Eagle* in 1959 and this affected the standard of

Figure 4.2 Ingersoll
Dan Dare pocket watch,
1953. The *Dan Dare*
stories spawned a range
of toys and collectibles.
The face of this child's
watch shows a helmeted
Dan Dare with ray gun
as a monster approaches.
Source: Courtesy of Science
Museum/Science and
Society Picture Library,
London.

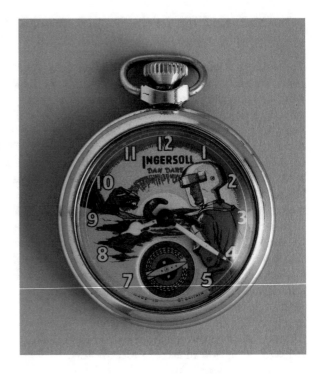

the *Dan Dare* stories. But by this time the changing cultural expectations of British youth perhaps had already been rendering *Dan Dare* increasingly out-dated. For prospective young readers in the United Kingdom, the imagination of outer space in such comics had now to compete with the (vicarious) realization of outer space in the newspapers, magazines and on television as the space shots of the 1960s gathered pace. *Dan Dare*'s harking back to the Second World War and the resiting of the United Kingdom's Fighter Command in the arena of space was now up against an actual exploration of space by the Soviet Union and the United States. As Gagarin breached the space frontier on television sets across Britain on 12 April 1961, soon followed by Shepard and Glenn, there was little room left for the fading empire of Colonel Dare. The progress and optimism recalled by Hennessy at the Science Museum was now overtaken by the realities of the Space Race.

Elsewhere this fading of the national imaginaries of outer space had long been underway. *Journey into Space* was a science-fiction radio series that ran on the BBC Light Programme, between 1953 and 1956. Its crew of characters – British, American and Australian, led by Captain Andrew 'Jet' Morgan – played out in melodramatic prose their adventures to the moon,

Mars and beyond. The opening installment of the first series, entitled 'Operation *Luna*,' attracted an impressive five million listeners, but numbers tailed off as people lost interest with a storyline that hesitated, literally, to get off the ground. But once the rocket *Luna* was spacebound, audience figures started to rise and by the final episode *Journey into Space* was attracting over 18 million listeners. Toward the end of the second series – 'The Red Planet,' the show's producers could claim, for the very last time in British broadcasting history, a radio program that was attracting a larger audience for its scheduling than the TV competition.[48]

Journey into Space's characters and settings carried resonances with the United Kingdom and its place in a postwar, Cold War, end-of-empire world. Where *Dan Dare* seemed a throwback to the gutsy aviators of the Second World War, 'Jet' Morgan and his team appeared closer to the present. Some of their adventures were 'launched' from a 'spaceport' in the dusty outback of South Australia where also, in 1946, the Anglo-Australian Joint Project had been established to allow the United Kingdom extensive, sparsely populated terrain for early rocket and missile program testing. The Joint Project range at Woomera, South Australia, went on to be used as the launch site for Britain's large rocket programs of the 1950s, 1960s and 1970s: Skylark (upper atmosphere and space sounding), Black Knight (test vehicle for the Blue Streak ballistic missile), the repurposed Blue Streak rockets of the European Launcher Development Organization and the Black Arrow satellite-launching rocket.[49] Although *Journey into Space* starts at a rocket research station in Nevada, the action then shifts to Australia and the secret Operation *Luna* base whence the crew blast off on their planetary odyssey, their dialogue a mix of eager and plucky BBC English, Hollywood American and cheeky Australian. Geppert's observation on the evolution of the geography of outer space is apt for these cosmic companions aboard the rocket *Luna*, 'An entire geography of outer space developed that presented itself as a continuation, if not a logical extension, of earlier geographies of imperial expansion and colonial domination.'[50]

Journey into Space, like *Dan Dare*, influenced many, including Alan Bond, Managing Director of Reaction Engines Ltd, the UK-based space propulsion company. In 2011 he accompanied the author to the Science Museum's store to inspect its collection of early British high-altitude pressure suits; he was trying to locate one purportedly used in the publicity and promotion of *Journey into Space*. When *Journey into Space* folded in 1956 there was no significant space radio fiction on British radio for another two decades and the arrival in 1978 of *The Hitchhiker's Guide to the Galaxy*.[51] But in this popular program the space narrative was constructed around comedy and parody, albeit hugely successfully so.

IV   Astroculture on television: *The Sky at Night* and *Doctor Who*

Radio's near monopoly in British broadcasting while television moved from the rudimentary to the more sophisticated began to be challenged when BBC TV was itself opened up to competition from commercial broadcasting in 1955. BBC television producers had now to become aware of audience ratings against the independent television sector in order to help justify their programing. They began to balance the existing, somewhat staid scheduling with a more populist and popular output. The developing space technoscience of the 1950s 'provided television broadcasters with a flexible tool that allowed producers to respond to changing demands: it had educational, informative and entertaining aspects – especially following Sputnik – that could be emphasized in different proportions.'[52] The astroculture of televised space was proving more beguiling than that of radio.

The technological and electronic surge from the Second World War had ushered in a new age of astronomy. The development of radar in particular had provided the necessary techniques and equipment with which to pursue radio astronomy. The United Kingdom became a global leader in this new discipline with research centers set up at the universities of Cambridge and at Manchester at the Jodrell Bank observatory in Cheshire.[53] Infrared, ultraviolet and X-ray observation were now becoming a reality, too, with clusters of spectrum-specific scientists all eager to prize grants from the national and international funding agencies. Much of this work, while adding by orders of magnitude to the professional scientists' perception and understanding of the universe, remained out of reach to the lay astronomer and more remote still to the general public.

Nevertheless, 'amateur' astronomy continued to prosper – the second most popular leisure activity in the United Kingdom after fishing, it has been said – and it was with this tradition and community in mind that the BBC launched a new television program in April 1957 that would merge the new with the old and with a regular astronomy and space update for television audiences. Just six months after its start, the program would report the launching of Sputnik and its ushering in of the new Space Age. The program was called *The Sky at Night* and ran monthly for 15 minutes and very late in the evening. It provided news and expert advice on current astronomical phenomena and viewing opportunities, very much continuing traditions established in print (by popular journals and magazines) and pursued through the activities of the many astronomy societies long established across the country.

Sky at Night became what some might describe as a national institution. It still runs today following the same basic format of short features on contemporary celestial subjects with imported expert advice provided by a relevant astronomer or space scientist. Until 2012, the program continued to be fronted by its original presenter, Patrick Moore (1923–2012).[54] Moore had been approached to front *Sky at Night* following his robust performance dismissing the existence of UFOs and the views of their learned proponents. He was an author, writer, practicing amateur astronomer and active member of the British Interplanetary Society. Over the next five decades, Moore's pithy, distinctive style – much aped by television impressionists – both informed and entertained millions, and he did as much as anyone to communicate and enthuse the country of the new astroculture.

Sky at Night was the United Kingdom's and certainly the BBC's main outlet for matters concerning space.[55] Even during the 1960s there was no other regular slot for a spaceflight-orientated factual series. Documentaries on the accelerating Space Race between the United States and the Soviet Union were few and far between, just one or two appearing in the BBC's flagship science documentary series 'Horizon.'[56] However, both the BBC and Independent Television invested greatly in their 'news and features' coverage of the American Apollo program. Presenters familiar to viewers from their work on other programs – like James Burke from *Tomorrow's World*, the BBC's science magazine program – became synonymous with Apollo, much as Patrick Moore had with astronomy – both becoming 'space *personae*' of the television age. When Apollo ceased in the early 1970s, so too did the newsworthiness of space on British television. The occasional documentary continued to appear, but the more involved treatments and series had to wait for the development of the sophisticated television technology of the digital age when computer graphics would augment footage with spectacular if mostly speculative imagery.

In 1970 the nation's most popular science-fiction television serial – *Doctor Who* – had returned to the screens after an unusually long break. For seven years previously the series' eponymous hero, the Doctor, had been traveling through space and time – and throughout the BBC year – in his 'Tardis' time machine, seeing off monsters, alien adversaries and sundry cosmic catastrophes.[57] *Doctor Who* had become one of the BBC's most successful programs with an entire generation of youngsters charging around school playgrounds on Monday lunchtimes screeching 'exterminate,' aping the killing orders of Who's most notorious villains, the Daleks, from the previous Saturday's episode. But now the Doctor had been exiled to earth by his masters, no longer allowed to explore the cosmos in his Tardis. The alien confrontations continued but in very terrestrial settings. In reality, the Doctor had been grounded because actors had made clear how exhausting the near year-round rehearsing and filming had become, and the BBC itself could no longer afford the escalating costs of set builds during what was a period of major expense as the Corporation introduced color TV broadcasting (Figure 4.3).

In some ways *Doctor Who*'s fall to earth in 1970 was a return to form for UK television and cinema's depiction of space science fiction. Neither had embraced the genre as American cinema and TV had. *Quatermass* (1955), for example – a hugely influential representation of science fiction in the United Kingdom on the small and big screen – was in the main set firmly on terra firma, even when dealing with alien life forms. The seminal *2001: A Space Odyssey,*

Figure 4.3 *Doctor Who* cast members Jon Pertwee (left), Katy Manning and Nicholas Courtney (right, rear) in character at the opening of the BBC TV Visual Effects exhibition at the Science Museum, London, in December 1972. Jon Pertwee was the first Doctor to be filmed in color.

Source: Courtesy of Science Museum/Science and Society Picture Library, London.

despite its very British credentials of cast and crew was, like the accent of its screen writer Arthur C. Clarke, a mid-Atlantic production that drew on American funding and the expertise of its national space-technology sector.[58] Critic John Oliver bemoans it 'as a defining moment in the diminution of any identifiable British characteristics that American-financed films shot in the United Kingdom, such as *MGM*'s own *Village of the Damned*, may have once possessed.'[59]

Nevertheless, *2001* influenced a subsequent British TV saga, *Space: 1999*, which ran from 1975 to 1977 and had itself evolved from the canceled second series of another title, *U.F.O.*[60] *Space: 1999* boasted high production values, not least because of the production personnel who had come from working on *2001*. But as art historian Henry Keazor has argued, it is unclear whether *Space: 1999*, with its minimalist, 1960s cut of dress and set, was a 'nostalgic look back or an encouraging look into the future.'[61] It was, perhaps, an exceptional excursion into imagined outer space on television and one informed by influences far broader in scope than any specifically British ones. From 1969, UK viewers had started to devour the first series of *Star Trek*, which had been bought by the BBC to fill the extended summer break of *Doctor Who*. The American hegemony of space science fiction continued to go boldly where British broadcasters feared to tread.

V   A grounding in space?

If we consider Geppert's 'heterogeneous array of images and artifacts, media and practices' as key components of an astroculture, then it is reasonable to group them into categories of the imaginary and of the actual – or technoscientific. Either is capable of imparting 'meaning to outer space,' and either can stimulate 'the individual and collective imagination.' It follows, however, that the proportions of each will vary in time as the nature and quantity of the components themselves change. The components of British astroculture during the 1950s and, for the sake of chronological simplicity, before the launch of Sputnik 1, included, inevitably, perhaps, a greater proportion of the imaginary. They assumed the forms of the BIS's activities and outputs, the printed adventures of comic book heroes like *Dan Dare* and the radio broadcasts of 'Jet' Morgan's *Journey into Space*. But with Sputnik came a sudden expansion in the technoscientific, and television, in particular, embraced this astrocultural category. This relatively new medium – its own technology evolving quickly, and the monopoly of the BBC challenged by newly enfranchised independent broadcasters – seized space from the radio, its producers regarding 'this, space technoscientific material as possessing intrinsic cultural importance [with] spectacular, dramatic and, above all, visual qualities.'[62]

Space radio had given way to Space TV – mostly the factual of the technoscientific, live by satellite from Cape Kennedy. Even the Andersons' imaginary *Thunderbirds* had been informed by the real, that is the Mercury 7 astronauts. The son of *Doctor Who* actor Patrick Troughton recalls that:

By Christmas [1968] my focus had turned from [the] science fiction of Doctor Who to science fact. Manned exploration of the Moon seemed a lot more exciting. Late in the evening on Christmas Eve, my father and I watched in silent amazement as live transmissions were beamed to Earth of spectacular views of the lunar surface. I think for Dad and myself, this was a time of change and awakening. For me, the lure of science fact began to offer far more excitement and interest than the false promises in a world of science fiction. For Dad, he had come to realize that time was precious. He needed to move on in his life and that meant leaving Doctor Who as soon as possible.[63]

UK television's pursuit of the technoscientific was matched elsewhere, the BIS complimenting television's reporting of the space shots of the 1960s with its own coverage in print and event, including that of British activity. By the 1970s the Society was expressing a wish to resuscitate the imaginary. Reality was now dominant and constraining; technoscience was eclipsing the imaginary.

In seeking an answer to the question of whether the 1970s was a period of transition in the United Kingdom's exploration of outer space, a broad form of enquiry has been adopted in which the exploration is considered in terms of astroculture and technoscience. Following Geppert's emphasis on the heterogeneity of astrocultural forms, this chapter suggests that astroculture may indeed include elements and representations of technoscience as well as those of the imaginary. In other words, the astrocultural comprises both the imaginary and the technoscientific, the respective proportion of each likely to shift over time. In three of the four astrocultural areas perused in this chapter – British Interplanetary Society activity; popular comic-book literature; and radio drama – the imaginary was well represented during the early to mid-1950s but less so at and following the launch of Sputnik 1. Before Sputnik, the imaginary worked, unbound by the constraints of the real. With the opening of the so-called Space Age, the imaginary was increasingly informed by the reality of what was now actually happening in space. UK television maintained a modicum of the imaginary over the years with science-fiction drama and children's programs, yet in the 1970s this genre too increasingly drew on the actuality of ongoing space activity in set designs and narratives, while the formerly space and time-traveling Dr. Who abandoned his cosmic travels altogether in his exile on earth. Television, however, embraced the technoscientific with alacrity in the late 1950s as it began its coverage of the details and drama of the space missions that followed through the 1960s. Yet, this withered too in the 1970s and the end of the Apollo missions to the moon; Skylab and the Apollo-Soyuz Test Project failed to recapture the spectacle and suspense of the Space Race.

UK space activity, seemingly technoscientific, had always struggled as a distinct policy area. Successive British governments through the 1960s had little interest in assigning space its own coordinating council or agency. Rather, space activities – inherited or accumulated from the interests of other parties: defense, science, telecommunications – were considered piecemeal, as extensions and responsibilities of terrestrial departments and ministries with their respective heads and ministers always mindful of their own priorities. In this manner, UK space policy retained an element of consistency throughout these years, the 1970s witnessing little change in this trait.

In short, this brief examination through astrocultural and technoscientific lenses of how Britain was exploring outer space in the mid-twentieth century suggests a 'grounding' well before the 1970s. The imaginary that had been building through the 1950s (and for many years before) was limited by the realities of Sputnik and what followed into space. The mind's eye was now constrained by what the physical eye was seeing. The change occurred with the opening of the Space Age and not with its 'closure' post-Apollo. In turn, actual UK space activity was never deemed a discrete entity by the polity, and the 1970s saw little change in this judgment. While this seemingly challenges the notion of outer space being 'key to the self-image of public, governmental and technical elites, and to modernist narratives of progress,' one might also suggest that other nations' more ambitious explorations of space were similarly grounded by very terrestrial priorities and that the difference to the United Kingdom was one of degree rather than type.[64]

Notes

1. As remembered by a member of the launch team in C.N. Hill, *A Vertical Empire: History of the British Rocketry Programme, 1950–1971*, 2nd edn, London: Imperial College Press, 2012, 310. Black Arrow was a small satellite-launching rocket derived from the Black Knight ballistic missile test vehicle.

2. Hill juxtaposes the two terms 'space program' and 'rocketry program,' implying the two are interchangeable; ibid., 1.

3. John Krige, Angelina Long Callahan and Ashok Maharaj, *NASA in the World: Fifty Years of International Collaboration in Space*, Basingstoke: Palgrave Macmillan, 2013, 25.

4. The United Kingdom's Skylark sounding rocket had been developed to return data on the upper atmosphere in support of the country's ballistic missile program but was made available for civilian scientific research. The first data set was returned in November 1957, representing also some of the earliest space data returned during the International Geophysical Year. For a comprehensive history of the rocket and its use in the early European space science effort, see Matthew Godwin, *The Skylark Rocket: British Space Science and the European Space Research Organisation, 1957–1972*, Paris: Beauchesne, 2007. The first UK satellite payload comprised six experiments on Ariel 1, built and launched by the United States, on 26 April 1962. Over the next three years two more satellites in the six-satellite Ariel program were launched, the third designed and built entirely in the United Kingdom. For a comprehensive history of Ariel and the early UK space-science program, see Harrie Massey and Malcolm O. Robins, *History of British Space Science*, Cambridge: Cambridge University Press, 1986, and for its international origins, see Krige et al., *NASA in the World*.

5. The civil-orientated 'Space Policy Review Committee: Report,' chaired by P. Rogers, Deputy Secretary of the Cabinet, in TNA/PRO CAB 134/2461: RSP(65)1, and the 'Space Policy Review: Military Interests in Space,' in TNA/PRO CAB 134/2545; T(O)(65)20, chaired by mathematician and cosmologist Hermann Bondi, both reported in 1965. In 1967 Parliament investigated UK space activities in the 'Thirteenth Report from the Estimates Committee,' while two years later the Cabinet Communications, Electronics and Space Committee delivered its own report; see 'The United Kingdom Space Policy: 1969 Review,' in TNA/PRO CAB 134/2649: CSC(69)9 (Final). Two years after that came the House of Commons Select Committee on Science and Technology's Fifth Report: 'United Kingdom Space Activities.'

6. Alexander C.T. Geppert, 'Rethinking the Space Age: Astroculture and Technoscience,' *History and Technology* 28.3 (September 2012), 219–23, here 219.

7. Geppert defines astroculture as a 'heterogeneous array of images and artifacts, media and practices that all aim to ascribe meaning to outer space while stirring both the individual and the collective imagination.' In seeking to define technoscience, he cites John Pickstone's description of 'technological projects which are heavily dependent on science (or vice versa),' explicitly including 'high-tech projects funded and directed by state agencies, or even by state-backed academic consortia.' Ibid., 220.

8. Black Arrow was the launch vehicle for the satellites of the United Kingdom's National Space Technology Program, designed to test key satellite technologies and components in space. For a firsthand account of these including Prospero, see B.P. Day and R.H. Gooding, *RAE Farnborough Space Department: A History*, London: British Interplanetary Society, 2015, chapter 8, 'Black Arrow Satellites,' 76–87.

9. For an in-depth discussion of why the Black Arrow program was canceled, see House of Commons, 'United Kingdom Space Activities,' Fifth Report from the Select Committee on Science and Technology, London: HMSO, 1971.

10. See Douglas Millard, *The Black Arrow Rocket: A History of a Satellite Launch Vehicle and Its Engines*, London: NMSI, 2001.

11. For an overview of the research performed under the Black Knight program, see Douglas Millard, *Discriminating Research: The Black Knight Research Programme*, MSc thesis, Centre for the History of Science, Technology and Medicine, Imperial College, London, 2005.

12. See D.G. King-Hele and D.M. Gilmore, *Technical Report G.W. 455: The Use of Blue Streak with Black Knight in a Satellite Missile*, May 1957, TNA: PRO AVIA 6/19852.

13. Krige et al., *NASA in the World*, 25.

14. Ibid.

15. Letter from Richard Porter to Professor van de Hulst, 14 March 1959, in John M. Logsdon, ed., *Exploring the Unknown: Selected Documents in the History of the U.S. Civil Space Program*, vol. 2: *External Relationships*, Washington, DC: NASA, 1996, 18.

16. See Godwin, *Skylark Rocket*.

17. For a comprehensive history of ESRO and ELDO, see John Krige and Arturo Russo, 'The Story of ESRO and ELDO, 1968 to 1973,' *A History of the European Space Agency, 1958–1987*, vol. 1, Noordwijk: ESA, 2000.

18. Krige et al., *NASA in the World*, 52.

19. Ibid., 59.

20. See Note 5 above.

21. 'Ministerial responsibility for space should not be vested in a single Minister and that each aspect of space should be the responsibility of the minister most closely concerned with the corresponding terrestrial activities.' 'Thirteenth report from the Estimates Committee, Session 1966–67, 27 July 1967,' London: MSO, 1967, viii. See Godwin, *The Skylark Rocket*, 249.

22. Prime Minister Edward Heath took the United Kingdom into the European Common Market on 1 January 1973.

23. See John Krige, *Fifty Years of European Cooperation in Space: Building on Its Past ESA, Shapes the Future*, Paris: Beauchesne, 2014, 166.

24. Robert A. Harris, ed., *The History of the European Space Agency: Proceedings of an International Symposium, 11–13 November 1998, The Science Museum, London*, Noordwijk: ESA, 1999, 26.

25. The United Kingdom's first space agency was established on 1 April 2010.

26. Geppert, 'Rethinking the Space Age,' 220.

27. For a detailed appraisal of the Society's influence on the culture of 'British outer space,' see Oliver Dunnett, *The British Interplanetary Society and Cultures of Outer Space*, PhD thesis, University of Nottingham, 2011.

28. For a full history of the V-2 and associated rocket program, see Michael J. Neufeld, *The Rocket and the Reich: Peenemünde and the Coming of the Ballistic Missile Era*, New York: Free Press, 1995.

29. See Andrew Chatwin, ed., *Val Cleaver: A Very English Rocketeer*, London: British Interplanetary Society, 2015, and Oliver Dunnett, 'Patrick Moore, Arthur C. Clarke and "British Outer Space" in the Mid Twentieth Century,' *Cultural Geographies* 19.4 (2012), 505–22. The Society's *Journal*, first published in 1934, was supplemented with the *Spaceflight* monthly magazine from 1956.

30. Leonard J. Carter, ed., *The Artificial Satellite: Proceedings of the 2nd International Congress on Astronautics*, London: British Interplanetary Society, 1952.

31. Ibid.

32. US military thinking on artificial satellites can be traced back to 1945. See Robert L. Perry, *Origins of the USAF Space Program, 1945–1956*, originally printed as *Volume V (Space Systems Division Supplement), History of Deputy Commander (AFSC) for Aerospace Systems, 1961*, reprinted by History Office, Space and Missile Systems Center, 1997; available at http://fas.org/spp/eprint/origins/index.html (accessed 1 October 2017).

33. Thore Bjørnvig, 'Transcendence of Gravity: Arthur C. Clarke and the Apocalypse of Weightlessness,' in Alexander C.T. Geppert, ed., *Imagining Outer Space: European Astroculture in the Twentieth Century*, Basingstoke: Palgrave Macmillan, 2012, 127–46, here 129 (= *European Astroculture*, vol. 1). See also Bjørnvig's contribution, Chapter 7 in this volume.

34. Clarke's 'The Sentinel' was first published in book form in the *Expedition to Earth* collection of short stories, London: Sidgwick & Jackson, 1954. *2001: A Space Odyssey*, directed by Stanley Kubrick, USA 1968 (Metro-Goldwyn-Mayer). See also Robert Poole's contribution, Chapter 5 in this volume.

35. Sounding rockets carry scientific instruments to high altitude for the study of space, the upper atmosphere or for photographing the earth and its surfaces. They follow high, slender parabolic trajectories. Early US sounding rockets employed surplus V-2 missiles as their first stages. The UK Skylark sounding rocket had been developed to return data on the upper atmosphere in support of the United Kingdom's nascent ballistic missile program. S. Fred Singer was aware of this and contacted Professor Harrie Massey (1908–83) of University College London, suggesting that the university might be interested with others in talking to the design authority of Skylark – the Royal Aircraft Establishment – with a view to using it for launching university experiments into space.

36. 'Space Flight Chronology, 1890–1956,' in Robert L. Perry, *Origins of the USAF Space Program 1945–1956*, n.p., 1961, here vi–viii; available at http://www.pax-americana.com/documents/origins.pdf (accessed 1 October 2017). 'Korolev and Freedom of Space: February 14, 1955–October 4, 1957,' in

David S.F. Portree, *NASA's Origins and the Dawn of the Space Age*, Washington, DC: NASA, 1998; available at http://history.nasa.gov/monograph10/korspace. html (accessed 1 October 2017). On the IGY more generally, see Roger D. Launius, James Rodger Fleming and David H. DeVorkin, eds, *Globalizing Polar Science: Reconsidering the International Polar and Geophysical Years*, Basingstoke: Palgrave Macmillan, 2010, and Hervé Moulin, 'The International Geophysical Year: Its Influence on the Beginning of the French Space Program,' *Acta Astronautica* 66.5/6 (March 2010), 688–92.

37. Geppert, 'Rethinking the Space Age,' 219–20.
38. See, for example, S.W. Smith, ed., *Teachers Handbook of Astronautics*, London: British Interplanetary Society, 1963; M.S.S. Hunt, *The History of Astronautics Since the War*, London: British Interplanetary Society, 1963.
39. See, for example, Kenneth W. Gatland, ed., *Spaceflight Today*, published in conjunction with the British Interplanetary Society, London: Liffe Books, 1963.
40. See, for example, Dominic Phelan, *Cold War Space Sleuths: The Untold Secrets of the Soviet Space Program*, Dordrecht: Springer, 2012, 3–4.
41. See Krige et al., *NASA in the World*, 148–9.
42. The Society's motto 'From Imagination to Reality' was adopted in 1987 and inspired by the subtitle of the Society's book *High Road to the Moon: From Imagination to Reality*. See R.A. Smith and Bob Parkinson, *High Road to the Moon: From Imagination to Reality*, Woking: British Interplanetary Society, 1979, 60, 71.
43. Bob Parkinson, 'The Daedalus Years 1970–1981,' in idem, ed., *Interplanetary: A History of the British Interplanetary Society*, London: British Interplanetary Society, 2008, here 54; Alan Bond, ed., *Project Daedalus: The Final Report on the BIS Starship Study*, London: British Interplanetary Society, 1978 (= *Journal of the British Interplanetary Society Supplement*).
44. See, for example, Kevin F. Long, Richard. K. Obousy and Andreas M. Hein, 'Project Icarus: Optimisation of Nuclear Fusion Propulsion for Interstellar Missions,' *Acta Astronautica* 68.11–12 (2011), 1820–9.
45. *Dan Dare and the Birth of Hi-Tech Britain*, exhibition at the Science Museum, London, May–November 2008.
46. Oliver Dunnett, 'The British Interplanetary Society and Cultures of Outer Space,' 254.
47. Dunnett recounts Morris's admiration of C.S. Lewis's Christian values but his viewing of 'the modernity of space exploration in a positive light, as opposed to Lewis, who considered it an ominous prospect.' Ibid., 244.
48. James Farry and David A. Kirby, 'The Universe will be Televised: Space, Science, Satellites and British Television Production, 1946–1969,' *History and Technology* 28.3 (September 2012), 311–33, here 314.
49. The first satellite to be launched successfully from Woomera, four years prior to Black Arrow's Prospero, was Australia's WRESAT. For a comprehensive history of the Joint Project, see Peter Morton, *Fire Across the Desert: Woomera and the Joint Project, 1946–1980*, Canberra: Australian Government Publishing Service, 1989.
50. Alexander C.T. Geppert, 'European Astrofuturism, Cosmic Provincialism: Historicizing the Space Age,' in idem, *Imagining Outer Space*, 3–24, here 3.

51. *The Hitchhiker's Guide to the Galaxy* was a 1978 BBC radio series later adapted by its author Douglas Adams into a series of books, a stage play, and a computer game.
52. Farry and Kirby, 'The Universe will be Televised,' 312.
53. Jon Agar, *Science and Spectacle: The Work of Jodrell Bank in Post-War British Culture*, Amsterdam: Harwood Academic Publishers, 1998.
54. His death brought with it a period of uncertainty as the BBC decided how and indeed whether to continue with a program that had been and is likely to remain the world's longest running with the same presenter.
55. For a comprehensive analysis of UK space television broadcasting and its evolution in the postwar years, see Farry and Kirby, 'The Universe will be Televised.'
56. For a history of UK science documentary that suggests a relative paucity of space coverage, see Tim Boon, *Films of Fact: A History of Science in Documentary Films and Television*, London: Wallflower Press, 2008.
57. Tardis is an acronym formed from 'Time And Relative Dimension In Space.' It is a time-traveling spaceship whose internal dimensions exceed its external ones. Externally it resembles a 1960s police telephone call box, its jammed 'chameleon circuit' preventing it adapting to the appearance of its surroundings.
58. See Robert Poole's contribution, Chapter 5 in this volume, for a listing of the principal secondary sources on the film, its making and significance.
59. John Oliver, 'Science Fiction: Britain's Distinctive Take on the Future,' in *British Film Institute Screenonline: The Definitive Guide to Britain's Film and Television History*, available at http://www.screenonline.org.uk/film/id/446205/index.html (accessed 1 October 2017).
60. Both series were produced by husband and wife team Gerry and Sylvia Anderson who had achieved tremendous success on UK and American television with children's series like *Fireball XL5* and *Thunderbirds*.
61. Henry Keazor, 'A Stumble in the Dark: Contextualizing Gerry and Sylvia Anderson's *Space: 1999*,' in Geppert, *Imagining Outer Space*, 189–207, here 191.
62. Farry and Kirby, 'The Universe will be Televised,' 313.
63. Michael Troughton, *Patrick Troughton: The Biography of the Second Doctor Who*, Andover, Hants: Hirst, 2012, 245.
64. Geppert, 'Rethinking the Space Age,' 219.

PART II

Reconfiguring Imaginaries

CHAPTER 5

The Myth of Progress: *2001 – A Space Odyssey*

Robert Poole

The 1968 film and novel *2001: A Space Odyssey* marked the cultural apex of the Space Age.[1] It was an Anglo–American project, the joint creation of the leading film director Stanley Kubrick (1928–99), famous for the nuclear war satire *Dr. Strangelove*, and the leading science-fiction and popular science writer Arthur C. Clarke (1917–2008; Figure 5.1), famous for his 1945 prediction of the communications satellite.[2] *2001* was the first science-fiction film to have genuinely plausible special effects. Its making was a miniature space program in its own right, breaking all previous records for production costs and generating considerable public interest; it was, in relation to the science of its day, the most scientifically accurate feature film ever made.[3] It was made in the mid-1960s at a time when space programs were accelerating on all fronts. The early human-cannonball style flights of the Mercury and Vostok programs were over and the more complex Gemini and Voskhod missions saw the first astronauts performing orbital maneuvers, walking in space and sending back stunning pictures of the blue planet below. The first space probes had recently reached Venus and Mars, their senders still optimistic about detecting signs of life. *2001* opened across the United States in the spring of 1968, and across the rest of the world in the spring and summer, during a lull in the US manned space program caused by the Apollo fire disaster of January 1967. *2001* was for a time the biggest show in space; it attracted record audiences, and was still running in many cities more than a year later as the Apollo 11 astronauts landed on the moon.[4] Arthur C. Clarke was then on the television commentary team for CBS, a prophet in his own

Robert Poole (✉)
University of Central Lancashire, Preston, UK
e-mail: RPoole@uclan.ac.uk

© The Author(s) 2018
Alexander C.T. Geppert (ed.), *Limiting Outer Space*
European Astroculture, vol. 2
https://doi.org/10.1057/978-1-137-36916-1_5

Figure 5.1   British science-fiction author Arthur C. Clarke on 9 June 1952, reading from his latest bestseller *The Exploration of Space*. Photograph by Bill Allen.
Source: Courtesy of Picture Alliance.

future, his words broadcast round the world on the communications satellites whose advent he had predicted. Never have fiction and reality run so close; there is a case for saying that the real winner of the Space Race was *2001: A Space Odyssey*.[5]

At its inception in 1964, *2001* had promised to be a kind of propagandist docudrama on the model of *Destination Moon* (1950) and *The Conquest of Space* (1955), designed with the best available scientific advice to make the case for space travel in a popular form.[6] It was at first trailed as *Journey Beyond the Stars*, an exciting saga of the exploration of the solar system, and it was the prospect of realistic scenes of space travel, as depicted on posters by the NASA artist Robert McCall (1919–2010), that initially brought audiences flooding in (Figure 5.2). Keen to stay ahead of present-day reality however,

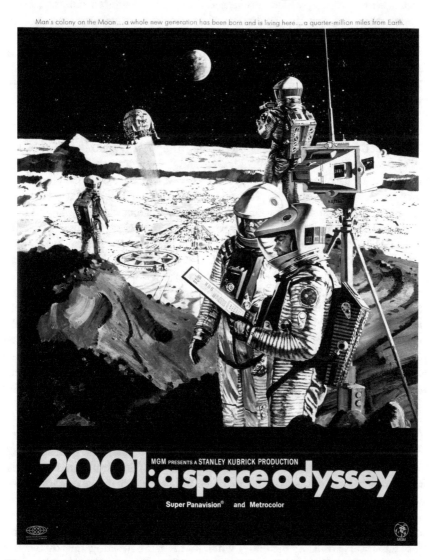

Figure 5.2   A 1968 poster for *2001: A Space Odyssey*. Audiences who came expecting a 1950s-style space adventure encountered something very different.
Source: Courtesy of Warner Brothers.

Kubrick and Clarke had raised their sights further into the future – to the moment of first contact with extraterrestrial intelligence, which at that time was widely anticipated – particularly in Europe – as a consequence of space travel. They explained that their aim was to prepare the public for the impact of such an encounter, which threatened shock but promised enlightenment. Kubrick went as far as filming dozens of interviews with leading scientists and philosophers on the subject of extraterrestrial intelligence for a documentary prologue, although the plan was in the end scrapped, leaving a legacy of interviews of considerable historical interest.[7]

Also scrapped, to the dismay of the serious space community, were almost all the traditional explanations about the technology of space travel. The 1940s and 1950s had been the heyday of 'hard' science fiction and the gadget story, and the much publicized attention to technological detail in the production of *2001* had led the public to expect something similar, but there were to be no 'now tell me, Professor' moments in Kubrick's film.[8] Even the narrative voiceovers, regarded as essential by Kubrick's advisors in order to explain the obscure plot, were abandoned. 'This isn't a proper science fiction movie at all,' thought one leading science-fiction writer, Lester del Rey (1915–93). 'A wasted opportunity,' complained another, Frederick Pohl (1919–2013). Kubrick described such reactions as 'obtuse.'[9] His aim, concealed from his eager technical staff, was to achieve a future environment plausible enough to allow the viewers to take it for granted and concentrate on the real story. An early section depicting lunar exploration in the near future came closest to satisfying expectations, with its glorious space ballet to the soundtrack of Johann Strauss II's 'Blue Danube' waltz and its realistically styled account of the discovery of an alien artifact on the moon. The subsequent long voyage to Jupiter (in the novel, Saturn) to investigate the apparent source of the alien artifact also contained plenty of conventional space interest, although the plot's central conflict was not between humans and aliens but between the astronauts and their rogue computer.

The recognizably science-fiction elements of the film were framed by a 20-minute wordless prologue entitled 'The Dawn of Man,' set at the dawn of human evolution in Africa, and by a mysterious, psychedelic final sequence which came to be known as 'the ultimate trip.' The film ended with the image of a human fetus floating in space above the earth, gazing down on the entire globe as the eyes of humanity, in the persons of the Apollo 8 astronauts, would do for the first time in December 1968. The film as a whole was slow, hypnotic and enigmatic. In interviews Kubrick encouraged philosophical speculations about its meaning, while all the time insisting that he did not give explanations. This was Kubrick's explanation, 'on the *lowest* level,' of the final stages of the film:

> When the surviving astronaut, Bowman, ultimately reaches Jupiter, this [alien] artifact sweeps him into a field of force or star gate that hurls

him on a journey through inner and outer space and finally transports him to another part of the galaxy, where he's placed in a human zoo approximating a hospital terrestrial environment drawn out of his own dreams and imagination. In a timeless state, his life passes from middle age to senescence to death. He is reborn, an enhanced being, a star child, an angel, a superman, if you like, and returns to Earth prepared for the next leap forward of man's evolutionary destiny.[10]

In the long run, it was its mystical aspect which accounted for the cult status of *2001: A Space Odyssey.*

2001 offered a vision of progress on the grandest scale, in which space travel is a natural extension of the earliest human technology, and contact with extraterrestrial intelligence triggers a step change in human evolution. On closer examination, however, there is a darker side to both film and novel: the hints that preparations for nuclear war are more advanced than those for first contact; the emotionless and deceitful human characters; the equally emotionless and deceitful supercomputer HAL 9000 who derails the mission to Jupiter; and the dark fable of the origin of humanity through weaponry and violence in the 'Dawn of Man' sequence. If *2001* is about transcendence, it is equally about the limits that have to be transcended. Sandwiched between the atomic black comedy *Dr. Strangelove* (1964) and the dystopian *A Clockwork Orange* (1972), a film that was simply a peon to progress would have been an anomaly in the career of a director whose *œuvre* was dominated by films exploring the human capacity for violence and deception. Much of the interest of *2001* for the historian lies in the contrast between its secular and progressive outward message and its more philosophical and pessimistic inner core. This yields insights into both the tensions within Western astroculture and the sources of its decline in the 1970s.

The long-running critical debate about the film has received extensive attention; much of it has been published.[11] More recently, however, the opening of both the Stanley Kubrick Archive at the University of Arts London and the Arthur C. Clarke Collection at the National Air and Space Museum in Washington, DC has allowed a more thorough historically based exploration of the making of *2001: A Space Odyssey.*[12] Using draft novel texts, scripts and screenplays, together with correspondence, research and production materials and other works by the film's creators, this chapter explores two elements of the film where the ambiguous attitude toward human progress is most marked: on the one hand the threat of nuclear war, and on the other the 'killer ape' model of human origins portrayed in the prologue of 'The Dawn of Man.' *2001* posed questions about the limits to the progress of a nuclear-armed primate on a single planet; it was a film equally about human origins and the human future.

Figure 5.3 Desolation and Progress, the two potential roads of atomic energy as portrayed in the British Atomic Scientists Association exhibition 'Atom Train' in Dorland Hall, London, 23 January 1947. The exhibition was the first to present an overall view of atomic energy to the public. Photo by Reg Birkett.

Source: Courtesy of Getty Images Warner Brothers.

I   The atomic crossroads

The idea that the invention of the nuclear bomb had brought mankind to a crossroads was part of the cultural architecture of the Cold War (Figure 5.3). The first atom bomb tests at Bikini Island in 1946 were code-named 'Operation Crossroads.' In the 1930s, English writer H.G. Wells (1866–1946) had characterized contemporary history as 'a race between education and catastrophe,' and with the atomic bombing of Japan in August 1945 the moment of choice seemed to have arrived. An editorial in *The Times* of London on 16 August 1945 warned: 'It must be made impossible for war to begin, or else mankind perishes.'[13] 'We are now faced with the naked choice between world cooperation and world destruction,' Britain's Prime Minister Clement Attlee told the House of Commons. The US Secretary of State for War Henry Stimson wrote in September 1945 that the present time marked 'the climax of the race between man's growing technical power for destructiveness and [...] his moral power.' Vannevar Bush, President Truman's science advisor, wrote: 'Down one path lies a secret arms race, down the other international collaboration and possibly ultimate control.' Journalist Norman Cousins, in a far-reaching early response, argued that the coming of the atomic bomb 'marked the violent death of one stage in man's history and the beginning of another.' The challenge to abolish war was 'the largest order man has had to meet in his 50,000-odd years on Earth.'[14] The wave of public anxiety about the future of civilization in the atomic age was led by the atomic scientists themselves. Their manifesto was entitled *One World or None?*, the question mark indicating an inescapable choice.[15] Britain's atomic scientists produced a similar volume entitled *Atomic Challenge*; their leading spokesman was the philosopher Bertrand Russell (1872–1970). Einstein led the American atomic scientists in promoting international control of the bomb and world government as the solution to the dilemma. There was a widespread consensus that human nature had to catch up with technology if it were to have a future.[16]

This sense of mankind at an atomic crossroads prompted an anxious search for alternative scenarios to destruction. The most popular of these, after world government, was spaceflight. 'Man will soon reach the planets unless we first destroy ourselves,' the science-fiction writer and philosopher Olaf Stapledon (1886–1950) told Clarke and other members of the British Interplanetary Society (BIS) in 1948.[17] It was widely expected that serious space travel would require atomic and not merely liquid-fuelled rockets, making the issue of atomic energy especially sharp. American astronomer and space advocate Carl Sagan (1934–96), who was approached – unsuccessfully – to be scientific advisor to *2001: A Space Odyssey*, wrote: 'The modern rocket [...] will prove to be either the means of mass annihilation through a global thermonuclear war or the means that will carry us to the planets and the stars.'[18] Kubrick and Clarke shared this set of assumptions. 'Space exploration is just about the only activity the human race will turn to as an alternative to blowing itself up,' declared Kubrick on the jacket of the novel of *2001: A Space Odyssey*. The statement echoed Clarke's claim in 1948 that with the advent of nuclear weapons 'interplanetary travel is now the only form of "conquest and empire" compatible with civilization.'[19] The leap into space was also a leap to safety.

Concerns about nuclear war lay behind the rapid growth in the late 1950s and early 1960s of the scientific version of the belief in extraterrestrial intelligence. The first serious paper on the subject in a mainstream scientific journal was published in 1959, and the first attempt at detecting alien radio signals was in 1961.[20] Both were founded on the assumption that communicating civilizations were necessarily long-lived, which in turn was based upon the assumption that atomic war was a hazard of the 'technological adolescence' of intelligent species rather than an enduring characteristic of advanced societies. If the search for extraterrestrial intelligence failed to make contact, argued Sagan, 'the most likely explanation [...] is that societies commonly destroy themselves before they are advanced enough to establish a high-power radio-transmitting service.'[21] The dangerous period was between the discovery of atomic energy and the migration into space, a period when a culturally immature civilization with the propensity to blow itself up was confined to a single planet. Once this phase was passed there was no limit to the power and longevity of civilizations. Indeed, the immense distances between the stars mean that proponents of extraterrestrial intelligence were positively required to build into their calculations an average lifespan for technological civilizations extending to tens of millions of years if there were to be any reasonable chance of their overlapping in both time and space. Allied civilizations would by definition be both very peaceful and highly advanced. When *homo sapiens* made contact it was likely to be welcomed into a 'galactic club' – provided that it posed no threat to the cosmic peace.[22] The vision of an Olympian galactic civilization, free from the vice and violence of present-day Western civilization, was an essential way in which the serious and scientific style of belief in extraterrestrial life sought to distance itself from the 'bug-eyed monster' school of pulp fiction and B-movies and the wilder speculations of ufologists.[23]

In an interview with *Playboy* magazine after the film had come out, Kubrick argued that, while first contact carried a risk of 'severe psychological dislocations,' it would most likely enrich society immeasurably. He went on to explain the connection between avoiding nuclear war and alien encounter:

Another positive point is that it's a virtual certainty that all intelligent life must at one stage in its development have discovered nuclear energy. This is obviously the watershed of any civilization; does it find a way to use nuclear power without destruction and harness it for peaceful purposes, or does it annihilate itself?

Any detectable aliens would be bound to have negotiated that watershed, showing that survival was possible; they might even 'give us specific guidelines for our own survival.'[24] The solution offered by *2001* was an optimistic one: salvation from the human propensity for war through contact with extraterrestrial intelligence. The outward optimism of the first Space Age was intimately bound up with the darkest fears of the twentieth century, and this was as true for Western society as a whole as it was for the makers of *2001*.

Kubrick's previous film had been the dark atomic satire *Dr. Strangelove*, made in 1962–63 when the Cold War theology of deterrence was at its height. In it, a false alert of a Soviet attack is set off by a mentally unbalanced American base commander, and the United States embark on a path to nuclear war, propelled not by any great forces of history but by accidents, misjudgements and personality disorders. In the end the bomb drops and mankind is wiped out. *2001* grew out of *Dr. Strangelove*, as Peter Krämer has explained in a succinct discussion of its origins. In one version of the script, Kubrick had experimented with the framing device of a commentary from the point of view of an alien civilization, explaining the end of human civilization as the result of failure to form a world government in time to prevent nuclear war.[25] With the 1962 Cuban Missile Crisis successfully negotiated and the Partial Test Ban Treaty put in place just before the film's release, Kubrick naturally sought to explore the way out of the atomic crossroads. Rejecting invitations to get involved in the movement for peace and nuclear disarmament, he ordered a year's back issues of *Sky and Telescope* magazine and made contact with Arthur C. Clarke, now living in Sri Lanka.[26]

Clarke too had wrestled with the problem of the atomic bomb. He had been in London with colleagues in the British Interplanetary Society when it was under assault from V-2 missiles engineered by former members of the BIS's German counterpart, the Berlin-based Verein für Raumschiffahrt (VfR). He was thus in a position to appreciate the contradictions posed by a version of progress based on rocket technology. He had discussed the issues soberly in his 1946 essay 'The Rocket and the Future of Warfare,' which was followed a few months later by his manifesto 'The Challenge of the Spaceship.'[27] The resurgence of science fiction in the United States and Britain in the postwar years saw a plethora of fiction which sought to come to terms with the threat of atomic technology, in much the same way as Clarke had sought to come to terms with rocket technology. Clarke's own early stories belonged to this period as he experienced a rush of inspiration that would provide him with the grand themes that he would rework for years to come. Early on in the preparations for *2001*, Kubrick's assistant Roger Caras provided summaries of eleven of Clarke's early short stories from the collection *Expedition to Earth* for possible use in the film (Figures 5.4 and 5.5). Six of these contemplate the destruction of a civilization by war, usually atomic.[28] In 'If I Forget Thee, O Earth,' colonists on the moon look back at their home planet: 'Across a quarter of

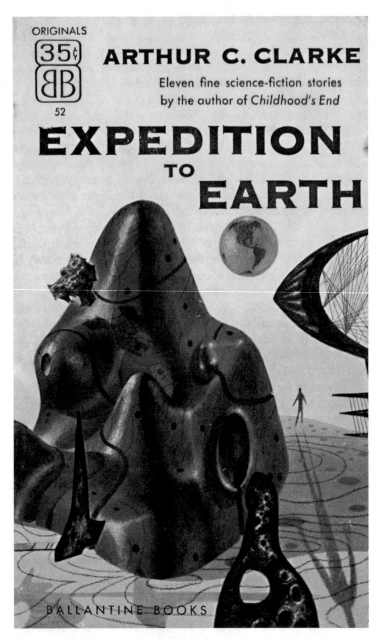

Figure 5.4 Arthur C. Clarke's 1953 story 'Expedition to Earth' imagined an Alien expedition to assist the evolution of intelligent life on earth. All eleven stories in the collection were summarized for Kubrick, who bought the rights to 'The Sentinel' to use in *2001: A Space Odyssey*. This early US cover emphasizes other-worldliness rather than space hardware.
Source: Courtesy of Penguin Random House.

Figure 5.5 The cover for the 1968 UK paperback edition of *Expedition to Earth* traded heavily on space hardware and the connection with *2001*. The promotional text flags 'The Sentinel' as the basis for the space travel part of the film, but fails to identify the title story ('Expedition to Earth' in the US, 'Encounter in the Dawn' in the UK) as the inspiration for the 'Dawn of Man' section.
Source: Courtesy of Penguin Random House.

a million miles of space, the glow of dying atoms was still visible, a perennial reminder of the ruined past.'[29] In 'Second Dawn' an alien race prematurely acquire the secret of atomic power before attaining the mental maturity to become telepathic and so overcome conflict: 'Alone, perhaps of all the races in the Universe, her people had reached the second crossroads – and had never passed the first. Now they must go along the road that they had missed, and must face the challenge at its end.' It is implied that the road for them will end in atomic war.[30] The earliest of these stories, 'Loophole,' from April 1946, offers a reversal of the scenario of H.G. Wells's *War of the Worlds*. A Martian civilization detects the atomic explosions on earth at the end of the Second World War and decides to set up a monitoring post on the moon. Fifteen years later, mankind develops atomic rockets, at which point a Martian battle fleet assembles to warn earth: '*These experiments must cease.* Our study of your race has convinced us that you are not fitted to leave your planet in the present state of your civilization.' Earth appears to comply but then secretly develops teleportation and wipes the Martians out in a surprise atomic attack.[31]

Among the short stories summarized for Kubrick was 'The Sentinel' – written in 1948 and published three years later – which provided *2001: A Space Odyssey* with its pivotal event: the discovery of an alien artifact on the moon. In the film, the excavation of the artifact exposes it to the sun, triggering a signal to its makers that man has reached the moon and is ready for some assistance to develop further. The original scenario in 'The Sentinel' is altogether darker than *2001*, owing more to the Nuclear Age than to the Space Age. The alien artifact resists all attempts to open it, until 'what we could not understand, we broke at last with the savage might of atomic power.' The scientist-narrator realizes that 'we have broken the glass of the fire alarm' and its makers will soon arrive to attend to the blaze. Humankind has reached the atomic crossroads; space travel 'is a double challenge, for it depends in turn upon the conquest of atomic energy and the last choice between life and death.' The story ends as the narrator wonders whether the alien fire brigade will want to assist earth or to cull its destructive civilization: 'I do not think we will have to wait for long.'[32] Caras marked the story: 'Read this.' Kubrick read it and bought the rights. In the spring of 1964, Clarke traveled to New York and met Kubrick. The two men immediately hit it off and set about co-writing a novel, provisionally entitled *Journey Beyond the Stars*, with which to secure the film deal that produced *2001: A Space Odyssey*.

The events of *2001* take place in a world where atomic power has been applied to rocketry, but the nuclear stand-off has not yet been resolved. In the early drafts of *Journey Beyond the Stars* the lunar base is described as 'a closed system, like a tiny working model of Earth itself,' redeploying Cold War technology:

Any man who had ever worked in a hardened missile site would have felt at home in Tycho. Here on the Moon were the same arts and hardware of underground living, and protection against a hostile environment; but here at last they had been turned to the purposes of peace. Tycho Base was more expensive than a Titan site, and this represented something new in the history of Man. After ten thousand years, he had at last found something as interesting as war.[33]

When the chief scientist at the base discusses why the aliens buried their artifact (named 'TMA One') on the moon rather than on earth, he makes the connection between atomic power and space travel: 'Perhaps the creatures who built TMA One wanted to make contact only with space-faring societies. They weren't interested in primitive, pre-atomic cultures.'[34] In the film itself the space sequences open with the orbital ballet, but an early draft of the screenplay makes clear that the first spacecraft that we see are in fact American and Russian 1,000-megaton bombs. 'Hundreds of giant nuclear bombs had been placed in perpetual orbit above the Earth,' explains a narrator: 'They were capable of incinerating the entire Earth's surface from an altitude of 100 miles.' Such weapons have brought a sinister peace to a rapidly advancing world, 'but no one could expect it to last forever.'[35]

The rest of the film outlines how the stalemate ends. The spaceship *Discovery* voyages to disaster in its search for the beings who have placed the sentinel on the moon. David Bowman, the last astronaut, like Odysseus, returns home as the sole survivor of his expedition, transformed by the experience of exile. In the film he is symbolically reborn as a 'space baby' or star child, but the book adds another crucial detail: he detonates the orbiting nuclear arsenal because 'he preferred a cleaner sky.' Clarke later wrote that some readers had interpreted this to mean that human civilization was destroyed, although he had intended to indicate that the detonation was harmless and mankind had been saved. 'But now,' he added, in 1972, 'I am not so sure.' Thinking of how Odysseus took revenge on those who had taken over and plundered his estates, Clarke wrote: 'We have wasted and despoiled our own estate, the beautiful planet Earth. Why should we expect any mercy from a returning Star Child? He may judge us all as ruthlessly as Odysseus.'[36]

Even without an apocalyptic ending it is possible to detect a strand of nuclear pessimism just below the surface of *2001*'s allegory of human progress. When an academic fan of the film sent Kubrick a study of the psychological responses of the American public to the October 1962 Cuban Missile Crisis, which appeared to bring the northern hemisphere close to nuclear war, Kubrick replied with a thoughtful letter: 'One of the more alarming features of the Cuban crisis, to my eyes anyway, was the lack of concern on the part of most of the people [...] who thought of it either as a bluff which they could not take seriously, or else they had an incredible kind of denial-resignation which allowed them to say if it happens it happens.'[37] In *Dr. Strangelove* the great crisis of human history is navigated by human sleepwalkers, unaware of what is happening, unequipped to cope with the consequences of their own technology. This could equally well describe *2001: A Space Odyssey*, except that this time the crisis is the discovery of extraterrestrial intelligence. The human characters in the film are tightly controlled and deceitful – as is the computer HAL, but there are suggestions throughout that more primitive tensions are not far below the surface – in the wary way the American and Russian officials interact at the space station, for example, and in the last astronaut's vengeful advance upon the computer which has just tried to kill him. *Homo sapiens* may have survived the atomic crossroads to make it into space, but his ancient, two-edged human nature has come with him. In the end, man is rescued from his primitive self through alien intervention. For those who have watched carefully the prologue to the film, 'The Dawn of Man,' however, this will come as no surprise, for this was how it all began.

II   Apeman, spaceman

'The Dawn of Man' sequence which opens *2001: A Space Odyssey* provides an account of the evolutionary origins of mankind that stakes out the ground for what will happen in the year 2001. In the film it takes the form of a word-less fable, but events were explained in detail in the novel when it appeared. The sequence opens two million years ago on the African savannah. A tribe of hominids is on the verge of dying out in a long drought, kept away from the water hole by a rival tribe and preyed upon by leopards. They live amongst other animals, but have no way of exploiting them: 'In the midst of plenty they were slowly starving to death.'[38] One day an alien monolith appears and by some mysterious process teaches them to use bone weapons. They attack and kill the leopard, kill the leader of the rival tribe of hominids and learn to hunt and butcher their prey, setting them on the evolutionary path to humanity with all the ingenious technology of violence. In the novel a link-ing narrative sketches out the next two million years of human history. After outlining the rise of language, agriculture, technology and civilization it con-cludes: 'The spear, the bow, the gun and finally the guided missile had given him [man] weapons of infinite range and all but infinite power [...] for ages they had served him well. But now, as long as they existed, he was living on borrowed time.'[39]

One famous scene in *2001* symbolized human progress. Having learnt to use bone weapons to hunt and to kill his rival, the chief hominid – Moon-watcher by name – hurls a bone into the air in triumph, where it spins over and over before the film cuts to the year 2000 and a bone-white orbit-ing spaceship. One simple and widespread response to this scene is awe at the scale of technological progress achieved by a simple primate, and won-der at the power of evolution. But how far has the man-ape himself evolved? An alternative reading is that technology has outrun human nature – the spaceship is, in both book and screenplay, a nuclear weapon. More pessimistically still, human history was formed in violence and might yet end in it. The big questions about the human future were bound up with those about human origins. Kubrick and Clarke, moreover, were not the only commentators to connect apeman and spaceman.

From the day in 1961 when NASA sent a chimpanzee into space ahead of any of the famous 'Mercury Seven' astronauts, the figure of the astronaut was mocked by the figure of the ape. The year 1968 saw the release not only of *2001: A Space Odyssey* but also of *Planet of the Apes* and a collection of science-fiction stories entitled *Apeman, Spaceman*.[40] The common theme was that *homo sapiens* had not evolved to keep up with his own technology and faced disaster if he could not adapt – a theme familiar from the response to the atomic bomb a generation before. As long ago as 1930 Olaf Stapledon had written in his novel of future history *Last and First Men*:

> His [man's] primitive nature could no longer cope with the complexity of his environment. Animals that were fashioned for hunting and fighting in the wild were suddenly called upon to be citizens, and moreover citizens of a world-community. At the same time they found themselves possessed of certain very dangerous powers which their petty minds were not fit to use.[41]

In the 1960 edition of his book *Adam's Ancestors*, British archeologist Louis Leakey (1903–72) worried that 'the over-specialization of our brain power made us capable of inventing the means of destruction of our species by atom bombs.'[42] A speaker at a well-publicized conference on aggression and war held by the American Anthropological Association in Washington, DC in 1967 warned: 'Hiroshima and Nagasaki and the nuclear age have changed for all time the relations of warfare to biology and the very existence of man.'[43] In the Nuclear Age the search for the sources of human nature suddenly began to look urgent.

The Space Age of 1957–72 also saw a surge in studies of fossil man, generating a wave of interest in human origins that almost rivaled interest in the human future. Until the mid-twentieth century it had been expected that the earliest human remains would be found in Eurasia or the Far East, but in the 1950s and early 1960s discoveries of fossil remains of early human ancestors by Louis and Mary Leakey in eastern Africa and by Raymond Dart in southern Africa established the view that the human species began in Africa. Discoveries of tools in association with the bones fostered the view that human origins had something to do with the discovery of technology, a view first set out in detail by the British Museum's paleontologist Kenneth Oakley (1911–81) in *Man the Tool-Maker*, published in 1950. The 'new physical anthropology' at the University of Chicago sought to unite studies of human behavior, primate behavior and fossil human remains, and developed the 'man the hunter' thesis, arguing that humankind grew from the social behavior that accompanied the use of tools and weapons.[44] More controversially, the South African paleoanthropologist Raymond Dart (1893–1988) claimed to have found evidence of the use of weapons by the distant ancestor *australopithecus* some two million years ago, long before the growth of the human brain. 'The blood-bespattered, slaughter-gutted archives of human history,' he argued, showed that violence was hardwired into human nature. Dart's views were popularized in 1961 in the bestselling book *African Genesis* by the writer Robert Ardrey (1908–80), the best-known example of a genre that later became known as 'savannah sociology.' Ardrey's core message was simple: 'Man had emerged from the anthropoid background for one reason only: because he was a killer. […] Far from the truth lay the antique assumption that man had fathered the weapon. The weapon, instead, had fathered the man.'[45] Man was both tool-maker and weapon-maker: 'from handaxe to hydrogen bomb his best efforts have been spent on the weapon's perfection.'[46] While Leakey and the academic professionals rejected Dart, Ardrey was his popular champion, offering an eloquent presentation of the 'killer ape' hypothesis which sustained a long-running public controversy over the human propensity for violence. What all these related hypotheses about human origins had in common was a belief that a formative phase in the forging of human nature had taken place on the African savannah in the distant evolutionary past. In 'The Dawn of Man,' Kubrick and Clarke were dramatizing an issue that was at the heart of anthropological debate in the 1960s.

As Kubrick and Clarke hammered out the storyline for *2001* through successive drafts they read both Leakey's *Adam's Ancestors* and Ardrey's *African Genesis*.[47] They also discussed plot outlines involving contact between intelligent extraterrestrials and primitive humans beginning in prehistoric Africa, drawing on one of Clarke's early stories, 'Encounter in the Dawn,' which had not been amongst those first purchased by Kubrick.[48] Clarke, together with the film's scientific advisor Frederick Ordway, had dinner with Leakey and his son Richard in New York in November 1965, and Richard visited the film set at Shepperton near London a couple of months later. Louis Leakey was later invited (but declined) to contribute an interview to the prospective documentary prologue.[49] When the 'Dawn of Man' scenes came to be filmed, Kubrick and his colleagues and actors went to enormous lengths to make the hominids appear realistic, visiting apes in the zoo and the displays on early man in London's Natural History Museum and consulting books, films and even seminars on the latest findings in primatology and ape social behavior. As much effort went into the realism of the prehistoric scenes as into that of the space scenes, for their message was of equal importance.

'The Dawn of Man' began life as a flashback, inserted late in the 1964 draft novel *Journey Beyond the Stars* to explain the backstory of the aliens which the astronauts are about to meet: they are the descendants of the visitors who started the man-apes on the road to humanity two million years before. In this form, it would have provided a dramatic revelation about the beginning of human progress and prepared the viewer for another assisted evolutionary leap. In the earlier drafts of the novel the man-apes are already violent in an ineffective way before the aliens arrive to teach them how to use stones and bones to hunt and butcher their prey. Success improves their diet, leaving them time to socialize and ultimately become human; technology and enterprise bring progress.[50] In the final scene, preserved in the published novel though not in the film, Moonwatcher's tribe prevails over its rivals not by crude violence but through ritual: they parade the leopard's head on a stick, like a trophy, which so overawes the other man-apes that they are easily routed. It is the combined triumph of ritual and power, not just a tale of animal violence but the first episode in human history.[51]

Clarke had developed his long view of human history in the immediate postwar years under the influence of the bestselling book *A Study of History* by the British historian Arnold J. Toynbee (1889–1975). Clarke had read Toynbee's books during the war, and attended his 1947 public lecture in London on 'The Unification of the World.' He embraced Toynbee's emphasis on the role of creative minorities in driving civilization forwards at crucial points in historical cycles of 'challenge and response,' and saw the role of rocket pioneers in his own age in the same light. Just as the renaissance age of exploration had rescued medieval Europe from stagnation and collapse, so the coming Space Age might rescue Western civilization from the cycles of war and self-destruction.[52] Seen in this light, Moonwatcher's tribe of man-apes is the original creative minority. There was, however, a crucial difference between Toynbee and Clarke. For Toynbee, the cycles of history were essentially a process of spiritual degeneration and renewal, whereas for Clarke civilization was driven by technology, animated by human creativity and enterprise.

Evolution and history are linked in a similar way in the sixth chapter of the novel, 'The Ascent of Man,' which bridges the gap between 'The Dawn of Man' and the Space Age. In the novel, Clarke describes how the intervention of the aliens is followed by a long period of technological stasis during which the man-apes' reliance on tools rather than physical strength allows them to evolve into a more dexterous and big-brained human form. They become capable of using language and therefore of acquiring culture and transmitting through the generations their successive discoveries of fire, metal, agriculture, settlement, writing and religion. By the end of this process 'the toolmakers had been remade by their own tools.' Here Clarke picks up Ardrey's phrase that 'the weapon had fathered the man,' a statement about violence, and turns it into a statement about technology and culture.[53]

Kubrick was not happy with Clarke's account of human origins, and in the spring of 1966 he returned to 'The Dawn of Man' section of the draft novel. He objected to the 'silly simplicity' of Clarke's account of how the aliens taught the man-apes through a kind of educational TV: 'it takes away all the magic.'[54] As to the 'Ascent of Man' chapter, Kubrick wrote: 'I think this is a very bad chapter and should not be in the book. It is pedantic, undramatic and destroys the beautiful transition from man-ape to 2001.' He was particularly anxious to remove any suggestion that the man-apes had been violent before the arrival of the aliens: 'they do not fight with each other [...] though they are strongly territorial, fighting and killing has not entered significantly into their lives.' Violence and tool-weapons had to arrive together.[55] Kubrick delayed the publication of the novel, facing down threats of legal action from Clarke's agent, while he worked out a cinematic way of conveying the ape-man transition and integrating it with the rest of the film.

Kubrick's solution was to use his own directorial imagination. The whole 'Ascent of Man' narrative was summed up in the single cut from bone to spaceship. The film was given a thematic unity by repeat appearances of the monolith – on the moon, in space and during the astronaut's final transformation to 'star child' – accompanied by repeated images and theme music to convey a transcendent sense of evolutionary progress. In May 1967, as he at last prepared to film the 'Dawn of Man' sequence, Kubrick cut Clarke's complex attempts to narrate the rise of civilization and produced a simpler screenplay which reproduced large chunks of Ardrey's *African Genesis*: 'Man had emerged from the anthropoid background for one reason only: because he was a killer,' ran the narrative voiceover.[56] Afterwards, explaining his next film, *A Clockwork Orange*, Kubrick continued to press the killer ape hypothesis. 'It's simply an observable fact that all of man's technology grew out of his discovery of the tool-weapon,' he told one interviewer.[57] 'I'm interested in the brutal and violent nature of man because it's a true picture of him,' he explained to the *New York Times*, 'and any attempt to create social institutions on a false view of the nature of man is probably doomed to failure.'[58]

In *2001: A Space Odyssey*, then, while Kubrick was offering enthralling visions of space travel and evolutionary progress as a way out of the Nuclear Age, he was also still worrying away at the issues of war and violence that had animated *Dr. Strangelove* and which fed through into his later work. For Clarke, *2001* was mainly about human progress; for Kubrick, it grew to be equally about human nature.

III   The myth of progress

The mid-twentieth-century belief that the human future lay in space travel has been identified by De Witt Douglas Kilgore as 'astrofuturism,' the belief that space travel was the key to a future of unlimited human progress and fulfillment. Kilgore contextualized and qualified astrofuturism by demonstrating that it was founded on a color-blind, commonwealth-style outlook whose appeal was less universal than its language. *2001: A Space Odyssey* was the cultural peak of astrofuturism, with its transcendent vision of an evolutionary leap into space. But this vision was bound up with anxieties about the limits faced by a hi-tech nuclear-armed primate confined to a single planet, illustrating further the historical context of astrofuturism.

The authors of *2001: A Space Odyssey* were both aware of the connections and tensions between the Space Age and the Nuclear Age. Clarke was always concerned about the possibility that the leap into space would be delayed by a failure of political imagination or even cut off by nuclear war, as expressed in his early stories. Clarke also perceived that even if man made it to the stars, nostalgia for the home planet would remain a powerful force.[59] One of the ironies of astrofuturism was that by measuring *homo sapiens* against a far-off peak of technological progress it made him appear to be still in infancy. 'What is the ape to man?' asked Friedrich Nietzsche in *Thus Spake Zarathustra*, and answered himself: 'A laughing-stock, a thing of shame. And just the same shall man be to the Superman [...] man is more of an ape than any of the apes.'[60] In a promotional essay for *2001* Clarke compared the future impact of the discovery that *homo sapiens* was only one junior intelligent species among many in the cosmos to the shock of the nineteenth-century discovery that he was only part of the animal kingdom.[61] In his long interview with *Playboy* magazine Kubrick affirmed: 'In the deepest sense, I believe in man's potential and in his capacity for progress.' But he went on to speak of the need to 'cure' the deep irrationality in man, as identified in *Dr. Strangelove*, in order to allow progress to happen. He also acknowledged the concerns of physicist Freeman Dyson, author of an early paper on the detection of extra-terrestrial intelligence, that the safest response to any alien communication would be to ignore it: 'Intelligence may be a cancer of purposeless technological exploitation, sweeping across a galaxy as irresistibly as it has swept across our own planet.'[62] The faith in progress was haunted by an awareness of how easily it could all go horribly wrong.

Clarke and Kubrick, for all their different emphases, shared a vision of evolutionary progress. Crucially, this vision was grounded in specific arguments about human nature and history. They believed in the mediating role of human agency, and they worked to induce the current generation of humans to sense their place in history and realize their evolutionary potential. This might seem like a modest dash of humanism in an otherwise grandiose project, but it has to be compared with the hardline astrofuturism propounded by the German-American community of former V-2 engineers such as Krafft A. Ehricke, Ernst Stuhlinger and Wernher von Braun.[63] Ehricke, for instance, formulated 'three fundamental laws of astronautics' (the formulation echoing Asimov's three laws of robotics and Newton's laws of motion) which essentially claimed that space was human destiny and there was no limit to human expansion. 'The concept of space travel disregards national borders, refuses to recognize differences of historical or ethnological origin, and penetrates the fiber of one sociological or political creed as fast as that of the next,' wrote Ehricke in the wake of Sputnik. 'Nobody and nothing under the natural laws of this universe can impose any limitations on man except man himself.'[64]

In this totalizing vision space travel existed beyond history and unhampered by any contradictions of human nature; lesser species, one senses, had better watch out. In *2001: A Space Odyssey* space travel was placed firmly within human history, albeit in a perspective so long that it was not easy to recognize, and equally firmly in juxtaposition to a human nature that by no means guaranteed *homo sapiens* a future, whether on earth or in space.

While Clarke and Kubrick's understanding of the leap into space was historically based, in the film it was conveyed using the structures and techniques of myth.[65] The title of *2001: A Space Odyssey* was a direct reference to the myth of Odysseus, the exiled warrior, returning home transformed by his experiences to set his homeland to rights. On the long voyage to Jupiter, the astronaut Bowman reads accounts of past explorations by Magellan and Cook, 'and he began to read the Odyssey, which of all the books spoke to him most vividly across the gulfs of time.'[66] Clarke and Kubrick had read Joseph Campbell's treatise on myth *The Hero With a Thousand Faces* while preparing the novel. 'We set out with the very deliberate intention of creating a myth,' explained Clarke afterwards. 'The Odyssean parallel was clear in our minds from the very beginning, long before the title of the film was chosen. [...] All the mythical elements in the film – intentional and otherwise – help to explain the extraordinarily powerful responses that it has evoked from audiences and reviewers.'[67] Clarke had used the metaphor of Odysseus in several earlier works on space, and told a press conference that the word 'odyssey' was used in the title to avoid its being pigeon-holed as science fiction.[68] Elsewhere, Clarke described the film as 'a realistic myth,' while for Kubrick it was 'a mythological documentary': 'If 2001 has stirred your emotions, your subconscious, your mythological yearnings, then it has succeeded.'[69] As well as Odysseus, the returning wanderer, there are strong echoes (in 'The Dawn of Man') of the biblical myth of Cain, son of Adam, who slays his brother Abel, and (in the transformation of spaceman into 'star child') of Nietzsche's theme of man and superman in *Thus Spake Zarathustra* – the inspiration for the Strauss fanfare of the same name which Kubrick chose as the *2001* theme music.[70]

The film's quasi-religious invocation of space travel as a kind of evolutionary progress was welcomed by most astrofuturists, and popularized by Clarke in a slew of promotional articles.[71] By refusing to show actual aliens it distanced itself effectively from UFO panics and science-fiction horror movies alike and aligned itself with the newly respectable scientific perspective on the search for extraterrestrial intelligence. It also aligned itself with the emphasis in Space-Age Christianity on the transcendence of God, free of any particular time, place or physical form.[72] The Roman Catholic Church in the United States conferred its 1968 award for 'Best Film of Educational Value' on *2001: A Space Odyssey*, citing 'the scope of its imaginative vision of man – his origins, his creative encounter with the universe, and his unfathomed potential for the future.'[73] Kubrick and Clarke, atheists both, joked that the studio had unwittingly funded 'the first $10m religious movie'; one critic described it as 'a shaggy God story.'[74] Among those offering favorable commentaries on the film were a Jesuit priest and a senior rabbi. *2001* provides evidence for the developing argument that astrofuturism has many of the characteristics of religious faith.[75]

The fit between progress, religion and myth in *2001* was facilitated by the model of externally assisted evolution adopted by Clarke and Kubrick. The concept of evolution as an internally driven forward tide of progress had been consistently rejected by religious thinkers since Darwin's day.[76] The preferred alternative, in both history and evolution, was a cyclical model of progress in which material forces can produce changes within limits but are not sufficient to achieve truly transformative change. In the cyclical model, as the historian of evolution Peter Bowler has explained, 'the origin of each new phase of human development – and hence the origin of the human race itself – was not the predictable outcome of a universal trend running through the whole sequence. [...] The origin of nations, like the origin of species, represented an essentially mysterious injection of creative power.'[77] A 'mysterious injection of creative power' is exactly the role of the alien monolith in *2001: A Space Odyssey*. It intervenes two million years ago in Africa when the man-apes are threatened with tribal extinction, and again in the year 2001 when *homo sapiens* is sleepwalking toward nuclear extinction. In-between lies one vast cycle of history – vast on the human scale, if not on the cosmic scale – during which humans have achieved much but come to the end of their earthly road. They require assistance from more advanced species if they are safely to negotiate the atomic crossroads and move on to a higher plane of existence. What *2001: A Space Odyssey* offered in 1968, then, was not a law of progress but a myth of progress. This was evident in Kubrick's replacement of Clarke's 'Ascent of Man' narrative with the single cut from bone to spaceship. The film's technological realism gave it credibility with space buffs, but its core appeal was accounted for by its use of a mythic structure, harnessed to the cause of progress.

As a final point, *2001: A Space Odyssey* can be viewed in its historical context. Present-day periodization often telescopes the period 1945–89 into a single Cold War, dominated by superpower conflict on the grand scale in politics, war and space. There was, however, a middle period, between the Cuban Missile Crisis of late 1962 and the onset of the Second Cold War in 1979–80. During this there were serious hopes of East-West accommodation and even convergence, accompanied by large-scale projects which appeared to have the potential to change the conditions of human existence: treaties to regulate the nuclear arms race, space programs to escape and understand the earth and in the 1970s international measures to tackle global environmental problems.[78] *2001* belongs to this middle stage of the long Cold War. Its astrofuturist utopianism originated in the classic Cold War of 1945–63 as Western culture reached for alternatives to nuclear extinction: atoms for peace, space travel and contact with extraterrestrial intelligence. But behind these utopias lay the fear that, with the atomic bomb, technology had breached the limits of what could be safely entrusted to a semi-evolved primate on a small planet. This combination of anxiety and idealism had its roots in the Nuclear Age and took flight during the Space Age. The Apollo years of 1968–72 were also the years of the environmental renaissance, linked to the first photos of earth from space, and of the 'limits to growth' movement, linked to the advent of systems-based world modeling. Their solutions were earth-based rather than space-based, and often social rather than technological, but they shared with *2001* the diagnosis that, on earth at least, *homo sapiens* had reached its limits.

Notes

1. The principal printed sources are: Arthur C. Clarke, *2001: A Space Odyssey*, London: Hutchinson, 1968; Jerome Agel, ed., *The Making of Kubrick's* 2001, New York: Signet, 1970; Arthur C. Clarke, *The Lost Worlds of 2001*, London: Sidgwick and Jackson, 1972; Piers Bizony, *2001: Filming the Future*, London: Aurum, 1994; David G. Stork, ed., *HAL's Legacy: 2001's Computer as Dream and Reality*, Cambridge, MA: MIT Press, 1997; Alison Castle, *The Stanley Kubrick Archive*, London: Taschen, 2005; Anthony Frewin, *Are We Alone? The Stanley Kubrick Extraterrestrial Intelligence Interviews*, London: Elliott & Thompson, 2005; Robert Kolker, ed., *Stanley Kubrick's 2001: A Space Odyssey: New Essays*, Oxford: Oxford University Press, 2006; Peter Krämer, *2001: A Space Odyssey*, Basingstoke: Palgrave Macmillan, 2010; Piers Bizony, *The Making of Stanley Kubrick's 2001: A Space Odyssey*, Cologne: Taschen, 2014; Christopher Frayling, *The 2001 File: Harry Lange and the Design of the Landmark Science Fiction Film*, London: Reel Art Press, 2015.
2. John Baxter, *Stanley Kubrick: A Biography*, London: HarperCollins, 1997; Arthur C. Clarke, *Greetings, Carbon-Based Bipeds!*, London: HarperCollins, 1999; Neil McAleer, *Odyssey: The Authorized Biography of Arthur C. Clarke*, London: Gollancz, 1992; Arthur C. Clarke, 'Extra-Terrestrial Relays: Can Rocket Stations Give World-Wide Radio Coverage?,' *Wireless World* 51.10 (October 1945), 305–8.
3. David A. Kirby, *Lab Coats in Hollywood: Science, Scientists and Cinema*, Cambridge, MA: MIT Press, 2011, 1–8.
4. Krämer, *2001: A Space Odyssey*, 90–2.
5. Howard E. McCurdy, *Space and the American Imagination*, Washington, DC: Smithsonian Institution Press, 1997.
6. Ibid., 45–8, 67–8, 193–5.
7. The interviews are collected in Frewin, *Are We Alone?*, and the originals are in the Kubrick Archive, University of the Arts, London.
8. Brian Aldiss, *Billion-Year-Spree*, London: Corgi, 1973, 244–325.
9. Lester del Rey, *Galaxy* 26.6 (July 1968), 193–4, and *Galaxy's* editor Frederick Pohl in *Film Society Review* 5.2 (1970), 23–7. Critical responses are collected in Agel, *Making of 2001*, and Stephanie Schwam, *The Making of 2001: A Space Odyssey*, New York: Random House, 2000; several hostile critics afterwards changed their minds.

10. Stanley Kubrick, 1970 interview with Joseph Gelmis, in Gene D. Phillips, ed., *Stanley Kubrick Interviews*, Jackson: University of Mississippi Press, 2001, 80–104, here 91.

11. Agel, *Making of 2001*; Schwam, *Making of 2001*; Michael Chion, *Kubrick's Cinema Odyssey*, London: BFI Publishing, 2001; Mark Crispin Miller, '2001: A Cold Descent,' *Sight and Sound* 4.1 (January 1994), 18–25; Kolker, *Stanley Kubrick's 2001*; Krämer, *2001: A Space Odyssey*.

12. Tatjana Ljujic, Peter Krämer and Richard Daniels, eds, *Stanley Kubrick: New Perspectives*, London: Black Dog, 2015, presents research based on the Kubrick archive, including Robert Poole, '*2001: A Space Odyssey* and the Dawn of Man,' 174–97, which is drawn upon for part of the present article.

13. 'Fulfilment,' *Times* (16 August 1945), 5.

14. Lawrence S. Wittner, *One World or None? A History of the World Nuclear Disarmament Movement Through 1945*, Stanford: Stanford University Press, 1993, 248–9; Norman Cousins, 'Modern Man is Obsolete,' *New York Times Saturday Review* (18 August 1945), 5–9; Scott C. Zeman, '"To See [...] Things Dangerous to Come to"; Life Magazine and the Atomic Age in the United States, 1945–1965,' in Dick van Lente, ed., *The Nuclear Age in Popular Media: A Transnational History, 1945–1965*, Basingstoke: Palgrave Macmillan, 2012, 53–78; Dolores L. Augustine, '"Learning from War": Media Coverage of the Nuclear Age in the Two Germanies,' ibid., 79–118; van Lente, 'Nuclear Power, Politics, and a Small Nation,' ibid., 149–74.

15. Albert Einstein, 'Atomic War or Peace?,' *Atlantic Monthly* (November 1945), reprinted in *Out of My Later Years*, New York: Philosophical Library, 1950, 185–99; Dexter Masters and Katharine Way, *One World or None?*, New York: Federation of Atomic Scientists, 1946. Alice Kimball Smith, *A Peril and a Hope: The Scientists' Movement in America, 1945–47*, Chicago: Chicago University Press, 1965; Paul Boyer, *By the Bomb's Early Light: American Thought and Culture at the Dawn of the Atomic Age*, Chapel Hill: University of North Carolina Press, 1985, paperback edn, 1994; Spencer R. Weart, *Nuclear Fear: A History of Images*, Cambridge, MA: Harvard University Press, 1988; Wittner, *One World or None*.

16. John D. Cockroft, ed., *Atomic Challenge: A Symposium*, London: Winchester, 1948, xx. Jessica Wang, *American Science in an Age of Anxiety*, Chapel Hill, University of North Carolina Press, 1999, 25–37. On the displacement of atomic anxieties in other directions, see Joachim Radkau, *Nature and Power: A Global History of the Environment* [2002], Cambridge: Cambridge University Press, 2008, 265–72; Robert R. MacGregor, 'Imagining an Aerospace Agency in the Atomic Age,' in Stephen J. Dick, ed., *Remembering the Space Age*, Washington, DC: NASA, 2008, 55–70.

17. Olaf Stapledon, 'Interplanetary Man?,' *Journal of the British Interplanetary Society* 7.6 (November 1948), 213–33.

18. Carl Sagan, quoted in De Witt Douglas Kilgore, *Astrofuturism: Science, Race, and Visions of Utopia in Space*, Philadelphia: University of Pennsylvania Press, 2003, 56.

19. Copy for cover of the first UK edition of *2001: A Space Odyssey*, 1968; Arthur C. Clarke, 'The Challenge of the Spaceship: Astronautics and its Impact upon Human Society,' *Journal of the British Interplanetary Society* 6.3 (December 1946), 66–81, here 72–3.

20. Giuseppe Cocconi and Philip Morrison, 'Searching for Interstellar Communications,' *Nature* 184 (19 September 1959), 844–6; Alastair G.W. Cameron, ed., *Interstellar Communication: A Collection of Reprints and Original Contributions*, New York: W.A. Benjamin, 1963.

21. Carl Sagan, 'The Quest for Extraterrestrial Intelligence,' in idem, *Broca's Brain*, New York: Coronet, 1980, 324–5.

22. Ronald Bracewell, *The Galactic Club: Intelligent Life in Outer Space*, San Francisco: Freeman, 1975; Robert Poole, 'ET and the Bomb,' unpublished paper, British Society for the History of Science conference, Exeter, July 2011.

23. Alexander C.T. Geppert, 'Extraterrestrial Encounters: UFOs, Science and the Quest for Transcendence, 1947–1972,' *History and Technology* 28.3 (September 2012), 335–62; Thore Bjørnvig, 'Transcendence of Gravity: Arthur C. Clarke and the Apocalypse of Weightlessness,' in Alexander C.T. Geppert, ed., *Imagining Outer Space: European Astroculture in the Twentieth Century*, Basingstoke: Palgrave Macmillan, 2012, 127–46 (= *European Astroculture*, vol. 1).

24. Phillips, *Stanley Kubrick Interviews*, 52–3.

25. Krämer, *2001: A Space Odyssey*, 18–19.

26. Stanley Kubrick Archive, University of the Arts, London (henceforth SKA), SK11/9/97.

27. Arthur C. Clarke, 'The Rocket and the Future of Warfare,' *RAF Quarterly* 17.2 (March 1946), 61–9; and idem, 'Challenge of the Spaceship'; Robert Poole, 'The Challenge of the Spaceship: Arthur C. Clarke and the History of the Future, 1930–1970,' *History and Technology* 28.3 (September 2012), 255–80.

28. SKA, SK/12/8/1/12; Clarke, *Expedition to Earth* [1953], London: Sphere, 1968. The six stories of extinction are 'Second Dawn,' 'If I Forget Thee, O Earth,' 'History Lesson' (UK title 'Expedition to Earth'), 'Superiority,' 'Loophole,' and 'The Sentinel.'

29. Clarke, *Expedition to Earth*, 45.

30. Ibid., 41.

31. See ibid., 146–52; 150: 'When mankind eventually develops atomic rockets, a Martian battle fleet assembles [...].'

32. Ibid., 172–4.

33. SKA, SK12/1/1/2; incomplete novel text (the 'Lucifer text'), 4.1–3.

34. Ibid., 7.12–13.

35. SKA, SK12/1/2/1, 'The Athena Screenplay,' 26.

36. Clarke, *Lost Worlds of 2001*, 239–40.

37. SKA, SK11/9/100 (correspondence between Kubrick and Irvin Doress).

38. Clarke, *2001: A Space Odyssey*, chapter 1. Chapter references are given as editions vary.

39. Ibid., chapter 6.

40. Tom Wolfe, *The Right Stuff* [1979], London: Picador, 1991, 212–37. Sherryl Vint, 'Simians, Subjectivity and Sociality: *2001: A Space Odyssey* and Two Versions of *Planet of the Apes*,' *Science Fiction Film and Television* 2.2 (Fall 2009), 225–50; Harry Harrison and Leon Stovers, eds, *Apeman, Spaceman*, New York: Doubleday, 1968.

41. Olaf Stapledon, *Last and First Men* [1930], Harmondsworth: Penguin, 1972, 21–3.

42. Louis Leakey, *Adam's Ancestors*, 4th edn, London: Methuen, 1960, 217–18.

43. Morton Fried, Marvin Harris and Robert Murphy, eds, *War: The Anthropology of Armed Conflict and Aggression*, New York: Natural History Press, 1968, 16–21. The conference took place in November 1967, after a controversy over the profession's stance on the Vietnam War had erupted the previous year.

44. Kenneth Oakley, *Man the Tool-Maker*, London: British Museum, 1950; Donna Haraway, *Primate Visions: Gender and Race in the World of Modern Science*, London: Routledge, 1989, chapters 6–8, 116–230; Anne Roe and George Gaylord Simpson, eds, *Behavior and Evolution*, New Haven: Yale University Press, 1958; see the 'Human Species' issue, *Scientific American* 203.3 (September 1960); Richard Lee and Irven DeVore, *Man the Hunter*, Chicago: Chicago University Press, 1968; Bruce G. Trigger, 'Aims in Prehistoric Archaeology,' *Antiquity* 44 (March 1970), 26–37.

45. Robert Ardrey, *African Genesis*, London: William Collins, 1961, 29, 158; Nadine Weidman, 'Popularizing the Ancestry of Man: Robert Ardrey and the Killer Instinct,' *Isis* 102.2 (June 2011), 269–99.

46. Ardrey, *African Genesis*, 204–5.

47. Clarke, *2001*, Part 1; SKA, SK/12/1/1 (novel texts), 12/1/2 (screenplays), 12/8/1/21–12/8/1/23 (voice-over narratives).

48. Clarke, *Lost Worlds*, 50–2; SKA, SK12/8/1/12.

49. Clarke, *Lost Worlds*, 34–5, 39, 50–2; Frederick I. Ordway, '2001: A Space Odyssey in Retrospect,' in Eugene M. Emme, ed., *Science Fiction and Space Futures: Past and Present*, San Diego: American Astronautical Society, 1982, 47–105, here 56; SKA, SK/12/8/1/62, Leakey to Caras, 21 February 1966.

50. 'Patchwork text' SKA, SK12/1/1/4, chapters 2–6; SK12/1/1/2, part I.

51. The relationship between novel and screenplay is complex. In summary, Clarke and Kubrick agreed to co-write a novel, provisionally entitled *Journey Beyond the Stars*, to secure a deal with MGM, which was done between April and December 1964. The novel, retitled *2001: A Space Odyssey* was then greatly revised and extended, a complex process which ended in April 1966, by which time much of the screenplay had been written and many of the sequences set in space had been filmed. The novel was to have come out before the film, but in the event Kubrick held it back while further changes were made to the screenplay, particularly the 'Dawn of Man' section which was filmed last of all in the summer of 1967. These changes were mostly not reflected in the novel which finally appeared in the summer of 1968, several months after the release of the film. Sections of the discarded novel drafts were printed in Clarke, *Lost Worlds*. See Poole, 'Dawn of Man.'

52. Clarke, 'Challenge of the Spaceship'; Arnold J. Toynbee, 'The Unification of the World and the Change in Historical Perspective,' in idem, *Civilization on Trial*, Oxford: Oxford University Press, 1948, 63–96. This argument is developed in Poole, 'Challenge of the Spaceship.'

53. Clarke, *2001: A Space Odyssey*, chapter 6. Clarke's chapter title later appeared as the title of Jacob Bronowski's 1973 BBC TV history of science and civilization, *The Ascent of Man*.

54. Kubrick to Clarke, 11 April 1966, SKA, SK/12/8/1/12; Clarke, *Lost Worlds*, 47–9.

55. Clarke, *Lost Worlds*, 48; Krämer, *2001: A Space Odyssey*, 59–65.

56. Dawn of Man notes, 30 May 1967, SKA, SK12/8/1/23; Ardrey, *African Genesis*, 29.

57. Phillips, *Stanley Kubrick Interviews*, 152.

58. Craig McGregor, 'Nice Boy from the Bronx?,' *New York Times* (30 January 1972), D1, quoted in 'The Hechinger Debacle'; available at www.visual-memory.co.uk/amk/doc/0037.html (accessed 1 October 2017).

59. See, respectively, his short stories 'Against the Fall of Night,' 'The Star,' and 'If I Forget Thee, O Earth,' variously anthologized.

60. Friedrich Nietzsche, *Thus Spake Zarathustra* (1883–85), prologue, Sections 3 and 4; available at http://www.gutenberg.org/ebooks/1998 (accessed 1 October 2017).

61. Clarke, 'The Universe Around Us,' draft article (1966) in SKA, SK/12/8/1/14.

62. Phillips, *Stanley Kubrick Interviews*, 68, 53–4.

63. Kilgore, *Astrofuturism*.

64. Krafft A. Ehricke, 'The Anthropology of Astronautics,' *Astronautics* 2.4 (November 1957), 26–7, and 65–8, reprinted in Marsha Freeman, *Krafft Ehricke's Extraterrestrial Imperative*, Burlington: Apogee Books, 2009, 119–26.

65. David G. Hoch, 'Mythic Patterns in *2001: A Space Odyssey*,' *Journal of Popular Culture* 4.4 (Spring 1970/71), 961–5; Philip Kuberski, 'Kubrick's *Odyssey*: Myth, Technology, Gnosis,' *Arizona Quarterly* 64.3 (Fall 2008), 51–73; Joseph Campbell, 'The Moonwalk: The Outward Journey,' in idem, *Myths to Live By*, New York: Viking Press, 1972, 233–49.

66. Clarke, *2001: A Space Odyssey*, 115.

67. Idem, *Lost Worlds*, 34; idem, 'The Myth of 2001' [1969], in *Report on Planet Three*, 247–9; Joseph Campbell, *The Hero With a Thousand Faces*, New York: World Publishing, 1968.

68. Agel, *Making of 2001*, 164; McAleer, *Arthur C. Clarke*, 199; Clarke, *Prelude to Space* (1947/1953), New York: Gnome Press, NEL edn, 1968, 97 (chapter 19); Alexander Walker, *The Film Director as Superstar*, New York: Doubleday, 1970, 39.

69. Arthur C. Clarke, *Report on Planet Three*, London: Gollancz, 1972, 246; Walker, *Film Director as Superstar*, 241; Agel, *Making of 2001*, 161.

70. Herbert Glass, 'Also sprach Zarathustra' at the LA Phil web site; available at http://www.laphil.com/philpedia/music/also-sprach-zarathustra-richard-strauss (accessed 1 October 2017); Timothy E. Scheurer, 'Kubrick vs. North: The Score for "2001: A Space Odyssey,"' *Journal of Popular Film & Television* 25.4 (Winter 1998), 172–82; David W. Patterson, 'Music, Structure and Metaphor in Stanley Kubrick's *2001: A Space Odyssey*,' *American Music* 22.3 (Fall 2004), 444–74.

71. See, for example, Clarke's typescript of an article for *Vogue Magazine*, 'The World of 2001,' SKA, SK12/8/1/11.

72. Geppert, 'Extraterrestrial Encounters'; Kendrick Oliver, *To Touch the Face of God: The Sacred, the Profane, and the American Space Program, 1957–1975*, Baltimore: Johns Hopkins University Press, 2013, 44–70.

73. Agel, *Making of 2001*, 327.

74. Baxter, *Stanley Kubrick*, 210; Agel, *Making of 2001*, 327, 244.

75. Ibid., 31–5, 55–7; Frewin, *Are We Alone?*, 131–40; Gabriel McKee, *The Gospel According to Science Fiction: From the Twilight Zone to the Final Frontier*, Louisville: John Knox Press, 2007, 23–5, 39, 150, 178; Bjørnvig,

'Transcendence of Gravity'; Roger D. Launius, 'Escaping Earth: Human Spaceflight as Religion,' *Astropolitics* 11.1–2 (2013), 45–64.

76. Peter J. Bowler, *Theories of Human Evolution: A Century of Debate, 1844–1944*, Oxford: Oxford University Press, 1987. Neither model, of course, was properly Darwinian.

77. Idem, *The Invention of Progress: The Victorians and the Past*, Oxford: Oxford University Press, 1989, 8–9.

78. See Robert Poole, *Earthrise: How Man First Saw the Earth*, New Haven: Yale University Press, 2008, and the contributions in Simone Turchetti and Peder Roberts, eds, *The Surveillance Imperative: The Rise of the Geosciences during the Cold War*, Basingstoke: Palgrave Macmillan, 2014.

CHAPTER 6

The Earthward Gaze and Self-Reflexivity in Anglophone Novels of the 1970s

Florian Kläger

It has become a commonplace to say that in the wake of the Apollo program, human self-perception was changed radically through the reversal of perspective constituted by the 'Earthward gaze' from space. Initial enthusiasm for spaceflight in the 1960s rendered outer space one of the period's 'major sites of utopian thinking,' we are told, and as the excitement over the discoveries and promises cooled, attention returned to the point of origin that was earth.[1] The program that 'was meant to show mankind that its home was only its cradle,' Robert Poole observes, 'ended up showing that its cradle was its only home.'[2] This is epitomized not only in the iconic images of 'Earthrise' and the 'Blue Marble,' but also in reflections about the contents of the Voyager plates that were sent out as representations of terrestrial life. As astronomer and novelist Carl Sagan (1934–96) remarked at the end of the 1970s, the chance that this message would ever be found by an alien civilization was slim, but 'its receipt by the inhabitants of Earth was guaranteed.' Reaching out into space and confronting its profound emptiness, the human imagination was inspired to contemplate the only intelligent life there seemed to be: its own. However, this also meant to admit that the previous enthusiasm for the Apollo missions and the dawning Space Age might have been a little out of proportion. W.H. Auden famously demurred in his 1969 poem on the moon landing: 'Worth *going* to see? I can well believe it. / Worth *seeing*? Mneh!'[3] Still, while the *quest* felt as if it had been frustrated,

Florian Kläger (✉)
Universität Bayreuth, Bayreuth, Germany
e-mail: klaeger@uni-bayreuth.de

© The Author(s) 2018
Alexander C.T. Geppert (ed.), *Limiting Outer Space*
European Astroculture, vol. 2
https://doi.org/10.1057/978-1-137-36916-1_6

the *questions*, aspirations and expectations it had raised did not go away. It became evident that all along, many of these questions had not been about the moon or outer space at all, but rather about humankind and its 'place in the universe' in a metaphorical sense, about its purpose, potential and self-perception in relation to its cosmic environs.

Not least among the Apollo program's consequences, then, was a heightened sense of self-reflection that originated in the popular response to spaceflight but engendered anthropological and cultural self-questioning as well. In particular, this appears to have been the case in countries that observed an upsurge in American post-Apollo 'planetary' visions tacitly, arrogantly and perhaps unconsciously assuming that the American experience was universal. Symptomatically, the Voyager Golden Record, launched in 1977, showed a photograph of earth taken from space labeled 'United States of America, Planet Earth.' This chapter examines the impact of this phenomenon of post-Apollo self-reflexivity on a medium of cultural self-reflection that has a long history of negotiating questions about the place of humankind in the universe, that is, the novel. Specifically, I want to indicate by reference to three authors of the 1970s – Doris Lessing (1919–2013), A.S. Byatt (1936–) and John Banville (1945–) – how novelists from non-spacefaring nations critically and productively engaged with the Earthward gaze facilitated by the Apollo missions. It is important to note that none of the novels examined here as much as mentions the recent missions to the moon. As decidedly 'literary,' non-science-fiction texts, they focus on such topics as the life of Nicholas Copernicus, the intricacies of northern English provincial life in the mid-1950s, and on the medical treatment of a present-day patient in a mental institution. In this way, they explore the self-reflexivity associated with the Earthward gaze as a cultural phenomenon as far as possible divorced from its concrete technological and ideological contexts.[4] Thus, they direct attention to the interpretive response elicited by the view from space. Often defined as a genuinely self-reflective genre, the novel seems the perfect site for negotiating the issues raised in the wake of the 1970s missions to space. As cosmological self-reflection became a prominent cultural phenomenon, the authors discussed here explored its implicit conception of humankind as a questing, and (self-)questioning, species; and what is more, their novels appropriated that phenomenon for their own generic self-reflection. In order to introduce this argument, the chapter briefly sketches the ways in which post-Apollo self-reflection was articulated outside the novel; then it indicates the ways in which that genre engaged with it critically; and finally, it situates these literary voices in the context of cultural and literary criticism of the 1970s.

I   Nostos, or the longest detour

Austrian philosopher Günther Anders (1902–92) famously remarked in 1970 that the truly decisive event of spaceflight was not that humankind reached the remote regions of space or the moon, but that for the first time, earth had the chance to see itself in the way people were accustomed to seeing themselves in the mirror. As earth 'became reflexive,' it was 'not the destination that proved to be what excited people, but the point of departure.'[5] Alexander Geppert explains that this brought about 'a radical change in self-perception on a genuinely global level, literally resulting in a new *Weltanschauung*.'[6] It was, however, by no means self-evident or inevitable that the Apollo missions should have prompted such a fundamental change. During the 1960s, the possibilities promised by outer space still dominated discourse about spaceflight; but as the missions proceeded, concern about the place of departure moved increasingly center-stage.[7] Following on the enthusiasm for technology and progress, awareness grew of the political, ecological and moral obligations they entailed, and of the 'limits to growth.' The gaze first turned Earthward literally, but it immediately acquired metaphorical meaning as well. Two notions in particular became attached to the world seen from space: the idea of its precariousness, fragility and limitations; and the interdependency of cohabiting in this exceptional place. Under the rubrics of 'one world only' and 'one world to share,' this dual development fundamentally affected the political and moral sense of orientation for many.[8] Gazing at themselves and their home from outside they began to reflect about their relationship to their home planet and to the other people and species they shared it with.

As Rüdiger Zill has shown, however, contemporaneous European interpretations of the Earthward gaze varied.[9] Thinkers like Emmanuel Lévinas (1906–95) and Maurice Blanchot (1907–2003) celebrated the fact that astronauts and cosmonauts 'left the Place' and 'existed in the absolute of homogeneous space.'[10] Still, there were also those, such as Hannah Arendt (1906–75) and Hans Blumenberg (1920–96), for whom spaceflight proved once and for all that earth was, indeed, the 'ground' to which humans always must return and the object that really came into focus through spaceflight. Even where commentators emphasized the outward movement towards space, they also acknowledged the bonds that remained between space traveler and planet, and what those who stayed at home could learn from the events: Lévinas praised spaceflight for presenting the opportunity 'to perceive men outside the situation in which they are placed, and let the human face shine in all its nudity.'[11] Outside all conventional frames of reference, Lévinas trusted, we come to see what is essentially human – and thus, we find a basis for morality. Commenting on Gagarin's transmissions from orbit, Blanchot made a similar point:

> It is therefore necessary, up there, for the man from the Outside to speak, and to speak continually, not only to reassure us and to inform us, but because he has no other link with the old place than that unceasing word, which, accompanied by hissing and conflicting with all that harmony of the spheres, says, to whoever is unable to understand it, only some insignificant commonplace, but also says this to him who listens more carefully; that the truth is nomadic.[12]

Blanchot directed attention from what was seen – the photographs, much discussed by historians of astroculture – to what was spoken and heard, 'the unceasing word' that accompanies human beings wherever they go. Human nature will not be left behind, and this serves as a reminder that what matters about visual impressions and discoveries is their discursive interpretation and function. Along with excitement about 'the Outside,' Blanchot thus also registered an insight about the inhabitants of 'the old place'; both he and Lévinas stressed the universally human element that was brought into relief by spaceflight in the individual astronaut.

Among those philosophers more preoccupied with the look behind than with the outward movement, Günther Anders argued that the view from outside corroborated the Copernican world view in that humankind finally saw earth as the object in space that it is.[13] Hans Blumenberg insisted that the opposite was true. As he saw it, Copernicanism had been inverted as humans realized that their home was also the only solid ground they could ever return to. In pointing out this fact, Blumenberg not only referred to literal ground that ensured physical survival, but also the metaphorical ground that would provide intellectual nourishment: objects in space like the moon, he argued, were simply 'too dull, too compact, too nondescript, too sterile for the astuteness of the human senses and of language.'[14] Clearly, this echoed Blanchot's interest in what humankind can learn about itself by contemplating its own loneliness in space, and the feelings it may generate. However, Blumenberg's focus was not on the individual traveler in space, but on the collective of those left at home and on their relationship to this place. Blumenberg argued, through an analogy between the *Odyssey* and accounts of contemporary space exploration, that both had something to say about their respective homes: 'to return to Ithaca requires and merits the longest detour.'[15] This appreciative sense of belonging created by and dependent on the epic effort of spaceflight is the subject of the final chapter in Blumenberg's monumental *Genesis of the Copernican World*, which first appeared in German in 1975. He concluded this study with the following assessment:

> It is more than a triviality that the experience of returning to the Earth could not have been had except by leaving it. The cosmic oasis on which man lives – this miracle of an exception, our own blue planet in the midst of the disappointing celestial desert – is no longer 'also a star,' but rather the only one that seems to deserve this name. It is only as an experience of turning back that we shall accept that for man there are no alternatives to the Earth, just as for reason there are no alternatives to human reason.[16]

Blumenberg did not lament the disappointment of the 'celestial desert,' he welcomed and celebrated that its discovery should make us accept that we have 'no alternatives to the Earth.' This was an – albeit unexpected – contribution the Apollo missions made to a tradition in the history of ideas that the *Genesis of the Copernican World* traced from Greek antiquity to the moon landings in a comprehensive stocktaking of just those 'ways in which human beings attempt to come to terms with and make sense of the infinite universe that surrounds us' that are studied today under the rubric of 'astroculture.'[17] Examining what he called 'the anthropological semantics of cosmology,' Blumenberg investigated how various historical cultures had interpreted the place of humankind in the universe.

One of Blumenberg's most fundamental and consequential arguments in this and other works on the subject was that we tend to make the stars 'about us' in the vein suggested by Ernst Cassirer (1874–1945) in his 1944 *Essay on Man*:

> What man really sought in the heavens was his own reflection and the order of his human universe. He felt that his world was bound by innumerable visible and invisible ties to the general order of the universe – and he tried to penetrate into this mysterious connection.[18]

The desire to find such organizational correspondences was not simply eradicated by the early modern epistemic shift that dispelled the belief in connections between the macrocosm and the microcosm.[19] As Blumenberg showed in his study of reactions to the Copernican revolution, when Galileo characterized his astronomical insights as ennobling man by placing him on a star among other stars, or when Nietzsche saw humankind debased and humiliated by its 'eccentric' position in the cosmos, they still ascribed meaning to a simple fact of nature that might otherwise be viewed as irrelevant to human self-conceptions. In other words, they projected meaning onto astronomical phenomena that give no objective indication of holding such meaning.[20] For Blumenberg, these meaning-ascriptions rendered cosmology 'the ontological model of the modern age.'[21] As his treatment in the post-Apollo *Genesis* showed, it has been very difficult for humans *not* to ascribe ontological significance to the ('merely' epistemological) data of astronomy, or in other words, not to anthropicize their cosmic surroundings.[22] Indeed, even before the moon landings, Hannah Arendt anticipated that this mechanism would determine their effect on the imagination:

> [O]nce the limit is reached and the limitations established, the new world view that may conceivably grow out of it is likely to be once more geocentric and anthropomorphic, although not in the old sense of the earth being the center of the universe and of man being the highest being there is. It would be geocentric in the sense that the earth, and not the universe, is the center and the home of mortal men, and it would be anthropomorphic in the sense that man would count his own factual mortality among the elementary conditions under which his scientific efforts are possible at all.[23]

By the longest of detours, humankind would arrive again at its own home, both physically and intellectually. Whereas for Lévinas and Blanchot, the central insight of the Apollo missions was epitomized by the individual in space, for Blumenberg and Arendt it concerned the collective at home. In this assessment, they agreed with numerous statements to the same effect made by returning astronauts; with the interpretation of the Earthward gaze that led President Johnson to send copies of the Earthrise photograph to all heads of state in the world; and with the gushing response of the ecology and peace 'one earth' movements.[24]

II   Lunatics and narcissists

Unsurprisingly, the literature of the day engaged with this phenomenon as well. One of the central observations of Ronald Weber's early study of American literary responses to spaceflight is of the 'dominant use of space exploration' in fiction 'as a means of deflecting attention back to earth and earthly concerns.'[25] Outside the United States, too, novelists seized on this subject, and in the following, I examine some of the ways in which they adapted it to their purposes. Doris Lessing's *Briefing for a Descent into Hell* (1971) explicitly responded to the Earthward perspective introduced by the Apollo program, registering it even on the first edition's dust-jacket. It is designed to look like a polished, if worn, wooden door labeled 'Patients Day Room' with an ominous lock in the lower-right corner and a peephole in the upper part of the door, which suggests that the viewer is gazing into the room/book through its door/cover (Figure 6.1). Beyond the peephole, earth is visible from space, in an artistic rendering of a photograph taken two years earlier on the Apollo 11 mission.[26] Numerous stars were added to the original image, forming a backdrop in the otherwise black space around earth. To open the book is thus to open the door onto space containing a hospitalized planet Earth in need of help or at least rest. The back cover, in turn, shows a view from the other side of the door (Figure 6.2). There is a cover for the peephole, pulled aside to reveal the non-descript head of a man, probably a doctor. This startling design implies that the patient is, or is in some way like, earth, and that the novel will allow the reader to assume both an outside and an inside perspective of the patient's suffering. The object in the reader's hand is a door that marks the threshold between patient and medical practitioner, between cosmic loneliness and workaday routine. In view of this design, it is startling that the novel's first page identifies its 'category' as 'inner-space fiction. For there is never anywhere to go but in.'[27] Outer space, so prominently and enticingly depicted on the front cover, is immediately and comprehensively rejected as a destination for humankind, suggesting that it should be read metaphorically for 'inner space' when it does occur in the novel.

Briefing for a Descent into Hell undertakes to record the arduous medical treatment of Charles Watkins, a Cambridge professor of classics who has lost his memory, in a number of seemingly disconnected, generically disparate narrative forms including dream visions, letters and memos, songs, interior monologue and 'autobiography.' This formal chaos requires readers to

Figure 6.1 The front cover of Doris Lessing's *Briefing for a Descent into Hell* (1971), an artistic rendition of a recent photograph taken during the Apollo 11 mission. Through this design, the book is made to appear as a door and window offering insight into earth's predicament, which is presented as that of a patient in need of medical or psychiatric care.

Source: Alan Tunbridge, *Briefing for a Descent into Hell* by Doris Lessing, London: Jonathan Cape, 1971, front cover. Courtesy of the Random House Group Limited.

Figure 6.2 The back cover of Lessing's *Briefing for a Descent into Hell*. It effects a shift in perspective that identifies the viewer/reader with earth, seen from the outside on the front cover: at the end of the novel, the jacket design promises, readers will be able to reconcile the outside and inside perspectives.

Source: Alan Tunbridge, *Briefing for a Descent into Hell* by Doris Lessing, London: Jonathan Cape, 1971, back cover. Courtesy of the Random House Group Limited.

position themselves consciously at a distance to the narrative, seeking out a perspective from which meaning can be discerned in what increasingly seems like the representation of pathological hallucinations. In the sense that Watkins's literally 'lunatic' dream visions include a journey to the moon on the back of a bird and several other flings into outer space, the categorization as 'inner-space fiction' is, and is not, ironic. It identifies the desire to obtain 'real' insight into the self with the reflex to look for it outside the self, thus commenting on the similar enthusiasm that characterized the astrofuturism of earlier decades that had hoped to find the destiny of humankind outside its terrestrial home. Indeed, Watkins's intuitions on how to make sense of his situation initially reflect a naïve and irrational hope for some deep revelation originating from the moon:

> Now the moon was in its last quarter and making a triangle, sun, earth, moon, whereas when I had reached that coast it was full, and sitting on the plateau's edge and staring into the moon's round face I had my back to the sun, which was through the earth, and the sun stared with me at the moon. Then the pulls and antagonisms and tensions from the sun and moon had been in a straight line through the earth, which swelled, soil and seas, in large bulges of attraction as the earth rolled under the moon, the sun; but now the tension of sun and moon pulled in this triangle, and the tides of the ocean were low [...]. I did not know why I thought so, but I had come to believe that it was the next full moon that I was waiting for.[28]

The reader feels the same 'pulls and antagonisms and tensions' in the narrative of *Briefing*, and, like Watkins, they trust that some event or point in time may arrive that makes sense of the powerful but contrary forces that constitute the novel. However, the arrival of the full moon effects nothing; Watkins's excruciating soul-searching must go on. There is no answer to man's profound desire for self-knowledge in space, this suggests. As Watkins continues to pursue his quest, it assumes the form of (imagined) journeys into outer space that come to figure for his wrestling with the limitations of self and self-knowledge and the desire for their transcendence.[29] One such instance directly assumes the perspective of the Earthward gaze:

> The Earth hung in its weight, coloured and tinted here and there, for the most part with the bluish tint of water [...] and humanity and animal life and bird life and reptile life and insect life – all these were variations in a little crust on this globe. Motes, microbes. And yet – it was mostly here that the enclosing web of subtle light touched the earth globe. It was for the most part through the motes or mites of humanity.[30]

From this 'Apollonian' perspective that again clearly references the iconic photographs of the day, humanity becomes that feature of the planet which bestows it with a special shine of life. The planet is ennobled, as it were, by its human population, tiny as its individual specimen might seem. Indeed, under Watkins's Earthward gaze, the individual is entirely eclipsed by the collective:

> [...] and from this vantage point it was amusing to see how passionate hatred, rivalries and competitions disappeared altogether, for the atoms of each of these categories were one, and the minute fragments that composed each separate pulse or beat of light (colour, sound) were one, so there were no such things as judges, but only Judge, not soldiers, but Soldier, not artists, but Artist, no matter if they imagined themselves to be in violent disagreement.[31]

Watkins, whose pathological yearning for harmony and wholeness is the driving force behind his disjunctive narrative, projects his desire into a cosmic vision. Here, and in the numerous other passages in which the narrator does so, it is a gaze from outer space back onto earth that seems to facilitate the longed-for accord, as it makes divisions among humanity disappear. In this way, Lessing harnessed the potential of the self-reflective Earthward gaze for her own exploration of a fragile mind in search of human connection. Readers were invited to relate their own post-Apollo experiences – their various associations of the Earthward gaze with expectations, frustrations and a revaluation of the mundane – to that of Watkins as he stumbles through his worlds of vision, hallucination and memory, and to reflect on their own view of humankind from 'outside.' Is Watkins mistaken in his intimations of transcendence, or does the cosmic perspective offer actual insight? Would only a madman interpret the images of earth from space as he does, or is the desire to do so deeply human? Crucially, as Watkins recovers his mental stability again at the close of the novel, he turns back into an unremarkable, bland person, offering apologies for his earlier behavior.[32] His visionary interlude stands as a profound experience that readers may have found engaging but that is now irrecoverably past: just as the popular enthusiasm for space exploration faded, so this protagonist finds himself back in a mundane existence that contrasts starkly with the exhilarating promises of his earlier dreams. It is left to the reader to decide which Watkins they prefer, the lunatic or the bore.

The contrast between the mundane and the transcendent, the rational and the insane that is blurred to great effect in *Briefing for a Descent into Hell* is also employed in A.S. Byatt's *The Virgin in the Garden* (1978). This novel portrays provincial life at the time of Elizabeth II's coronation in 1952, as a northern small town puts on a play about Elizabeth I called *Astraea*. The eponymous virgin is, on the one hand, the adolescent Frederica Potter, cast in the role of young Elizabeth in the production. On the other hand, it is this queen herself, who in the iconography of her day was conventionally identified with the lunar deity Cynthia and thus, the moon, by virtue of her exulted femininity and virginity. Involving a large part of the local community, the production of *Astraea* causes subtle realignments in the town's social constellations, building up to Frederica's departure.

Local biology teacher Lucas Simmonds approaches Frederica's pubescent brother Marcus, his pupil, by introducing him to a pamphlet he wrote called *The Plan and the Pattern*. It presents an esoteric blend of pseudo-science, myth and religion committed to challenging 'our Megalanthropic, Anthropocentric belief that Man is the highest order of Being given to us in sense-perception.'[33] For Lucas, anthropocentrism of the kind that Lessing had represented in Watkins's visions is deplorable; Byatt's 1950s lunatic already loathes the narcissism that Christopher Lasch would soon diagnose, in its American brand, as the defining feature of the 1970s.[34] As indicated above, a kind of anthropocentrism that might be deemed narcissistic derived from the Earthward gaze as people realized that while humanity's home is not the center of the universe, humankind itself ought to be its own prime concern (if only for lack of alternatives). Lucas does not reject that perspective on earth entirely; indeed, his own alternative 'system' involves a similar Earthward gaze in that it looks at the planet from the outside, contemplating not only its physical make-up of 'Lithosphere,' 'Hydrosphere' and 'Atmosphere,' but also the 'layer of [human] Thought' that Lucas calls 'Noussphere, the Earth-Mind.'[35] As he yearns to reach the next stage of human development in the cosmos, Lucas's desire to transcend anthropocentrism is itself testament to his own kind of narcissism: taking a global, planetary or even cosmic perspective on life on earth is the expression of his belief that he possesses deeper insight than anyone else, which makes his mundane existence unbearable for him. He thus criticizes a collective narcissism (humankind exulting itself) and falls prey to individual narcissism (one man exulting himself). As the novel proceeds, Lucas increasingly loses touch with the community around him, focusing entirely on the pupil he shares his theories with, even abusing him sexually, and ultimately being hospitalized after a failed suicide attempt.

Lucas's personal narcissism is perhaps no worse than that of several other male characters in the novel's provincial setting, whose moral decrepitude is likewise expressed through their sexual depredations against minors. Their interests are emphatically and exclusively of this world, however, and, through this structural doubling, Byatt suggests that irrespective of whether one's fantasies are mundane or spiritual, it is dangerous to make them the measure of all things. To overcontemplate man's place, purpose and stature in the universe appears as a characteristically masculine impulse grounded in a sense of psychological, scientific and spiritual self-importance that is as pathological as it is pathetic. In this way, Byatt seems to pick up on Auden's criticism of spaceflight in 'Moon Landing,' which he began by stating: 'It's natural the Boys should whoop it up for / so huge a phallic triumph, an adventure / it would not have occurred to women / to think worth while.'[36] Although it is perhaps no less essentialist than Auden's, Byatt's exploration of this theme is afforded greater range, depth and complexity by her novelistic medium. In the world of *The Virgin in the Garden*, which is centered on the parochial but firmly assertive of a wider spatial, temporal and intellectual frame encompassing London and the history of European art, masculine behavior such as Lucas's emerges as just as inhibitive of social existence as it would be to disregard a larger perspective altogether. Thus, the novel indirectly satirized the anthropic enthusiasm that informed some of the more esoteric 1970s interpretations of the Earthward gaze as part of the 'one world' movement, placing it in a utopian tradition that had failed, and continued to fail, to improve anyone's lives. Pointedly, this satire did not address the material, technological or military implications of the Apollo experience at all; in the English hinterlands, this experience was to be articulated mostly through its repercussions on human self-perceptions. Like Lessing's novel, *The Virgin in the Garden* engaged the self-reflective Earthward gaze as a symptom of narcissism, but Byatt's sting was more clearly directed at its specifically English, and indeed, specifically masculine, dimension.

In *Doctor Copernicus*, a fictional retelling of the eponymous astronomer's life published two years before *The Virgin in the Garden* (in 1976), John Banville also explored the psychological implications of the Earthward gaze.[37] One of the novel's four parts is narrated by Georg Joachim Rheticus, Copernius's student, who is portrayed by Banville as devious, overambitious and resentful against his master. This character relates how, when reading his master's still unpublished book, *De Revolutionibus Orbium Coelestium*, he was filled with sadness as he saw a diagram illustrating the heliocentric system:

> It was sorrow! Sorrow that old Earth should be thus deposed, and cast out into the darkness of the firmament, there to prance and spin at the behest of a tyrannical, mute god of fire. I grieved, friends, for our diminishment! [...] Beloved Earth! He banished you forever into darkness. And yet, what does it matter? The sky shall be forever blue, and the earth shall forever blossom in spring, and this planet shall forever be the centre of all we know. I believe it.[38]

Once again, clearly summoning mental images based on 'Blue Marble'-type photographs in 1970s readers, Banville here attributes to Rheticus the world view that Nietzsche was to call the 'Copernican diminishment of man': earth's eccentric position in space entails, for Rheticus, an insult to human dignity and an exile from its righteous place. The consolation he finds is a perfect expression of the 1970s anthropic answer to post-Apollo *tristesse*. However, in a later conversation with his student, Banville has Copernicus reject the legitimacy of his Nietzschean interpretation:

> You imagine that my book is a kind of mirror in which the real world is reflected; but you are mistaken, you must realise that. In order to build such a mirror, I should need to be able to perceive the world whole, in its entirety and in its essence. But our lives are lived in such a tiny, confined space, and in such disorder, that this perception is not possible.[39]

Copernicus is convinced that a true perspective on earth from outside would have to offer not only what his famous diagram in *De Revolutionibus* depicts, nor what a 'Blue Marble'-style image shows, but 'the world [...] in its essence,' and that is not possible – at least not through mimetic media. Images and scientific treatises cannot reflect the world 'in its essence.' A visual representation of earth in space, however true it may be to astronomical fact, has no meaning to offer to the observer in and of itself. '*Seeing is not perceiving*,' Copernicus insists; the 'real world' is not visible to the eye.[40] Like Lessing and Byatt, then, Banville pondered the meaning-ascriptions to the view from space. The novelists highlight the process by which a 'mere' visual phenomenon is charged with much more significance than it would appear to merit. The 'lunatics' Lucas and Watkins and the anthropocentric Rheticus overtax the Earthward gaze with making a statement about the place, purpose and stature of humankind in the universe. In their different ways, all three novels suggest that to do so is less than wholesome. The self-reflective anthropocentrism of the Earthward gaze, celebrated by philosophers in the wake of Apollo, was viewed rather more critically by these novelists. One reason for this is that as novelists, they took a special attitude toward both self-reflexivity and anthropicity.

III   Anthropic worlds

In spite of their skepticism, Byatt, Lessing and Banville developed their critical attitudes toward utopian anthropocentrism into something more constructive. As they attacked overinterpretations of the Earthward gaze, they also recognized the motivation behind them: it was to place humanity in the center of the picture, and to make the world speak about humankind. They attested to a basic human desire for an 'anthropically' meaningful world, but they also admonished that looking for such meaning in the external world could lead to misapprehension, mental instability and social isolation. Still, they had an alternative to offer to a reality that resists such meaning-ascriptions: the world of the novel. The novel could function as a response to the questions that remained after Apollo because it was 'about' humankind in a way that the external world had seemingly proved not to be; it could suggest meaning and purpose to existence that had not been found in contingent reality.

This view of the genre as an apt medium for the kind of self-reflection that the Earthward gaze prompted is not without foundation in literary tradition. The novel has been described as the self-reflexive genre par excellence, and particularly the postmodern novel – in its heyday during the 1970s – is often characterized by reference to its self-reflexive aspects.[41] In the present context, suffice it to say that as a product of human culture, the novel is inescapably 'anthropic' in that its origin, subject and telos is humanity – unlike the potentially meaningless universe, a novel's world is indubitably 'about' humankind.[42] What renders the novel remarkable in this regard, particularly vis-à-vis other arts such as cinema, painting, architecture or music, is its exclusively narrative production of meaning. The novel does not (generally) rely on visual, auditory, haptic sense impressions to create its worlds; its effect is created by the exercise of the 'poetic imagination' on the basis of verbal narrative. In this respect, it closely reflects the ways in which humans make sense of their own lives: events in the past, present and future are made sense of and integrated into experience through narrative 'emplotment.'[43] As Walter L. Reed emphasized in 1976, the novel 'explores the difference between the fictions which are enshrined in the institution of literature and the fictions, more truthful historically or merely more familiar, by which we lead our daily lives.'[44] One might thus call the novel a self-reflexive medium for human self-reflection.

Thus, it comes as no surprise that in the wake of Apollo, novelists were quick to realize the potential that their genre held for negotiating the questions that arose from the Earthward gaze. Conversely, they also tried to harness the potential for self-reflection inherent in the Earthward gaze and its ubiquitous discussion for their own concerns with the novel form. This reciprocity is registered in the three novels discussed here in a way that is intimately bound up with their attitude to the Earthward gaze, and they partake of a more general trend in the period toward a pronounced emphasis on the genre's 'self-questioning' that one contemporary observer traced to a shift of mood since the 1960s that produced a watershed in novelistic production.[45]

In Lessing's *Briefing for a Descent into Hell*, hospitalized Watkins remarks after a dream passage in which he has spontaneously broken into song: '*I gotta use words when I talk to you.* Probably that sequence of words, "I've got to use words," is a definition of all literature, seen from a different perspective.'[46] It is exactly his erratic and probing attempts to 'use words' that constitute *Briefing for a Descent into Hell*, a confusing and seemingly disharmonious jangle of dream visions, memories and narratives that only merge into a meaningful whole when viewed as a 'complete' novel. The cacophony has a distancing effect as the contrast between ostensibly objective and wildly subjective narratives alerts readers to their difficult task of reconciling perspectives. Betsy Draine perceptively comments on 'the gradual gathering of a meaning that transcends the "main" plot' of this novel:

> That meaning inheres neither in the fate of Charles Watkins at the end of the linear action nor in his evaluation of his situation, but rather in the interconnections of situations, characters, and action among the several versions of his story. [...] If no single level of action can be relied on, then one begins to seek meaning not by determining the definitive meaning of any one of the parts, but by submitting to stages of narration, each of which opens up several lines of inquiry which may lead eventually to spiritual insight.[47]

Thus, while Watkins himself does not achieve any lasting insight, readers are challenged to find some unity in this baffling plurality of disparate narratives in order to make the reading experience worthwhile.[48] The difficulty they encounter in *Briefing* renders readers self-consciously aware of the hermeneutic processes by which they make sense of ordinary experience and presents the novel as a medium in which such processes can be exercised, as if in a sandbox. When, in the passage quoted above, Watkins hallucinates his trip to outer space and assumes the 'planetary' perspective of mankind as 'a minute grey crust here and there in the Earth,' this seems to elide all individuality and subjectivity: 'there were no such things as judges, but only Judge, not soldiers, but Soldier, not artists, but Artist, no matter if they imagined themselves to be in violent disagreement.'[49] This passage clearly foregrounds the relationship between collective and individual, or genre and specimen, acknowledging the need for such complexity-reducing external frames. As the novel demands to be read as 'inner-space fiction,' it invites readers to question their own preconceptions about what constitutes interiority, and to explore the role of fiction in mediating between individuals and larger groups. In this way, it anticipates an effect later described by Apollo 14 astronaut Edgar D. Mitchell (1930–2016), who, in 1986, spoke of the 'individual/universal dyad': 'On the spectrum of consciousness, points of view can be anywhere along that. Most of us are clustered down toward the individual point of view. The moon experience catapulted us toward the other end.'[50] Lessing reflected, in this novel, about the particular 'use of words' that is literature, and its potential for negotiating a similar shift of perspective from the individual to the universal. She used Watkins's views of space and from space to illustrate the universal appeal, the comprehensive ambition and the metaphysical desire that is behind hermeneutic endeavors. The effect Mitchell described of heightening a sense of collectivity of course contrasted with the pathological narcissism of the 1970s decried by Lasch and others, and Lessing suggested the novel genre to be a tool that allows readers to explore a more wholesome middle ground between these two models.

Lucas, in Byatt's *The Virgin in the Garden*, allows himself to be guided by individual instincts that lead him to abuse Marcus, signaling his narcissistic moral disconnection. The novel presents his yearning for a meaning and goal that transcends the mundane as a symptom of his mental illness, but it also registers that to wish for cosmic order and harmony is only human: the victim Marcus too is 'attracted' by Lucas's system and consoled by it.[51] Another character, the artist Wilkie, devises a 'cosmic' musical score for the performance of *Astraea*: it is 'the true music of the spheres' based on a medieval manuscript explaining the 'correspondences between Doric, Lydian, Phrygian and Mixolydian modes, the planets in the heavens, and the muses.' How, Frederica wryly asks Wilkie, will anyone realize what his music is about 'when nobody *knows* you've been fixing all these planetary octaves and transcendental notes'? His response is that 'people will intuit an order if an order is there, even if they can't name it or the principles it's derived from,' and the narrator comments that '[i]t was difficult to know how seriously he [Wilkie] took himself. Certainly he did like order, a plurality of perceptual orders, he was a fixer and an orchestrator.'[52] Byatt here presents another manipulative character using ideas of cosmic order for his own purposes, and she implies parallels between Wilkie's scheming and the work of the novelist: *The Virgin in the Garden* itself is self-consciously artificial and wrought in symbolic patterns, thus mirroring the musician's project. Not all patterns will be apparent to all readers, but some will be intuited and give pleasure; and thus, the novel's intricately meaningful world is held up as a more effective answer to yearnings such as Lucas's for a plan and a pattern to existence. This desire, it is suggested, could only ever be fulfilled in art, which is as highly anthropic as *Astraea* and as *The Virgin in the Garden*, with their epic scope, orderly arrangement and attention to detail exulting the human and the transitory over the realm of the cosmic immutable. Where Lucas fails in his metaphysical quest, the novel manages to integrate a plurality of plans and patterns that, in its ambiguity, is more successful in answering the human desire for order than Lucas's straining for monolithic truth from a cosmic perspective. Art and the anthropic world of the novel have more to say about mankind than the universe, Byatt proposed, thus providing her own answer to the post-Apollo frustrations of the Earthward gaze.

Lessing and Byatt emphasized that readers should view the plurality of perspectives presented in their novels by assuming a vantage point that surveys the whole rather than its parts, producing comprehension from comprehensiveness, as it were, just as the photographs of earth from space did for many viewers. While they were not unequivocally enthusiastic about visions of the 'one world' type, they did present them as salutary alternatives to the 'intense preoccupation with the self' that Lasch deemed definitive of 'the moral climate of contemporary society' in the 1970s.[53] *Doctor Copernicus*'s argument in favor of the novel genre was quite different. Banville's protagonist is a scientist who (not unlike Lucas and Watkins) is driven by a desire for knowledge of 'the thing itself, the vivid thing,' which as a child he intuits to be 'summon-able' through names (that is, words). Young Nicolas Koppernikg (as his name is spelled in the novel's first part) realizes 'that the birth of the new science must be preceded by a radical act of creation,' and it is in making his revolutionary discovery that he encounters this 'thing':

> What mattered was not the propositions, but the combining of them: *the act of creation*. He turned the solution this way and that, admiring it, as if he were turning in his fingers a flawless ravishing jewel. It was the thing itself, the vivid thing.[54]

The novel's central conceit is that the scientist's work is like that of the artist, and that the novel can recast our view of the world as significantly as a scientific treatise, if not more so. Having lost his mystical power of insight in adulthood, the dying Copernicus is granted peace again in an epiphanic culmination of his life in the novel's concluding pages. Praying for death, he suffers through conversations with Andreas Osiander, who has written a preface to his book. Copernicus is incensed at Osiander's cautious phrasing in the preface that 'it is not necessary that these hypotheses should be true, or even probable; it is enough if they provide a calculus which is consistent with the observations.'[55] In spite of what he told his pupil Rheticus about the disparateness between scientific description and 'truth,' Copernicus feels that the retraction of any truth claim beyond the merely descriptive betrays his quest. He still does pursue the 'thing itself,' and finding that his written work is deprived of this dimension, the moribund astronomer hallucinates Osiander to transform into his (long-dead) brother Andreas Koppernikg, who tells him that it is possible to intuit 'the other, the thing itself, the vivid thing, which is not to be found in any book, nor in the firmament, nor in the absolute forms':

> It is that thing, passionate and yet calm, fierce and coming from far away, fabulous and yet ordinary, that thing which is all that matters, which is the great miracle. You glimpsed it briefly in our father, in sister Barbara, in Fracastoro, in Anna Schillings, in all the others, and even, yes, in me, glimpsed it, and turned away, appalled and [...] embarrassed. Call it acceptance, call it love if you wish, but these are poor words, and express nothing of the enormity.[56]

In recalling the people who meant something to Copernicus, and who allowed him to feel some human contact beyond his obsession with knowledge, Andreas here summons a pageant of the characters the reader has encountered, reminding them at the very end of the text of all they have experienced during their reading and suggesting that this 'great miracle' of the novel cannot be rephrased in 'poor words.' It emerges that Copernicus's quest for 'the thing itself' does not reach its goal in his book in the sense that the *Commentariolus* or *De Revolutionibus* are 'his,' but that *Doctor Copernicus* as 'his' book realizes it. Banville powerfully affirms the epistemological potential of the novel genre: by appropriating the iconic figure of Copernicus, he invests his statement for the similarity between scientific and poetic epistemology with particular force. Commenting on novelistic engagements with the new physics of the twentieth century, Banville in 1985 claimed to perceive a development toward 'a new kind of novel, a new definition of fiction':

> The old certainties are going. In their place can come a new poetic intensity, once the form is freed of its obligations to psychologize, to spin yarns, to portray 'reality.' How the change will come about no one yet knows. But as science moves away from the search for blank certainties it takes on more and more of the character of poetic metaphor, and since fiction is moving, however sluggishly, in the same direction, perhaps a certain seepage between the two streams is inevitable.[57]

This notion of the genre's potential also constitutes a theme of *Doctor Copernicus*, where it is presented as the result of the quest of a scientific mind. Its protagonist advises his student not to misread the astronomical phenomena, but he also struggles with finding a 'correct' way of interpreting them himself. The 'Copernican shift' he finally makes is to comprehend that both scientific and literary works are only about themselves, but that it is enough to accept that 'all that can be said is the saying,' because behind that saying, something like a 'life' and a 'true world' might emerge with the help of the poetic imagination. That the novel should be an instrument for answering this question is a momentous claim, and it is one that Banville made in response to what he perceived as the exaggerated 'misreadings' of science by his contemporaries. Banville did not make as much of the power of the novel to create a 'reception community' as Lessing and Byatt did. Instead, his romantic faith in the power of the novel to communicate insights of the poetic imagination seems oriented toward the subjective experience of the reader.

IV   Homo significans

At the beginning of his *Genesis of the Copernican World*, Blumenberg gives an account of a meeting between Georg Wilhelm Friedrich Hegel (1770–1831) and Heinrich Heine (1797–1856) in 1829. One night, the poet visited the philosopher in Berlin and was made to wait outside his study. Looking out of the window at the night sky to pass the time, Heine

> was seized by a romantic mood, as so often in his youth, and he began, quietly at first but raising his voice unawares, to fantasy about the starry skies, divine charity and omnipotence coursing through them, etc. All of a sudden, and oblivious to where he was, he felt a hand on his shoulder, and heard the words, 'It's not the stars, but what man reads into them – that's what matters!' He turned and faced Hegel.[58]

Hegel's assessment resonates with the post-Apollo experience of those who were not taken in by the (mainly American) enthusiasm for 'global visions.' As it became clear that the Space Age would not keep all its promises, this prompted critical reflections about the 'readings' or meaning-ascriptions that had raised hopes in the first place. Thus, contrary to what one might expect, and contrary to the 'whole earth' and 'one world' spirit, critics such as Lessing, Byatt and Banville did not unequivocally celebrate the return to the terrestrial and human. Instead, they very much acknowledged that this regained anthropocentrism was informed by the self-centered pathology typical of the 'Me Decade' that projected meaning onto images and data that were not 'about' humankind in any specific way.[59] In different ways, they attacked the narcissism implicit in any such avid embrace of self-reflection, arguing that to take a cosmic perspective consciously and willingly centered on the earthly and human could be a symptom of much that was wrong with contemporary society. They denounced both the radical self-centering of the 'culture of narcissism' and the starry-eyed idealism of the 'whole earth' and 'one world' movements, at least insofar as these are based on an anthropicizing Earthward gaze. Where this gaze exalted earth and humankind as the only sources of ideals and meaning, the novels warn that 'mere' images should not be misread in this way.[60]

In her seminal 1971 essay 'On Photography,' Susan Sontag warned that among the prime narcissistic functions of photography was 'self-surveillance.' As Lasch elaborated at the end of the decade, this was so not only because the camera 'provides the technical means of ceaseless self-scrutiny but because it renders the sense of selfhood dependent on the consumption of images of the self, at the same time calling into question the reality of the external world.'[61] The narcissistic self only exists insofar as it is perceived (*esse est percipi*). Certainly, this is true of novels and their worlds; in this respect, they are not only essentially anthropic but also narcissistic. Much like the images of earth from space, their meaning is not something found, but something made by their readership; and as the images are made to 'speak' about mankind and are turned into icons of anthropic self-reflexivity, the novels denounce this process as originating not from some inherent quality of the images but from the desire to make them 'about us.'

In a way, the novels thus arrived at the conclusion Hans Blumenberg had also identified as 'the final Copernican consequence': the post-Copernican, non-geocentric world did not offer any reply to the question about mankind's place in it, and post-Copernicans felt pressed to address that question themselves.[62] Where 'readings into the stars' proved disappointing, Lessing, Byatt and Banville presented the novel as the more rewarding object and instrument for human sense-making. At their hands, the genre challenged readers to assume an 'Apollonian' perspective, not on earth but on fictional worlds, integrating a multiplicity of viewpoints into a whole that comprised more than the sum of its parts and hinted at a 'true essence' eluding visual and 'literal,' non-poetic representation. This represents a romantic aesthetics with an emphasis on reader participation that was closely in step with contemporaneous developments in literary theory. Here, too, the reader's involvement in the production of meaning rose to prominence in the same decade with the international success of reader-response criticism and reception aesthetics.[63] Just as the Earthward gaze had exposed the fundamentally human desire for the universe to be meaningful, literary critics, too, had concluded that man is 'not just *homo sapiens* but also *homo significans*: a creature who gives sense to things,' as Jonathan Culler emphasized in 1975.[64] The novels discussed in this chapter foregrounded this parallel between the Earthward gaze and novel reading: as the viewers of earth's photographs 'read into' these images by projecting meaning onto their 'text,' so the readers of novels 'produce' their meaning. The former, however, are misguided in believing that the pictures are 'about them'; the latter are entitled to do so, because the world of the novel, unlike the cosmos, is strongly and thoroughly anthropic. This world, and the novel at large, gains its anthropic meaning not least from the reader, who becomes a participating agent akin to the 'prosumer' Alvin Toffler saw as crucial to 1970s economics.[65] In this way, engagements with the Earthward gaze enabled novelists to address social developments on a much wider scale, reassessing the place of literature in culture just as viewers of the Apollo photographs were re-evaluating the place of mankind in the universe.

Notes

1. Alexander C.T. Geppert, 'Space *Personae*: Cosmopolitan Networks of Peripheral Knowledge, 1927–1957,' *Journal of Modern European History* 6.2 (2008), 262–86, here 262. Unless otherwise stated, all translations are my own. Discussing the impact of the photograph of 'Earthrise' from the moon and the 'cultural value of space exploration,' Valerie Neal in 1994 stated the received opinion concisely: 'That one image had profound impact. It provoked new perceptions and new emotions, jarring people out of complacency about our knowledge, our place in the universe, and our security.' While the moon may have lost some of its mystery through the landings, earth conversely 'became a thing of wonder and beauty, newly seen from space as a jewel, set in a vast black void. With this image, the literary imagination turned homeward and inward'; see Valerie Neal, 'What is the Cultural Value of Space Exploration?,' in National Geographic Society, ed., *What is the Value of Space Exploration? A Symposium*, Washington, DC: National Geographic Society, 1994, 14–15, here 15.
2. Robert Poole, *Earthrise: How Man First Saw the Earth*, New Haven: Yale University Press, 2008, 199. See also Robin Kelsey, 'Reverse Shot: *Earthrise* and *Blue Marble* in the American Imagination,' in Gareth Doherty and El Hadi Jazairy, eds, *Scales of the Earth*, Cambridge, MA: Harvard University Press, 2011, 10–16; Andrew Chaikin, 'Live from the Moon: The Societal Impact of Apollo,' in Steven J. Dick and Roger D. Launius, eds, *Societal Impact of Spaceflight*, Washington, DC: NASA, 2007, 53–66; Roger D. Launius, 'Perceptions of Apollo: Myth, Nostalgia, Memory or All of the Above?,' *Space Policy* 21.2 (May 2005), 129–39; idem, 'Interpreting the Moon Landings: Project Apollo and the Historians,' *History and Technology* 22.3 (September 2006), 225–55; Benjamin Lazier, 'Earthrise; or, The Globalization of the World Picture,' *American Historical Review* 116.3 (June 2011), 602–30; Denis E. Cosgrove, *Apollo's Eye: A Cartographic Genealogy of the Earth in the Western Imagination*, Baltimore: Johns Hopkins University Press, 2001, 257–67.
3. W.H. Auden, 'Moon Landing,' *New Yorker* (6 September 1969), 38.
4. For a similar reason, the multitude of science-fiction novels of the 1970s are not discussed here, since a generic context that renders spaceflight a regular part of the novel's world will tend to treat its implications differently.
5. Günther Anders, *Der Blick vom Mond: Reflexionen über Weltraumflüge* [1970], 2nd edn, Munich: C.H. Beck, 1994, 12, 89–90.
6. Alexander C.T. Geppert, 'Flights of Fancy: Outer Space and the European Imagination, 1923–1969,' in Steven J. Dick and Roger D. Launius, eds, *Societal Impact of Spaceflight*, Washington, DC: NASA, 2007, 585–99, here 594.

7. In 1971 the German magazine *Der Spiegel*, in an interview with Wernher von Braun, polemically spoke of man's 'crash landing on earth,' and von Braun agreed that the Earthward gaze from space had superseded the gaze toward the stars in importance; see Wernher von Braun, 'Rücksturz zur Erde,' *Der Spiegel* 25.7 (8 February 1971), 137–44, here 139. I am grateful to Alexander Geppert for bringing this interview to my attention.

8. This is the argument presented by David Kuchenbuch, '"Eine Welt": Globales Interdependenzbewusstsein und die Moralisierung des Alltags in den 1970er und 1980er Jahren,' *Geschichte und Gesellschaft* 38.1 (2012), 158–84, here 161; see Denis E. Cosgrove, 'Contested Global Visions: *One-World, Whole-Earth*, and the Apollo Space Photographs,' *Annals of the Association of American Geographers* 84.2 (June 1994), 270–94, for a different perspective. On the broader ideological context, see Matthew Connelly, 'Future Shock: The End of the World as They Knew It,' in Niall Ferguson, Charles S. Maier, Erez Manela and Daniel J. Sargent, eds, *The Shock of the Global: The 1970s in Perspective*, Cambridge, MA: Harvard University Press, 2010, 337–50; and Sam Binkley, *Getting Loose: Lifestyle Consumption in the 1970s*, Durham: Duke University Press, 2007, 247–9.

9. Rüdiger Zill, 'Zu den Sternen und zurück: Die Entstehung des Weltalls als Erfahrungsraum und die Inversion des menschlichen Erwartungshorizonts,' in Michael Moxter, ed., *Erinnerung an das Humane: Beiträge zur phänomenologischen Anthropologie Hans Blumenbergs*, Tübingen: Mohr Siebeck, 2011, 300–26. See also Lazier, 'Earthrise,' and Rüdiger Zill, '"Die Erforschung der Rückseite des Mondes durch reines Denken": Technikphilosophie zwischen *Sputnik 1* und *Apollo 11*,' in Igor J. Polianski and Matthias Schwartz, eds, *Die Spur des Sputnik: Kulturhistorische Expeditionen ins kosmische Zeitalter*, Frankfurt am Main: Campus, 2009, 332–49.

10. Emmanuel Lévinas, 'Heidegger, Gagarin and Us,' in idem, *Difficult Freedom: Essays on Judaism*, London: Athlone Press, 1990, 231–4, here 233.

11. Ibid.

12. Maurice Blanchot, 'The Conquest of Space' [1961], in Michael Holland, ed., *The Blanchot Reader*, Oxford: Blackwell, 1995, 269–71, here 270–1.

13. Anders, *Blick*, 96–7. See Frank White, *The Overview Effect: Space Exploration and Human Evolution*, 2nd edn, Reston: American Institute of Aeronautics and Astronautics, 1998, on the 'Copernican Perspective' corroborated by spaceflight; esp. 61.

14. Hans Blumenberg, *Die Vollzähligkeit der Sterne*, Frankfurt am Main: Suhrkamp, 1997, 482.

15. Ibid., 383.

16. Hans Blumenberg, *The Genesis of the Copernican World* [1975], Cambridge, MA: MIT Press, 1987, 685.

17. Alexander C.T. Geppert, 'European Astrofuturism, Cosmic Provincialism: Historicizing the Space Age,' in idem, ed., *Imagining Outer Space: European Astroculture in the Twentieth Century*, Basingstoke: Palgrave Macmillan, 2012, 3–24, here 9 (= *European Astroculture*, vol. 1).

18. Ernst Cassirer, *An Essay on Man: An Introduction to a Philosophy of Human Culture* [1944], New Haven: Yale University Press, 1992, 48.

19. See Michel Foucault, *The Order of Things: An Archaeology of the Human Sciences* [1966], Abingdon: Routledge, 2003.

20. See Hans Blumenberg, 'Kopernikus im Selbstverständnis der Neuzeit,' *Akademie der Wissenschaften und der Literatur in Mainz: Abhandlungen der geistes- und sozialwissenschaftlichen Klasse* 5 (1964), 339–68, here 367.
21. Hans Blumenberg, *Paradigms for a Metaphorology* [1960], Ithaca: Cornell University Press, 2010, 113.
22. I use the term 'anthropicization' in a broad sense to designate a process by which the universe is interpreted in a manner that posits the existence of mankind as its *telos* and mankind as the measure of all things in it. See the recent comprehensive account and critique of this foundational idea in Western philosophy by Wolfgang Welsch, *Homo mundanus: Jenseits der anthropischen Denkform der Moderne*, Weilerswist: Velbrück, 2012.
23. Hannah Arendt, 'The Conquest of Space and the Stature of Man' [1963], in idem, *Between Past and Future: Eight Exercises in Political Thought*, New York, 2006, 260–74, here 273.
24. 'In contrast to the bleak and lifeless Moonscape, Earth represented, in the words of Apollo 8 command module pilot Jim Lovell, "a grand oasis in the big vastness of space"' (Chaikin, 'Live,' 54). Astronaut Joseph Allen commented: 'With all the arguments, pro and con, for going to the Moon, no one suggested that we should do it to look at the Earth. But that may in fact be the one important reason' (quoted after Launius, 'Interpreting,' 245). See also the interviews included in White, *Overview Effect*, 179–282. Walter A. McDougall, *...The Heavens and the Earth: A Political History of the Space Age* [1985], Baltimore: Johns Hopkins University Press, 1997, 412; Cosgrove, 'Contested Global Visions'; and for a recent overview, Kuchenbuch, 'Eine Welt.'
25. Ronald Weber, *Seeing Earth: Literary Responses to Space Exploration*, Athens: Ohio University Press, 1985, xiii; see also the contributions in Laurence Goldstein, ed., *The Moon Landing and its Aftermath*, Ann Arbor: University of Michigan, 1979 (= *Michigan Quarterly Review* 18.2).
26. The source appears to be a photograph labeled AS11-36-5355 by NASA, taken from a distance of some 180,000 kilometers. The image shows Iran, Lessing's country of birth, Zimbabwe, where she had lived until 1949, and her current home, Britain. For further information on the photograph, see http://spaceflight.nasa.gov/gallery/images/apollo/apollo11/html/as11-36-5355.html (accessed 1 October 2017).
27. Doris Lessing, *Briefing for a Descent into Hell* [1971], London: Flamingo, 2002, 9. Note that the inward movement is thus identified with the 'descent into hell' from the novel's very beginning.
28. Ibid., 54–5.
29. Watkins also identifies with Ulysses adrift on the sea, with a man lost in a forest and a city, and a British soldier he once knew in the Second World War, among other things. His hallucinations thus feed on myth and memory as well as on metaphor. Readers versed in English literary history might find echoes of Geoffrey Chaucer's late-fourteenth-century *House of Fame* in Watkins's space excursions on the back of a bird. Lessing gestures towards these traditions to indicate that narrative has a strong tradition of responding to human self-questioning.
30. Ibid., 94.
31. Ibid., 95.
32. Ibid., 250.

33. A.S. Byatt, *The Virgin in the Garden* [1978], London: Vintage, 1994, 190.

34. Christopher Lasch, *The Culture of Narcissism: American Life in an Age of Diminishing Expectations* [1979], New York: Norton, 1991.

35. Byatt, *Virgin in the Garden*, 190–1. This is reminiscent of the theory of the noosphere in Pierre Teilhard de Chardin, *The Phenomenon of Man* [1955], New York: Harper Perennial, 2008. The novel is set before the publication of this work but was written later, when Teilhard was well known. Lucas's system also alludes, of course, to Lovelock's Gaia hypothesis, popularized since the early 1970s; Binkley, *Getting Loose*, 160. See also the theories on 'planetary conscience' propounded in White, *Overview Effect*.

36. Auden, 'Moon Landing,' 38.

37. John Banville, *Doctor Copernicus*, in idem, *The Revolutions Trilogy*, London: Picador, 2000.

38. Ibid., 208–9.

39. Ibid., 239.

40. Ibid.

41. For a recent restatement, see Christoph Bode, 'The English Novel as a Distinctly Modern Genre,' in Christoph Reinfandt, ed., *Handbook of the English Novel of the Twentieth and Twenty-first Centuries*, Berlin: de Gruyter, 2017, 23–41, esp. 31–4. In the 1970s, the argument was influentially presented in Robert Alter, *Partial Magic: The Novel as a Self-Conscious Genre*, Berkeley: University of California Press, 1975. At the close of the decade, further comprehensive treatments of self-reflexivity in the novel followed in Linda Hutcheon, *Narcissistic Narrative: The Metafictional Paradox*, Waterloo: Wilfrid Laurier University Press, 1980; as well as Walter L. Reed, *An Exemplary History of the Novel: The Quixotic versus the Picaresque*, Chicago: University of Chicago Press, 1981.

42. I develop this point at greater length in my forthcoming *Reading into the Stars: Cosmopoetics in the Contemporary Novel*, Heidelberg: Winter, 2018.

43. See Mark Turner, *The Literary Mind*, New York: Oxford University Press, 1996; Lisa Zunshine, *Why We Read Fiction: Theory of Mind and the Novel*, Columbus: Ohio State University Press, 2006; David Herman, *Storytelling and the Sciences of Mind*, Cambridge, MA: MIT Press, 2013. The notion of 'emplotment' is familiar to historians, of course, from Hayden White, *Metahistory: The Historical Imagination in Nineteenth-Century Europe*, Baltimore: Johns Hopkins University Press, 1973. The 1970s also witnessed an increased critical awareness of the importance of narrative and social constructions in the natural sciences, owing to the work of Bruno Latour, Thomas S. Kuhn and others.

44. Walter L. Reed, 'The Problem with a Poetics of the Novel,' *NOVEL: A Forum on Fiction* 9.2 (1976), 101–13, here 104. Reed specifically stresses the novel's exploration of 'those human fictions – economic, political, psychological, social, scientific, historical, even mythical – which lie beyond the boundaries of the prevailing literary canon' (104). Among those fictions, of course, the 'one world' and 'whole earth' interpretations of the Earthward gaze would have to be included.

45. Malcolm Bradbury, 'Introduction,' in idem, ed., *The Novel Today: Contemporary Writers on Modern Fiction*, London: Fontana, 1977, 7–21, here 8.

46. Lessing, *Briefing*, 105.

47. Betsy Draine, *Substance Under Pressure: Artistic Coherence and Evolving Form in the Novels of Doris Lessing*, Madison: University of Wisconsin Press, 1983, 92.
48. Incidentally, this effect of meaning production by the reader is one that was identified early on by readers of Brand's *Whole Earth Catalog*, which was said to prompt, through its catholic inclusiveness, a feeling that there was some hidden plan or structure to be discovered behind the apparently chaotic jumble of products it advertised. Binkley, *Getting Loose*, 152, quotes from a 1970 article in *Harper's* on the *Catalog* to this effect, commenting further that in addition to the publication's variety, its '[d]isorganization and typographic glitches [...] transformed passive readers into active interpreters of cryptic, if sincere, lifestyle discourses.'
49. Lessing, *Briefing*, 93, 95.
50. Quoted after White, *Overview Effect*, 203.
51. Byatt, *The Virgin in the Garden*, 191.
52. Ibid., 414.
53. Lasch, *Culture of Narcissism*, 25.
54. *Doctor Copernicus*, 101–2.
55. Ibid., 271.
56. Ibid., 277.
57. John Banville, 'Physics and Fiction: Order from Chaos,' *New York Times Book Review* (21 April 1985), 1, 41–2, here 42.
58. Blumenberg, *Genesis*, 69–70.
59. See Chaikin, 'Live,' 59.
60. Garb, 'Perspective,' 266–7.
61. Lasch, *Culture of Narcissism*, 48.
62. Blumenberg, 'Kopernikus,' 368.
63. Witness the publication during the 1970s of central works by Wolfgang Iser and Hans Robert Jauss of the 'Constance School,' as well as Norman N. Holland, Stanley Fish, Jonathan Culler and others in the United States, but also of Umberto Eco's *The Role of the Reader*. An assessment of reader-response criticism in the decade is included in James L. Machor and Philip Goldstein, 'Reception Study in the U.S., 1985–2012,' *Anglistik: International Journal of English Studies* 24.2 (2013), 17–30, here 17–20.
64. Jonathan Culler, *Structuralist Poetics: Structuralism, Linguistics and the Study of Literature* [1975], London: Routledge, 2002, 304.
65. Alvin Toffler, *The Third Wave*, London: Pan Books, 1980, esp. 275–99.

CHAPTER 7

Building Outer Space: LEGO and the Conquest of the Beyond in the 1970s

Thore Bjørnvig

I   Cosmic play: setting the scene

Studies of European history have tended to interpret the 1970s as a decade of both personal and global crisis. Characterized by a sense of 'no future' and lagging faith in economic growth, efficient political planning and the boons of rational thought, science and technology, the 1970s constituted a turning point for the worse. Studies of space history have offered similar bleak views of the 1970s, considering the period as a low point in terms of space exploration. The idea of 'limits to growth' contributed to a growing anti-expansionism while the trust in technology and interest in outer space waned.[1] The belief system of astrofuturists – that is, space enthusiasts who saw large-scale space exploration projects as the pathway to the completion of humankind's destiny in outer space – was undercut by an earth-centered mindset in which a rediscovery of the planetary home was paramount.[2] In December 1972 the Apollo program was terminated and, thus, the 'classic' space agenda, including orbital stations, interplanetary vehicles, a moon base, expeditions to Mars and establishment of a Mars colony, was set aside.[3] From an American perspective, space historian Roger D. Launius has stated that the 'latter half of the 1970s might best be viewed as a nadir in human space exploration, with the Apollo program gone and the Shuttle not yet flying.' In view of the United States' space activities, the 1970s have even been referred to as a 'lost decade.'[4] This indecisiveness regarding real-life space exploration was

Thore Bjørnvig (✉)
Copenhagen, Denmark
e-mail: thorebjoernvig@gmail.com

© The Author(s) 2018
Alexander C.T. Geppert (ed.), *Limiting Outer Space*
European Astroculture, vol. 2
https://doi.org/10.1057/978-1-137-36916-1_7

reflected in the science-fiction genre, which downplayed the unbridled astro-futuristic optimism in favor of a more skeptical approach, coinciding with stylistic self-rumination.[5]

However, as the Apollo adventure came to a premature end, a new level of space enthusiasm was on the rise in the United States, connected to the old Space Age verve, yet by necessity of a more utopian, fantasy-like character. As De Witt Douglas Kilgore has argued, in the 1970s a second generation of space futurists emerged which, forced by budget cuts and nudged by leftist political winds, aspired to look beyond national space programs to realize the conquest of space. Here, Princeton physicist and space activist Gerard K. O'Neill's (1927–92) vision of an American suburban utopia of free-floating space colonies presented itself as a new focal point that energized space enthusiasts. This contributed to a boom in space interest groups in the late 1970s and early 1980s. Simultaneously, the period was characterized by a revival of popular astroculture as exemplified by the popularity of *Star Wars* (1977). This suggests that the intensification of interest in popular astroculture, O'Neill's grand schemes and the growing number of space interest groups may have had a common root in the frustration with the demise of the classic (first-generation astrofuturist) agenda for manned spaceflight. If true, then not only activist projects – which aimed at realizing the classic space agenda – but also *Star Wars* and related phenomena fulfilled a need: the former, by promising the realization of a utopian space; the latter by producing a fantasy of outer space as an exciting place full of human (and alien) adventure available to the public.[6]

Thus, during the 1970s, the enthusiasm for the human exploration of space did not simply die out, but was reconfigured in outlets different from the initial national concerns with outer space; that is, in pro-space movements, in semi-private, academic space initiatives and in popular cultural depictions of outer space. A transition took place in the public imagination from the anticipation of imminently advancing humankind's destiny as a spacefaring species typical of the 1960s, to a nostalgic-utopian depiction in the 1970s of the promised high adventures and colonization plans lost with the termination of the Apollo program.

While this may have been the case in the United States – the heartland of the Western Space Age – what about the periphery of Space Age geography? In Europe, things looked somewhat different, as if the conquest of space were still something right around the corner. In 1973 the decision to form the European Space Agency (ESA) was taken, and NASA and Europe reached an agreement to collaborate on the Spacelab program to create a laboratory to be flown into space on the prospected US Space Shuttle. ESA was established in 1975, the first ESA astronaut selection took place in 1977 and in 1978 the French Centre National d'Etudes Spatiales (CNES) suggested the idea of the European space plane Hermes. In 1979 the first successful Ariane 1 launch was completed, and France and the Soviet Union signed an agreement to fly

French astronauts to the Russian space station Salyut. Thus, while the new wave of astrocultural enthusiasm also swept across Western Europe, resulting in, for example, the sudden enthusiasm for 'space disco,' there were other factors at work here, which may, in some instances, have added a different flavor to the phenomenon.[7]

This chapter offers a case study of the early period of Danish toy company LEGO's now famous space-themed toy series Legoland Space, dubbed 'Classic Space' by adult fans of LEGO, spanning the period between 1978 and 1987. The development of interest in Legoland Space reflected popular interest in outer space in Western Europe and the United States in the late 1970s. However, it differed from this general interest in small but significant ways. At a time when popular astroculture veered toward unbridled Flash Gordon-like science-fiction fantasy, as in *Star Wars*, Legoland Space signified a more modest extrapolation from known space technologies and embraced the classic space agenda along with a techno-utopian dimension.[8]

II   Commanding the future: moon colonies and utopian plains

Legoland Space was the culmination of slow but sure progress toward specialization and role-playing in the toy industry. In 1949 the LEGO Group introduced 'Automatic Binding Bricks,' followed in 1955 by a 'System of Play.' The latter consisted of a town plan, with bricks and accessories sold separately. The first foreign sales office opened in West Germany in 1956. By 1961 LEGO bricks were being exported to many other European countries and the first sets were introduced to the United States. During the 1960s models expanded beyond the Town concept and during the 1970s more models were introduced, for example, a Wild West and a harbor scene. By the mid-1970s two LEGO figures were introduced, facilitating role-play in a toy line otherwise belonging to the category of 'construction' toys.[9]

The first space sets were sold in 1978 and comprised buildings and vehicles made out of LEGO bricks, along with studded plates to play and build on, some with flat, decorated landing courses and some with molded crater ridges. The theme's chief designer Jens Nygaard Knudsen was predominantly inspired by the US space program and the Apollo missions.[10] This was reflected in how the mini-figures of the sets were called 'astronauts' in the company's internal papers, that there were plans to obtain illustrations depicting moon landings from the American Embassy in Copenhagen to use as sources of inspiration for designing the space boxes, and also in the theme's logo, which resembled the well-known NASA 'meatball' symbol (Figure 7.1).[11]

Advertisements showed a variety of space models put together in the LEGO system of play, depicting a colony on a moon-like planetary surface (Figures 7.2, 7.3 and 7.4). The box containing the set 'Space Command Center' (1979) displayed a photograph of the finished model in front of a

Figure 7.1 White astronaut mini-figure, introduced in 1978, with the Legoland Space logo on the torso.
Source: Courtesy of LEGO Group.

sculpted landscape and a backdrop with a painted starry night sky. The land-scape resembled a moon-like planetary surface with sand dunes rolling off into the horizon (Figure 7.5).[12]

Figure 7.2 'Take-off to new and exciting adventures with LEGOLAND Space,' this ad published in the Swiss journal *Rataplan* proposed in August 1980.
Source: Courtesy of LEGO Group.

The model and the dunes are in daylight, yet the sky is black with stars, suggesting a planet without atmosphere. A gas planet with orange and brownish hues like those of Jupiter floats in the sky and the model is placed on a gray, molded crater baseplate. To the left a white astronaut is

Figure 7.3 'LEGOLAND Space turns the journey through space into an exciting adventure,' this ad published in the *Schweizerische Schülerzeitung* promised in April 1979. *Source*: Courtesy of LEGO Group.

walking toward the middle of the picture, carrying a black radio-set. The left part of the Space Command Center is a control room. Here, another white astronaut is sitting in a chair, facing a standing red astronaut. The wall beyond holds, on the top part, a row of inverted slope bricks with

Figure 7.4   Published in the Danish comic *Anders And* in July 1979, this ad depicted Lego's Space Command Center, with another set added. 'Soar out into space. Watch out. The exciting space program is full of ideas,' it declared.
Source: Courtesy of LEGO Group.

computer screen patterns, yellow transparent windows in the middle and another row of sloped bricks with computer patterns. The wall behind the left chair consists almost solely of one big 'screen'-brick with a large image

Figure 7.5   The box containing set 926, Space Command Center (1979).
Source: Courtesy of LEGO Group.

of an astronaut suspended in space, repairing a satellite. A close-up reveals a small moon in the upper right corner of the screen and what might be a glimpse of the spaceship from which the astronaut has ventured forth in the lower right corner.[13]

On top of the control room is a satellite dish on the base of which the logo of Legoland Space is printed: a planet with a small spaceship circling it and blasting away into space. On the right side of the control room is a garage-like structure with another, smaller satellite dish and a four-wheeled vehicle parked in front. The vehicle has a sloped brick with a computer screen pattern and on top of it another brick, suggesting a camera, or some sort of measurement instrument. Another vehicle is in the mid-background, a sort of moon-buggy driven by a red astronaut holding an additional radio-set. The white astronaut outside seems to be walking toward the unmanned vehicle in front of the garage. Since both are holding radio-sets, they may be talking to each other. When not monitoring the satellite-repairing mission, the two astronauts in the control room might also be engaged in conversation, the red astronaut perhaps getting ready to go outside.

All in all, the picture creates the impression of a small colony outpost, bustling with as much activity as four astronauts can muster. Though devoid of linguistic narrative, there is no doubt: the great space adventure is under-way, the dream of leaving earth and exploring the universe has become reality. The astronauts, all wearing the exact same stylized smile on their faces, seem to be perpetually happy while they, like an ant colony of hardworking engi-neers, build on the barren sands. The blue and gray colors of the buildings give off a cool, controlled ambience and between them, transparent elements of yellow, blue, green and red emit a magic glow. There is neither a repressed indigenous race, nor any hostile aliens, just as there are no weapons in sight. Everything takes place in a calm, controlled and cheerful way, conducted by what seems to be a peaceful, egalitarian society with a clear utopian streak.[14]

III   The astrocultural topography of Legoland Space

Just as Legoland Space promised children a bright and limitless future in outer space at a time when the awareness of limits to growth was at its height, it also offered the LEGO Group a way out of economic crisis. Between 1976 and 1977, the LEGO Group reached an economic low point and in response developed a more differentiated range of products. Along with Space another Legoland theme was introduced, Castle, a medieval world of knights, jousts and citadels. Simultaneously, the now iconic LEGO 'mini-figure' was intro-duced, the two new themes centered around it (Figure 7.6). Both introduced in 1978, the two themes played a momentous role in bringing the company out of economic stagnation, Space being the more successful of the two.[15]

Legoland was the 'land of LEGO,' undefined in time and space and 'the land where anything can happen.' Themes could take place in various time periods, a concept offering 'infinite possibilities of development.'[16] However, one could imagine various periods for specific themes, such as the Viking Age in 980, the 'age of chivalry' (the Castle theme) in 1200, Town in 1980 and Space in the year 2200. A more solid pointer to the exact time in which Space took place is found on the brick of the Space Command Center depicting an astronaut in space, where the inscription 'L.L. 2079' is likely shorthand for 'Legoland 2079.' This placed the action a hundred years in the future, fit-ting for an outer space theme reflecting and extrapolating realistic and already existing space technologies.

The primary target group was children between the ages of 4–8 years and the secondary target group their parents. Pre-launch market analysis indicated a strong interest in the outer space theme, though boys were more interested

Figure 7.6   Selection of LEGO Classic Space-era astronaut mini-figures, 1978–87.
Source: Courtesy of LEGO Group.

than girls. In Europe there were approximately 3.9–4 million 'active' boys who might buy LEGO sets.[17] Estimated sales of the Legoland product lines were 15.2 million boxes with total turnover of about DKK304 million. Children became very engaged in the product and there was a high degree of repurchase.[18] Europe was responsible for 75 percent of the turnover in 1978 with Germany, England and France as paramount countries.[19] In 1979 LEGO presented its hitherto most comprehensive program of new products with a corresponding rise in turnover. Here, Space was a great success both in the United States and in Europe, and LEGO received the British Association of Toy Retailers Toy of the Year Award for the Space theme in 1979. There was a 23 percent increase in turnover on the European market compared to 1978 and LEGO was gaining in popularity and turnover on the American market as well.[20]

The earliest space set made by LEGO was 'Space Rocket,' a simple Apollo-like rocket exclusively released in the United States in 1964, probably due to the country's heavy involvement with space technology development. Legoland Space was created by Jens Nygaard Knudsen, who worked for LEGO from 1968 until retirement in 2000.[21] In the 1970s Knudsen mainly designed Town models; however, he also designed a few space-themed models, 'Rocket Base' in 1973, and, 'Space Module with Astronauts' in 1975, depicting a manned Apollo moon landing. He continued making outer space models that eventually transformed into Legoland Space, as well as several subsequent space themes. Despite being nearly ready by 1976, the world-wide release of Space was held back until 1979 to be the great surprise for the Nuremberg Toy Fair that year.[22]

In an interview, Knudsen pointed out some of the inspirations for Lego-land Space, such as the French science-fiction author Jules Verne (1828–1905), the comic book series *Tintin*, *Flash Gordon* serials and science-fiction movies of the 1950s, such as the BBC's *The Quatermass Experiment* (1953), and later movies such as *2001: A Space Odyssey* (1968). With respect to space toys, Knudsen cited Danish toy company Reisler's 1950s space series of plastic astronauts and Dinky's *Thunderbirds* toys as major inspirations.[23] Real-life space exploration and the first moon landing were important as well:

> Of course I was fascinated when they landed on the moon – I sat up all night. But it was science fiction, which I was mainly inspired by. The unknown in space, to land on a foreign planet: Who will you meet? Will they be peaceful? Will they want to fight? The unknown, that was what fascinated me, the black hole, around in space, that was what I was fascinated by and still am. Now new planets are being discovered, I hear about it and I'm fascinated by it. I'm sorry that we can't travel with the speed of light – because then we won't be able to reach very far out into the solar system. But still, we ought to be able to make use of a few more planets.[24]

Thus, Knudsen pointed to real-life space exploration and popular astro-culture of his childhood and youth as sources of inspiration rather than the vibrant popular astroculture of the 1970s. However, the latter primed children's interest in early Legoland Space sets. For example, in Europe the French comic series *Valérian et Laureline* depicted a universe teeming with strange and wondrous life, while the Belgian comic series *Yoko Tsuno* told of an alien race secretly inhabiting earth.[25] Central pioneers of electronic space music also came from Europe and shaped the musical dimension of astro-culture. Frenchman Jean-Michel Jarre's *Équinoxe* and German Kraftwerk's *Mensch-Maschine* were both issued in 1978. From 1975 to 1977 the science-fiction TV series *Space: 1999*, produced in England, broadcast the adventures of moon base Alpha, while in 1979 James Bond conquered outer space in *Moonraker*.[26] In the United States *Close Encounters of the Third Kind* and *Star Wars* both premiered in 1977, creating unforgettable images of aliens, spaceships and exotic planets. The trend continued in 1978 with the TV series *Battlestar Galactica* and *Star Trek: The Motion Picture*, while in 1979 the British American co-production *Alien* sent shock waves through the Western imagination. However, just as popular astroculture of the mid- to late-1970s does not seem to have influenced Knudsen's design of Legoland Space significantly, the more dystopian elements of the Danish science-fiction milieu of the 1970s did not either, despite the fact that Knudsen was an avid reader of one of the genre's prominent Danish authors, Niels E. Nielsen

(1924–93), well-known for his dystopian themes.[27] But the popular astroculture of the 1970s certainly spawned other space toys against which the LEGO Group had to compete.

IV   A dream come true: selling space to children

One of the LEGO Group's competitors was Tente, another construction toy-line based on plastic bricks and created by the Spanish company Exin-Lines Bros S.A. in 1972, and marketed in the United States and Japan by the company Hasbro. In the late 1970s, Tente was also developing an outer space theme and, consequently, a member of Knudsen's design team defected to Tente, bringing along some as yet undisclosed ideas of the new Legoland Space theme. Hasbro was ready to sell its first Tente outer space sets in the United States in 1978; this forced the LEGO Group to rush the introduction of a few Legoland Space models, such as the Space Command Center, to the US market the same year. However, there were many other contenders on the space toy market. Both *Star Trek*, which was aired in the 1960s but kept being rerun in the 1970s, and *Battlestar Galactica* spawned a deluge of toys.[28] *Space: 1999* also became a popular toy subject, with moon bases, vehicles and action figures.[29] Kenner's *Star Wars* toys, put on the market simultaneously with the movie, became immensely popular and affected the toy market profoundly. Due to the practical fitting of *Star Wars* figures into spacecrafts, the standard 'GI Joe'-sized type of action doll was downscaled to an average of about 8.5 cm.[30] At the same time, these toys were a consummation of the increasing narrativization of toys from the 1950s through the 1970s, for example by modeling them on films and TV series. Part of this development was the growing propensity to produce toy figures of specific characters such as 'Luke Skywalker' (from *Star Wars*) rather than, for example, generic 'Indians' or 'cowboys.'[31]

Legoland Space premiered at a time when the LEGO Group still had not developed fully fledged fictional backgrounds and storylines for their products.[32] Stig Hjarvard's assertion, however, that LEGO sets of the 1970s and 1980s came entirely without a storyline, that the only narrativization was the instruction manuals' scripting of the building processes and that the sets thus facilitated completely open-ended play, is not entirely correct. First of all, the technologized and futuristic outer space setting of Legoland Space drew on the already established science-fiction genre. Second, the space sets themselves framed play through design concepts such as 'astronauts,' 'command base,' 'exploratory vehicle,' 'spacecraft,' 'mineral-detecting vehicle,' 'rocket,' 'satellite' etc., which established certain parameters within which play would likely be carried out. And third, if one looks closely enough at material accompanying the products, along with material from the LEGO archives, a storyline for Legoland Space can be traced.[33] For example, in a 1979 US LEGO catalog an advertisement read:

Here's a whole new galaxy from LEGO! With the new sets [...] you can imagine you're exploring the exciting world of outer space. Each set is fun in itself – but look how all the sets work together for even more fun. There are outer-space Action Figures too, complete with removable astronaut helmets and life support packs. Make up your own outer space adventures! Visit a friendly planet [...] patrol for Negative Forces. After you play, take everything apart and build a universe. [...] Just the way you want it![34]

Thus, the text suggests that LEGO space products provide access to a 'new galaxy' in which the child can partake in the exploration of the 'exciting world of outer space.' One set is fun in itself; however, the more sets you have, the greater the fun will be. The sets provide the possibility of creating one's own adventures, yet two play scenarios are suggested, both pointing to the presence of other, possibly alien, factions in outer space: to visit a 'friendly' planet, or patrol for 'Negative Forces.'

Another example of narrativization can be found in a LEGO idea book from 1980, published in the United States and relating the adventures of two mini-figures named Mary and Bill. In the first part of the picture story they are in Legoland Town, where they enter a cinema showing a science-fiction movie. When they leave, a spaceship – recognizable as one of the marketed models – has landed outside. They put on helmets and oxygen tanks, take off in the spaceship and travel to a space colony. Here they are greeted by red and white astronauts who show them around. Then a distress signal (indicated by the letters 'SOS') is emitted by an alien spaceship that has emergency landed close-by, the aliens being Town mini-figures with small bricks instead of the conventional mini-figure heads. After having helped repair the spaceship, Mary and Bill follow it to the aliens' base and, finally, they travel to a Legoland Castle-type of world. Thus, the idea book provided a narrative about the human colonization of a moon-like planetary body, friendly interaction with aliens and a peaceful visit to an alien settlement.[35]

Later narrativizations of Legoland Space can be found in, for example, the unpublished internal prototype book *Trapped in Space* from 1984, illustrated with photographs of various LEGO space models. Here, the spaceship 'Spearhead' is on 'a mighty trail-blazing mission, which takes humans out among the stars to explore the far reaches of the galaxies.' The purpose is to search for planets that can support human life and with resources of value to earth, and to look for intelligent alien races. Commissioned by the LEGO Group and building on *Trapped in Space*, Danish comic author Frank Madsen (1962–) created a comic series relating the adventures of the young space cadet Jim Spaceborn, his meetings with aliens, robots, the tyrant Kazak and his horde of black astronauts. Both reiterate the basic narrative outlined above, though emphasizing encounters with aliens and hostile factions. Published later than the Space sets in focus here, they are examples of more fully developed narrativizations not present at the outset of Legoland Space. Internally in the LEGO Group and in communications to shopkeepers, however, basic narratives were developed early on.[36]

According to a marketing strategy paper, Space was to be 'a line children can collect to create their own adventures in the unknown space.' The concept was to be communicated through 'elements that help man to explore space.' Here, 'spacemen' were to be represented by mini-figures, 'space appliances' by radars and control stations, 'space traffic' by spacecraft and space rockets and a 'space planet' by building plates. The 'basic consumer benefit' of all this was crystallized in the sentence: 'Build your own LEGOLAND Space and play,' evoking the double meaning of space as both outer space and a space of children's play.[37] This generic slogan was mirrored in texts such as the advertisement cited above and, for example, in a shop exhibition brochure given to shopkeepers: 'You can now build your own planet. With the space ships, rockets and radar trucks you can go exploring far out in space.' The 'evidence' for this would be that the sets were related in scale, plates and figures linked the sets and provided a play setting, and, last but not least, versatility, thus referring to the LEGO bricks' combinatorial possibilities and systemic philosophy.[38]

The space models were to be marketed as having 'a lot of functions and play value' and these in combination would create 'an exciting play environment.' The astronauts' different colors – at first only white and red, subsequently also yellow, blue, black and more – represented different kinds of functions, such as pilots, engineers and scientists (Figure 7.6 above). Having differently colored figures with different functions was thought to stimulate play.[39] The theme's planetary surface was not to be understood as an 'existing planet' as this would tie the sets too narrowly to the images of the real moon. This was perhaps in admission to the negative impact real-life telecasting from the moon had had on sales of outer space toys: the stark reality of a barren moon, devoid of both drama and aliens, broadcast on earth made sales of space toys plummet. Perhaps therefore, LEGO preferred a 'fantasy milieu,' such as 'the Planet of Legorion' which might have been 'conquered by emissaries from Legoland Town.' A product program for 1979 states that Space is set 'on another planet inhabited by persons who have arrived from the Earth several years ago. Their equipment is used for exploration of the planet they live on and the closest Galaxies.'[40]

Thus, Legoland Space was not to be tied too closely to real-life space exploration, yet was meant to be realistic. 'From the beginning,' explained Knudsen, 'we were supposed to make it accurate. We were to keep it on a scientific background, that's how we started, we weren't meant to have space wars. I remember that was really important.' Clearly, then, the theme was intended to draw children into an exciting world of manned space exploration and colonization extrapolated from real-life space exploration, enabling them to participate in a dream of conquering space through playing with the space models of the LEGO system. Yet, they were also expected to have an interest in space prior to being exposed to Legoland Space. The LEGO Group was aware of the growing fascination with outer space mediated by 'movies, TV and diverse comics' in both Europe and the United States. A letter to LEGO stockists, promising that sales would 'reach the sky' with the new outer space line, stated that 'young customers are interested in Space and with the new [LEGO] boxes which contain a lot of new elements, they will make the models look real!' With the new line, LEGO confidently professed that they were 'entering what is now interesting your "little" customer: Space,' and that the range of new elements promised to 'make the models look like in reality.' Twenty years later, in 1999, the LEGO Group finally tapped directly into the *Star Wars* myth and began producing the *Star Wars* set, which to this day is one of the most profitable LEGO products, responsible, for example, for every fifth euro made by the LEGO Group in Germany.[41]

V   Imagination, realism and specialization

The realism of the new LEGO Space sets had a double meaning, both as the emulation of existing space technologies and as the making of believable, realistic-looking fantasy space toys. The second meaning of such 'realism' was especially strong in late 1970s astroculture. Research conducted by the LEGO Group during 1978–79 indicated a growing demand for realism, and whereas competitors such as Playmobil and Matchbox had already met this demand, LEGO was only beginning to do the same. LEGO began introducing more specialized elements, for example, a circular radar dish with a more limited use than the basic eight-stub brick, and launched the more realistic-looking mini-figure (Figure 7.1 above). Also, LEGO began emphasizing play value on behalf of the constitutive construction-value of the bricks, increasing it by focusing on role-play and accepting specialized components for specific themes, despite compromising with the flexibility of the individual brick. In accordance with LEGO's systemic concept, the play value of models could be increased 'by collecting the units in popular product lines built on a joint play theme.' For example, the play value of a car would be extended if it was part of the larger City theme. Yet, there was always the fear that overly specialized elements would limit children's imagination.[42]

As such, Space was a compromise between accommodating tendencies of the toy market (including realism, specialization, role-playing and science fiction), and the LEGO System's raison d'être as a construction toy. Likewise, the mini-figures were an answer to competitive pressures from action-figures and, for example, Playmobil figures, but differed from *Star Wars* figures, for example, in that they had no specific attached identities apart from colors. As such, they were mere bricks, so to speak, in the great machinery of space colonization as opposed to the *Star Wars* action-figures of the rebel faction, the technology of which 'offered a human "fit".'[43]

This anonymity was, to some extent, a result of LEGO's rootedness in Scandinavian design tradition. Despite the demands of an Americanized consumer culture, the peculiarity of Scandinavian design was still visible in Legoland Space of the late 1970s. The simplicity, the clearness of line and the controlled use of color characterizing the theme all added up to a subtle, yet significant, sense of proportion. While embodying a thirst for conquest and colonization, the aesthetic suggested a modest and balanced mindset in which materiality and physical space had found a pleasing equilibrium. True to the principle of modernism, Scandinavian design aesthetic seeks 'to strike the optimum balance between form, function, material, color, texture, durability and cost so as to create democratic design solutions.' Growing out of modernism and its obsession with machine aesthetics, it moved toward the promotion of organic design, sharing with modernism the democratic ideal of creating everyday objects of outstanding design. Thus, with a marked focus on humanism, Scandinavian designers created a less doctrinal form of modernism. This in turn was influenced by the Nordic adherence to social democratic liberalism, creating an aesthetic with a tendency to produce items that are deemed 'timeless.' In addition, LEGO bricks embody another central tenet of Scandinavian societies, namely their inclusive, child-centered nature, which also may explain the hesitancy of the LEGO Group to embrace war as a play theme.[44]

On the one hand, due to a combination of market pressures and Knudsen's astrofuturist leanings, the theme embodied a version of space exploration rooted in the classic space agenda and influenced by overseas toy design trends. On the other hand, the Scandinavian design ethos held sway. This becomes especially apparent when considering the LEGO mini-figures' eternally smiling faces and the absence of weapons in LEGO toys. In contrast to Kenner's *Star Wars* toys, early LEGO Space products depicted manned space exploration as a Scandinavian, social democratic, utopian adventure of very wholesome proportions. Combining the LEGO system's philosophy of play and construction, drenched in a Scandinavian design tradition, with the astrofuturist dream of colonizing outer space, Legoland Space paved a unique way out of the disturbing conundrum of the 1970s. Promising an unrestrained future in space for the human race, yet suggesting an almost subdued level of grandeur, it at once suggested something entirely new and seemingly timeless.[45]

VI   Timeless futures: reconnecting with the sacred

Created by members of the baby boomer Apollo generation, Legoland Space was an interpretation of how the human presence in outer space might have looked if the classic space agenda had been realized. According to Daniel Lipkowitz, it took place in 'a future that seemed just around the corner' and depicted technologies 'that weren't too far removed from the real space technology of the 1970s and 1980s.' It was so 'realistic' that 'it should come as no surprise if space technicians from NASA in Houston, USA, came on a study trip to Billund,' one LEGO employee remarked in praise of their own

efforts. In fact, at some point in the late 1970s a delegation from the European Space Agency visited Billund to see LEGO's Space models. Though the
planet 'Legorion' may not have been intended to look too much like earth's
moon, to many it surely did. As such, Legoland Space fitted the optimistic,
indeed utopian, notion of 'the endless Space Age' and the mindset identified as 'astrofuturism' by Kilgore.[46] As Louis J. Halle, inspired by Gerard K.
O'Neill, wrote in 1980, 'thriving and proliferating colonies in outer space'
would offer an 'unlimited future' for mankind. This, in turn, suited LEGO
well since a deep admiration of technology and an optimistic view of progress
and the future have been characteristics of the LEGO dynasty.[47]

It is a long established paradigm for LEGO to make toys that at once fit
into the time, yet do not limit the imagination; 'The limits are where our
imagination ends,' CEO Godtfred Kirk Christiansen (1920–95) once proclaimed. Around 1954 he formulated ten characteristics that LEGO toys
ought to possess, the first of which was 'Unlimited play potential' and
another 'Development, imagination, creativity.'[48] A construction toy, the
LEGO system's fundamental element is the brick, which links it to the sphere
of architecture and engineering. In 1959 Christiansen explained:

> Just as we as adults have our world and want to make, conceive and
> build everything ourselves, so would the child create his own world.
> Just as the architect – master builder – has his bricks and manufactures
> buildings ranging from the simplest, small houses to the most wonder-
> ful castles, so the child has in its LEGO bricks the elements of its world.
> All that which involves the child's imagination, all that the child sees
> around it, it can build with LEGO elements. It can build houses, air-
> planes, ships, mills, puppet theaters, doll furniture – it can build what it
> wants.[49]

The man-made, mechanical nature of many of Christiansen's examples points
to a deeper reason why Legoland Space became a success: its connection to
the idea of the machine.[50] Dan Fleming has identified four dominant themes
of modern toy culture, one of which is 'the theme of the machine' as embodied in toy cars and trucks, construction sets and model spacecraft. According
to Fleming, construction sets are about piecing things together until something more than the sum of the parts emerges, thus rehearsing 'the nature
of the machine in general' and toys such as tinplate cars and aircraft embody
'the power of technology' with which children try to come to terms. Traditional constructional toys were 'the chief manifestation of the mechanical
impulse' created by 'the excitement of industrial expansion, capitalist growth
and unproblematic technological progress,' and when the metal-based Mec-
cano construction sets lost momentum, plastic-based LEGO took over.[51]

Thus, the LEGO system worked synergistically with a space theme whose
ideology was rooted in Western astrofuturism. Though it is only one amongst
many other product lines, its constant success, numerous incarnations and

connections to real-life space endeavors points to the effectiveness of this syn-ergy.[52] But what role did children and their LEGO Space sets play in relation to real-life space exploration? In general, popular astroculture is not merely responding to space exploration but also constitutive of it. Space exploration is enacted through space toys in everyday contexts and is thus domesticated. Space toys may, through miniaturization and compression of complex tech-nological objects and maneuvers, confer a feeling of agency and control on children. In addition, children's play with space toys in imagined outer space settings may also be a precursor of real life space exploration; in this respect the influence of space toys is no different to that of science fiction. Through playing with space toys the future is at once invoked, negotiated and created. By taking on the role of space explorers, by constructing space vehicles and by enacting the human venture into outer space, children playing with Lego-land Space not only contributed to the naturalization of space exploration and came to terms with its anxiety-provoking aspects. They also participated in a dream so inspiring that they might themselves become its ambassadors and entrepreneurs. That this is what might happen when children play with LEGO's more realistic outer space themes, such as the current spaceport of LEGO City, is the premise that the LEGO Group builds on when entering into educational joint ventures with ESA and NASA. For example, approx-imately five kilograms of LEGO were recently sent up to the International Space Station as part of a cooperation between NASA and LEGO aimed at kindling children's interest in science and technology, thus creating the 'builders of tomorrow,' as the slogan goes, in the form of engineers, scientists and even astronauts.[53]

As a testimony to Legoland Space's success in the late 1970s and early 1980s the theme still holds sway over the imagination of the generation that grew up building and playing with it. As already mentioned, 'adult fans of LEGO,' so-called AFOLs, have labeled it LEGO 'Classic Space' and laud the ingenuity of the theme in a multitude of ways and places.[54] Gizmodo blogger Jesus Diaz's experience of seeing his childhood's LEGO 'Galaxy Explorer' among the first of the Legoland Space models may be representative of the feelings that are activated in space-oriented AFOLs when re-engaging with LEGO space models:

> Lift off. Godspeed. Boom. A wave of emotions took control hitting my head like a Lego Airbus 380. Dozens of images started to appear in my head, Polaroids of Xmas and birthdays that I thought were faded. [...] Feelings and moments from times when everything was innocent and your only concern was your bike, a big carpet full of Lego bricks, and the amount of cocoa in your cereals.[55]

Elsewhere, Diaz has stated that 'watching Star Wars for the first time and the Lego Galaxy Explorer is what made me want to be an astronaut,' thus con-firming the thesis that popular astroculture in the form of space toys may in fact shape children's dreams of adult vocations.[56] In fact, AFOLs often see their LEGO hobby as a means of reconnecting with their childhood, and in turn see their childhood experiences with LEGO as the reason for their fascination with LEGO as adults. A common trope among AFOLs, the period between playing with LEGO when children and taking up LEGO as a hobby as adults, is designated the 'Dark Age.' Just as AFOLs' feelings for their childhood experiences with LEGO are strong, so are their notions of 'love of the brick' and the sense of purity that surrounds it. Strong animosity toward other 'clone' brands (such as United States Mega Bloks) suffuses the AFOL community, suggesting an almost ritualistic partaking in activities regarded as sacred.[57]

The sacred dimension of consumption has been pointed out by researchers who suggest that in contemporary Western society, where religion has become secularized and the secular sacralized, consumption may be a principal locus of transcendent experience. For example, collecting the toys of one's childhood may express a longing to reconnect with childhood, which again may be construed as 'nostalgia for paradise,' where childhood serves as a sacred geography and a site of personal transcendence. In religion and mythology it is common to find notions of a primordial, sacred time outside normal, profane time. These notions may be connected to a paradisiacal time or state and reconnecting with it has transcendent, transformative power. Interestingly, many grown-ups connect their childhood with a notion of 'timelessness,' seeing it as a carefree space devoid of the burdens of adult life; in effect a kind of prelapsarian, sinless state where time is abolished and eternity glimpsed.[58]

In Western Europe, the idea of childhood as embodying an innocent and sinless state emerged with Romanticism and culminated at the beginning of the twentieth century, when childhood was sanctified, thereby legitimizing the withdrawal of children from the workplace and their enrollment in the educational system. This suggests that the notion of childhood and children's play as something timeless and sacred is a historically determined construction. This goes for the four modes of timelessness that merged in Legoland Space as well: the timelessness of design, childhood, play and the paradisiacal. Framed by the sacred dimensions of consumption and enhanced by utopian and religious dimensions of astrofuturism and pro-space rhetoric, this goes a long way to explain the constant appeal of Legoland Space. Astrofuturistic and pro-space discourse is often structured according to the Judaic-Christian apocalyptic narrative of fall and redemption, and in this mythological framework outer space attains salvational potential: the nostalgia for paradise is relieved when mankind conquer the heavens.

By the 1960s, children's toys had lost much of the appeal to parents' nostalgia, who could no longer identify with the increasingly strange fantasy worlds which they embodied. At the same time, toys began losing their educational value as means of preparation for the future life of the child. However, given Legoland Space's emulation and extrapolation of Apollo-era space technologies, this theme still satisfied parents' longing for both nostalgia, educational realism and intergenerational identification. Considering the power of the gift of toys as crossgenerational conveyors of ideology and values, space toys are not just peripheral ephemeral objects. Rather, they constitute instances of coded material culture with which children make sense of the world, carrying meanings across generations, reinforcing already established mythologies and narratives of the alluring and redemptive properties of outer space.[59]

VII   Conclusion: limits and inevitabilities

Legoland Space was a result of a complex process of demarcations and transgressions, imitations and innovations, and it embodied notions of limitlessness and timelessness in its pursuit of the future. It created a toy world of systemic unlimitedness by way of the combinatorial potential of its elements; it set the scene for unlimited play and the unhampered unfolding of imagination; and it held the promise of an unlimited future of expansion and prosperity for mankind in outer space. At the same time, it encapsulated a design tradition that evoked a timeless quality through its aesthetic; it presented a future with the potential to instill in the consumer child a sense of the timeless through play; and it offered parents and adult fans of LEGO a chance to reconnect with their own childhood's play, potentially entering a sacred space of transcendence. However, just as the notion of timelessness was – and still is – a product of history, so the promise of unlimitedness created limits and, also, was set against a range of limits of practical, ideological, commercial and societal nature. In order to reach a new level of realism, LEGO elements had to be specialized, thus limiting the construction-value of the sets. The ideology of endless construction and progress, while articulated as a limitlessness of imagination, could just as well be seen as setting up limits, delineating the notion of what imagination is, what it should be used for and what material is ideal for its expression. Politically correct deliberations impeded the implementation of conflict in the theme, thereby enhancing its utopian characteristics while accommodating to parental approval. Thus, while answering to market pressures and astrocultural tendencies emerging in the heartland of the Space Age, the United States, Legoland Space retained values and propensities of the Scandinavian periphery of the Space Age.

Finally, the societal and economic limits set up by the 'lost decade' of the 1970s, with its termination of the classic space agenda, seem to have fueled the desire for limitless spaces of the imagination. In an American context, space historian Roger D. Launius has discussed how the zealous utopianism of manned space exploration gathered new strength in the 1970s and found an outlet in pro-space movements.[60] Those who had experienced the thrill of children's astroculture in the 1950s and the excitement of the Apollo program in the 1960s reached the age of parenthood in the mid- to late-1970s. As recently suggested, pro-space leanings may be rooted in childhood daydreaming and memories and, as toy companies are well aware, the parental group is as important to target as children. The appeal of Legoland Space to parents of the late 1970s, in the United States as well as in Western Europe, may have lain in the fact that it presented an opportunity to give their children that of which they themselves had been bereft: a hopeful vision of a future of limitless progress and expansion into space.[61] Paradoxically, this vision was expressed in a medium whose aesthetics brought it into the sphere of the timeless. But perhaps that was part of what made that particular future so attractive: its very timelessness infused it with a distinct feeling of inevitability.

Notes

1. See, for example, Andreas Wirsching, Göran Therborn, Geoff Eley, Hartmut Kaelble and Philipp Chassaigne, 'The 1970s and 1980s as a Turning Point in European History?,' *Journal of Modern European History* 9.1 (2011), 8–26; Niall Ferguson, 'Introduction: Crisis, What Crisis? The 1970s and the Shock of the Global,' in idem, Charles S. Maier, Erez Manela and Daniel J. Sargent, eds, *The Shock of the Global: The 1970s in Perspective*, Cambridge, MA: Harvard University Press, 2010, 1–21; and Michael A.G. Michaud, *Reaching for the High Frontier: The American Pro-Space Movement, 1972–84*, New York: Praeger, 1986, 58. Alexander C.T. Geppert, 'European Astrofuturism, Cosmic Provincialism: Historicizing the Space Age,' in idem, ed., *Imagining Outer Space: European Astroculture in the Twentieth Century*, Basingstoke: Palgrave Macmillan, 2012, 3–24, here 15 (= *European Astroculture*, vol. 1); see also his introduction, Chapter 1 in this volume. Unless otherwise noted, all interviews were conducted in December 2011.
2. De Witt Douglas Kilgore, *Astrofuturism: Science, Race, and Visions of Utopia in Space*, Philadelphia: University of Pennsylvania Press, 2003; Robert Poole, *Earthrise: How Man First Saw the Earth*, New Haven: Yale University Press, 2008, esp. 199.
3. Michaud, *Reaching*, 7–8; Kilgore, *Astrofuturism*, 150–75.
4. Roger D. Launius, 'Compelling Rationales for Spaceflight? History and the Search for Relevance,' in Steven J. Dick and Roger D. Launius, eds, *Critical Issues in the History of Spaceflight*, Washington, DC: NASA, 2006, 37–70, here 14; see also Roger Launius's contribution, Chapter 3 in this volume. Michaud, *Reaching*, 19–21. See as well Walter A. McDougall, 'A Melancholic Space Age Anniversary,' in Steven J. Dick, ed., *Remembering the Space Age*, Washington, DC: NASA, 389–95.
5. Brian W. Aldiss and David Wingrove, *Trillion Year Spree: The History of Science Fiction* [1973], London: Paladin Grafton, 1988, 427–84.
6. On O'Neill, see Kilgore, *Astrofuturism*, 150–85; Howard E. McCurdy, *Space and the American Imagination*, Washington, DC: Smithsonian Institution Press, 1997, 139–61, and the contributions by Roger Launius and Peter Westwick, Chapters 3 and 12 in this volume. On the growth of space interest groups, see Michaud, *Reaching*, 137–61, and on the significance of *Star Wars*, ibid., 130.
7. Brian Harvey, *Europe's Space Programme: To Ariane and Beyond*, London: Springer, 2003, 162, 271, 274, 294–5, 250; after many delays the first Spacelab mission was flown in 1983. On space disco, see Jussi Kantonen, 'Space Disco'; available at http://www.discostyle.com/space_disco.html (accessed 1 October 2017).

8. For an overview of the Legoland Space theme, see, for example, Sarah Herman, *Building a History: The LEGO Group*, Barnsley: Pen & Swords, 2012, 86–8. For a definition of 'Classic Space' LEGO, see the Brickipedia entry 'Classic Space,' according to which 'Classic Space' is defined by containing 'no truly different factions as in the time after 1987. All astronauts of Classic Space were peaceful explorers, and there were no antagonistic groups.' See http://lego.wikia.com/wiki/Classic_Space (accessed 1 October 2017).

9. Spanning almost a century, from 1932 to the present day, the history of the Danish company LEGO has been retold in numerous publications, for instance in Herman, *Building*, 13–71.

10. Jens Nygaard Knudsen, Interview, Holsterbro, Denmark, 10 January 2012. Archival material in the LEGO Group Archives indicates that the common name used for the Legoland Space series was simply 'Space.'

11. The Space mini-figures were called 'astronauts' in, for example, Jørgen Exner, 'Minutes of Meetings in Enfield on April 25–26,' 1979, Memo/Report, stamped 'Interlego 10 May 1979,' LEGO Group Archives, LEGO System A/S, box 1998/121 IMK, folder 'Mødereferater 1979,' 6. The LEGO Group Archives are situated in Billund, Denmark and will hereafter be referred to as 'LGA,' citing the box number/name before the folder, if there is a folder, or 'stamkort' [master record], if there is such. Plans for contacting the American Embassy in Copenhagen were mentioned in Plan Design, 'Space,' Konferencerapport [Conference Report], 5 December 1977, LGA, box 1996/162, 3. All translations are my own.

12. In analyzing the materiality of this set, I have been inspired by Jules David Prown's methodology, put forth in 'Mind in Matter: An Introduction to Material Culture Theory and Method,' *Winterthur Portfolio* 17.1 (1982), 1–19, where Prown suggests three successive levels of analysis: description, deduction and speculation when dealing with the study of material culture. The Space Command Center was first put out on the US market in 1978 as set number 493 without a 'proper' crater base plate; it was subsequently released in Europe as set number 926 with molded crater plates; see Sebastian Eggers, Christian Horstkötter and Tobias Kaminsky, eds, *LEGO Collector: Collector's Guide/Sammler Katalog*, Dreieich: Fantasia, 2008, 124, 140. See also a video posted on YouTube which shows the original box and model, available at http://www.youtube.com/watch?v=8-t7U44eVwM (accessed 1 October 2017). By virtue of being a LEGO product, the Space Command Center in a sense consisted of more than one model: Children were encouraged to build their own models, additional models were suggested on the outside of the box and an internal LEGO document suggested as many as twelve 'bi-modeller' [secondary models]; see Produktudviklingsafdelingen, 'Kommando Central [Command Center],' 5 April 1978, LGA, stamkort [master record] 493. All of these secondary models resembled realistic technologies or scientific experiments, such as a 'star telescope' and a 'radar station with rotating tower.'

13. A close-up photograph of this brick is available at http://peeron.com/inv/parts/3754p01?img=9297 (accessed 1 October 2017).

14. At the time, LEGO's policy was that its toys should be devoid of conflict, warfare and weapons. To this day no contemporary weapons are created by LEGO, reflecting the need to preserve the image of LEGO as a pedagogical, educational

toy; see Knudsen, Interview, and also Jonathan Bender, *LEGO: A Love Story*, New York: John Wiley, 2010, 61.

15. On the more varied range of products, see Per Thygesen Poulsen, *LEGO – en virksomhed og dens sjæl*, Albertslund: Schultz, 1993, 130. The idea with the mini-figures was that children would be able to 'collect one or several naturalistic play lines where everything within your theme is designed to fit around the mini-fig-ures'; see Kjeld Kirk Kristiansen, Bent Rotne, Finn Mørkenborg, Bent Clausen, Jørgen Klausen, and N.C. Jensen, '1979 Product Programme,' 1978, LGA, box A 113, folder A113/1, 10–11. See also Nevin Martell, *Standing Small: A Cel-ebration of 30 Years of the LEGO Minifigure*, London: Dorling Kindersley, 2009, 10. On Space as the most successful product line, see, for example, Cortzen, *LEGO*, 223; Knudsen, Interview. Other new product lines featured around this time were Technic in 1977 and Fabuland and Scala in 1978, see V. Petersen, 'Nyt fra legetøjsbranchen i USA,' Memo/Report, with letter from John M. Sullivan to Kjeld Kirk Kristiansen, 29 May 1979, stamped Interlego A/S 6 June 1979, LGA, box 1998/121 IMK, 1979 Div. memo/rapp., 1979 Mødereferater, 1979 From to LEGO System, folder 1998/121 IMK 'From/to LEGO Systems 1979.' See also Bent Rotne, 'Endnu et stort udviklings-år,' *Klodshans* 7 (May 1979), 13 (= *LEGO System A/S – i året 1978*).

16. Niels Thøgersen, Christian Majgaard, Bent Krog and Niels Christian Jensen, 'Marketing Strategy for LEGOLAND,' 22 May 1978, stamped Interlego A/S 14 June 1978, LGA, box 1998/121 IMK, 1977 LEGO Aktivitetsplan, 1978 Pro-jekt-stillings Vurdering, 1978–80 Produktstrategier, folder 1998/121 'Produkt-strategier 1978–80,' 26.

17. Thøgersen et al., 'Marketing,' 16–17, 26, 29, 33–4. For more musings on the possible reconstruction of an overall fictional timeline for all the LEGO Space themes, see 'The Space Timeline'; available at http://www.eurobricks.com/forum/index.php?showtopic=76493 (accessed 1 October 2017); see also PER, 'Summary af: Research gennemført fra 1. januar 1978 til 1. april 1979: Resul-tater, konklusioner og hypoteser,' 5 June 1979, LGA, box 1998/121 IMK, 1979 Div. memo/rapp., 1979 Mødereferater, 1979 From to LEGO System, folder 2817 'Summary af: Research gennemført fra 1. januar 1978 til 1. april 1979. hypoteser. konklusioner,' 13.

18. Thøgersen et al., 'Marketing,' 35; PER, 'Summary,' 23–4.

19. Klaus Rasmussen, 'Europa er vores hjemme-marked,' *Klodshans* 7 (May 1979), 6–7, here 6 (= *LEGO System A/S – i året 1978*).

20. Godtfred Kirk Christiansen and Kjeld Kirk Kristiansen, '1979 – et travlt år,' *Klods-hans* 8 (May 1980), 4 (= *LEGO System A/S 1979*); Verner Petersen, 'De ameri-kanske, canadiske og japanske markeder,' ibid., 29; on LEGO as toy of the year, see: 'Årets legetøjspris 79,' ibid., 20; Hans Hyldelund, 'Det europæiske marked,' ibid., 5; turnover in America: Petersen, 'De amerikanske,' 20.

21. On the 'Space Rocket,' see Eggers et al., *LEGO Collector*, 39 and on why the set was introduced in the United States, see Herman, *Building*, 88; on the 'Rocket Base' and 'Space Module With Astronauts' (the European version of the 1976's set 'Moon Landing'), see Gerhard R. Istok, *Unofficial LEGO Sets/Parts Collectors (1949–1990s)*, here vol. 2, chapter 29, 2012. On Knudsen and Legoland Space, see Mark Stafford, 'The Truth About Space!,' *Brick Journal* 6.2 (Summer 2009), 38–43, Knudsen, Interview; see also Bjarne Panduro Tveskov, Interview.

22. Knudsen also directed the creation of other paramount themes such as LEGO Castle, Wild West, Aquazone and Pirates themes. On Knudsen's importance in the creation of the mini-figure, see Stafford, 'Truth.' For the complicated design process all LEGO products go through, see Poulsen, *LEGO*, 142–9. In 1978 Knudsen was appointed chief designer because of the Space theme's success.

23. According to Knudsen's memory, these were, along with Japanese toy robots, the only space toys available in Europe at the time. However, there were both *Space: 1999* and *Star Wars* toys in Denmark in the late 1970s; see Torben Rølmer Bille, 'Rendyrket nostalgi,' *Kulturkapellet* (11 November 2008), available at http://www.kulturkapellet.dk/filmanmeldelse.php?id=319. For examples of Dinky *Thunderbirds* toys, see http://www.thunderbirds.free.fr/english/Dinky-toys.html; for Reisler's space figures, see http://tohan.dk/reisler_55mm_rumforskere.htm (all accessed 1 October 2017).

24. Knudsen, Interview.

25. Frenchmen Christin and Mézières's comic series *Valérian et Laureline* (begun in 1967) greatly influenced visualizations of outer space, with *On the False Earths* (French: *Sur les terres truquées*) coming out in 1977. The first album in Belgian Roger Leloup's series *Yoko Tsuno*, *Le trio de l'étrange*, related the discovery of an alien underground civilization on earth and subsequent albums revisited the extraterrestrial theme, with, for instance, *Les titans* published in 1978; see the fan-web site http://yoko.tsuno.free.fr (accessed 1 October 2017).

26. On *Space: 1999*, see Henry Keazor, 'A Stumble in the Dark: Contextualizing Gerry and Sylvia Anderson's *Space: 1999*,' in Geppert, *Imagining Outer Space*, 189–207.

27. Knudsen, Interview; Niels Dalgaard, *Den gode gamle fremtid: Modernitetskritik og genrevalg i halvfjerdernes danske science fiction-roman*, Copenhagen: Museum Tusculanum, 1997.

28. On Tente in relation to LEGO, see Knudsen, Interview; Stafford, 'Truth'; Eggers et al., *LEGO*, 124; for early Tente space sets see http://www.latenteteca.com/modelos/?serie=Astro and on Tente, see http://en.wikipedia.org/wiki/Tente_(toy). For *Star Trek* toys of the 1970s, see, for instance, http://www.plaidstallions.com/startrektoys.html and for *Battlestar Galactica* toys, see http://www.lurexlounge.com/bsg/vehicles01.php (all accessed 1 October 2017).

29. See http://catacombs.space1999.net/main/merc/vmmalpha.html and http://catacombs.space1999.net/main/merc/vmmfigure.html (both accessed 1 October 2017).

30. However, *Space: 1999* also had figures at about this height.

31. On the role of TV and movies in the narrativization of toys, see Gary Cross, *Kid's Stuff: Toys and the Changing World of American Childhood* [1997], Cambridge, MA: Harvard University Press, 2001, 147–87, and Mark S. Young, 'Creating a Sense of Wonder: The Glorious Legacy of Space Opera Toys of the 1950s,' in Cynthia J. Miller and A. Bowdoin Van Riper, eds, *1950s 'Rocketman' TV Series and Their Fans: Cadets, Rangers, and Junior Space Men*, Basingstoke: Palgrave Macmillan, 2012, 149–62. On the narrativization of toys in relation to LEGO, see Stig Hjarvard, 'From Bricks to Bytes: The Mediatization of a Global Toy Industry,' in Ib Bondebjerg and Peter Golding, eds, *European Culture and the*

Media, Exeter: Intellect, 2004, 43–63, and Aaron Smith, 'Beyond the Brick: Narrativizing LEGO in the Digital Age,' Conference Paper MiT7 Unstable Platforms – The Promise and Peril of Transition, Massachusetts Institute of Technology, Boston, 2011, available at http://web.mit.edu/comm-forum/mit7/papers/LEGO_AaronSmith_2011.pdf (accessed 1 October 2017). On Kenner's *Star Wars* toys, see Dan Fleming, *Powerplay: Toys as Popular Culture*, Manchester: Manchester University Press, 1996, here 93–102.

32. Daniel Lipkowitz, *The LEGO Book*, London: Dorling Kindersley, 2009, 72. Though the surrounding toy culture increasingly built their play universes on narratives derived from, for instance, cinematic universes LEGO held back in this area, see Hjarvard, 'From Bricks to Bytes,' 51–2, and Smith, 'Beyond the Brick,' 4.

33. Hjarvard, 'From Bricks to Bytes,' 56. Hjarvard, however, also states that during the 1970s and 1980s, LEGO sets were increasingly organized according to thematic universes built on already existing fictional worlds, ibid., 54–5. As for the 'design concepts' these were also explicitly given as names of models, for example in LEGO catalogs.

34. In the British catalog from 1979 there is no mention of either 'Negative Forces' or 'Action Figures,' which probably reflects differences between the European and US toy markets.

35. *Legoland Idea Book*, Enfield: LEGO Systems, 1980.

36. Douglas Hill, *Trapped in Space*, London: Pumpkin, 1984; Frank Madsen, *Jim Spaceborn: The Unknown Galaxy*, Billund: LEGO Publishing, 1986.

37. On children's popular culture as 'spaces,' see Claudia Mitchell and Jacqueline Reid-Walsh, *Researching Children's Popular Culture: The Cultural Spaces of Childhood* [2002], London: Routledge, 2010.

38. 'LEGOLAND,' [probably exhibition brochure given to owners of toyshops, ca. 1978], LGA, box 2000/50 SEU. Thøgersen et al., 'Marketing,' 30; 'evidence' presumably refers to how consumers should be convinced of the reality of the possibility to build this space.

39. Ibid., 19. On the color of the space mini-figures, see Plan Design, 'Space,' 2, 5. Their specific functions, however, were not disclosed to customers, yet in *Trapped in Space* red astronauts were defined as 'pilots and navigators,' yellow as 'the experts who look after all the machinery and technology, and often invent special new items' (that is, engineers) and the white as 'scientists,' see Hill, *Trapped*, 2.

40. James H. Gillam, *Space Toys of the 60s: An Illustrated Guide To Major Matt Mason, Zeroid Robots and Star Team, and Colorforms Outer Space Men*, London: Collector's Guide Publishing, 1999, 16, 95, 102, 124, 141. 'Legorion' is mentioned in Plan Design, 'Space,' 2; on the setting of Space, see BC, NCJ, JK, and FMR, 'Product Programme 1979: Assortment Strategy,' 3 March 1978, LGA, box A 113, folder A 113/1 'Produktprogram 1979,' 36; see also Thøgersen et al., 'Marketing,' 18–19. On realism of early LEGO space sets, see also Peter Thomas, 'Bricks in Space: Drei Jahrzehnte Raumfahrt im Spielzeugformat,' *Frankfurter Allgemeine Sonntagszeitung* 8 (22 February 2009), V14.

41. Knudsen, Interview; awareness of outer space fascination, see Plan Design, 'Space,' 2; letter to stockists, see Ian Coutts, 'September/October Release: Legoland Space Range,' letter to LEGO Stockists in South Africa, September 1979, LGA, box 2000/50 SEU, 1; on what interested the 'little customer,' see 'September Novelties 1979: Legoland Space,' presumably draft of letter for South African stockists, no date, LGA, box 2000/50 SEU. On the sales of *Star Wars* in

Germany, see Timo Kotowski, 'LEGO überwindet die magische Grenze,' *Frankfurter Allgemeine Zeitung* 27 (1 February 2012), 15 (earnings on LEGO sets in Germany in 2011 were €297 million); see also Thomas, *Bricks*.

42. On the growing demand for realism, see PER, 'Summary,' 17; on the creation of specialized elements, see Kristiansen et al., '1979 Product,' 16; on increasing play-value: BR [Bent Rotne], 'Speech/March Conference 1978,' LGA, box A113 (box one), folder A 113/1 '1979 Produktprogram,' 4–5; fear of limiting imagination: Poulsen, *LEGO*, 128–9. That this fear was well grounded became clear during the economic crisis of the LEGO Group around 2003–04; see Maaike Lauwaert, 'Playing Outside the Box: On LEGO Toys and the Changing World of Construction Play,' *History and Technology* 24.3 (September 2008) 221–37; and David Robertson, *Brick by Brick: How LEGO Rewrote the Rules of Innovation and Conquered the Global Toy Industry*, New York: Random House, 2013.

43. Fleming, *Powerplay*, 99.

44. Charlotte and Peter Fiell, *Scandinavian Design*, Cologne: Taschen, 2000, 13–14, 406–7. While a concept of a 'timeless' quality of design is difficult to pin down, in relation to Scandinavian design it is, according to Fiell and Fiell, defined as 'simplification without upsetting the balance of form and function; truth to materials; and a deep respect for the task at hand, while facilitating meaningful connections between the user, their tools for living, and the environment.' Ibid., 32.

45. For a similar line of thought, see Askild Matre Aasarød, 'Mannen som solgte Skandinavia,' *Plot* 15 (August/September 2013), 72–92.

46. Lipkowitz, LEGO, 72; Olaf Thygesen Damm, 'Det største "nyhedsår" nogensinde – vi kigger på årets mest iøjnefaldende og spændende produktnyheder,' *Klodshans* 7.1 (March 1979), 9–11, here 11; on the visit from ESA, see Knudsen, Interview. Unfortunately, I have been unable to procure more information on this incident. 'Endless Space Age': Peter Nicholls quoted in Michaud, *Reaching*, 125; Kilgore, *Astrofuturism*; see also Launius, 'Compelling,' 38.

47. Louis J. Halle, 'A Hopeful Future for Mankind,' *Foreign Affairs* (Summer 1980), 1129–36, here 1133; Launius, 'Compelling,' 41–3. Though Halle is thinking of O'Neill-like colonies in outer space rather than surface colonies, that does not reduce the relevance in relation to Legoland Space, especially not since part of O'Neill's scheme was to mine the moon; see Michaud, *Reaching*, 65. Furthermore, Jens Nygaard Knudsen and his team experimented with models of 'free floating' space habitats. These involved base plates fastened to anglepoise lamps, thus adding a new dimension to play. These were never realized, though; see Knudsen, Interview. On optimistic view of progress, see Poulsen, *LEGO*, 95; Wiencek, *World*, 39.

48. Cortzen, *LEGO*, 173, 221; Wiencek, *World*, 48.

49. Cortzen, *LEGO*, 137.

50. Plants and animals can, of course, be built with LEGO bricks and such models have been a standard part of LEGO assortments since the beginning. Moreover, it should be noted that, for example, the Pirate theme, introduced in 1989 and also a brainchild of Nygaard Knudsen, was an even greater success than the Legoland Space theme, see Poulsen, *LEGO*, 148–9. For the great range of building possibilities offered by LEGO bricks, see, for instance, John Baital and Joe Meno, *The Cult of LEGO*, San Francisco: Starch Press, 2011.

51. Fleming, *Powerplay*, 41, 43, 85, 90–1.

52. Other LEGO themes have enjoyed comparable success and durability, such as the Castle theme. For a brief overview of LEGO themes depicting realistic contemporary space exploration technology, see Thomas, *Bricks*.

53. Fraser MacDonald, 'Space and the Atom: On the Popular Geopolitics of Cold War Rocketry,' *Geopolitics* 13.4 (Winter 2008), 611–34; see also Michaud, *Reaching*, 123 on science fiction and pro-spacers; and James S. Ormrod, 'Pro-Space Activism and Narcissistic Phantasy,' *Psychoanalysis, Culture and Society* 12.3 (2007), 260–78 on the power of children's daydreaming in shaping their grown-up lives. On the LEGO-NASA educational collaboration, see, for example, http://aboutus. lego.com/fi-fi/news-room/2011/april/lego-education-goes-into-orbit-with-space-shuttle-endeavor. On the use of the concept 'Builders of Tomorrow,' see the LEGO homepage, http://parents.lego.com/en-us/lego-and-society/builders-of-tomorrow-2, see also interviews with designer of LEGO City Space Daire McCabe, Lotte Sørensen (marketing, LEGO Education) and Vicki Stolz (marketing, LEGO City) (interviews conducted on 13 December 2011). Children, however, are also able to create their own meanings differing from those provided by the toy industry, see Mitchell and Reid-Walsh, *Researching*, 27–9. For some examples of recent LEGO cooperations with real life space programs, see 'A Lego Space Probe,' 27 September 2010; available at http://www.enjoyspace.com/en/news/a-lego-space-probe and Robert Z. Pearlman, 'LEGO Figures Flying on NASA Jupiter Probe,' 4 August 2011; http://www.space.com/12546-lego-figures-jupiter-juno-spacecraft.html (all accessed 1 October 2017).

54. For instance, http://neoclassicspace.com/ (accessed 1 October 2017) where AFOLs celebrate LEGO 'Classic Space,' and also Peter Reid and Tim Goddard, *LEGO Space: Building the Future*, San Francisco: No Starch Press, 2013, which develops a storyline alluding to LEGO's different space series, especially Classic Space, by means of both text and photographs of homemade LEGO models. According to one estimate, there may be as many as 20,000 active AFOLs worldwide; see Bender, *LEGO*, 186. The great popularity of the animated classic space minifigure 'Benny' in *LEGO: The Movie*, directed by Phil Lord and Christopher Miller, USA 2017 (Warner Brothers) also points to the attraction of the theme.

55. Jesus Diaz, 'Lego Secret Vault Contains All Sets In History,' 23 June 2008, http://lego.gizmodo.com/5018990/lego-secret-vault-contains-all-sets-in-history/all; see also the comments to this post confirming that Diaz is not alone in having such sentiments. See also Jeremy Reimer, 'Lego Space Ships UPDATED!,' 29 September 2011, http://jeremyreimer.com/monarch/m-item?i=98; 'This is Galaxy Explorer LL928 calling your Childhood,' 26 July 2009, http://knightwise.com/this-is-galaxy-explorer-ll928-calling-your-childhood/ and 'All your Space LEGO base are belong to us, or, Memories of sparkly legs and being baked by the space heater,' 23 January 2013, http://craicmonkeysdelight.wordpress.com/2013/01/23/all-your-space-lego-base-are-belong-to-us-or-memories-of-sparkly-legs-and-being-baked-by-the-space-heater (all accessed 1 October 2017).

56. Jesus Diaz, 'Lego Space Timeline Brings Back My Best Childhood Memories,' 6 May 2009, http://gizmodo.com/5241294/lego-space-timeline-brings-back-my-best-childhood-memories (accessed 1 October 2017). For a general discussion of the advantages and disadvantages of using adults' memories in the study of children's popular culture, see Mitchell and Reid-Walsh, *Researching*, 55–62.

57. Yun Mi Antorini, *Brand Community Innovation: An Intrinsic Case Study of the Adult Fans of LEGO Community*, PhD thesis, Department of Intercultural Communication and Management, Copenhagen Business School, 2007, 178–9, 184, 234, 237; available at http://pure.au.dk/portal/files/8530/Brand_Community_Innovation_by_Yun_Mi_Antorini.pdf (accessed 1 October 2017).

58. On consumption as locus of transcendence and nostalgia for childhood as nostalgia for paradise, see Russell W. Belk, Melanie Wallendorf and John F. Sherry Jr., 'The Sacred and the Profane in Consumer Behavior: Theodicy on the Odyssey,' *Journal of Consumer Culture* 16.1 (June 1989), 1–38, here 2, 8–10, 12. Further on 'nostalgia for paradise,' see Mircea Eliade, *Myths, Dreams, and Mysteries: The Encounter Between Contemporary Faiths and Archaic Realities*, New York: Harper & Row, 1975, 59–72; originally published as *Mythes, rêves et mystères*, Paris: Libraire Gallimard, 1957; and on connections between timelessness and myth William E. Paden, *Religious Worlds: The Comparative Study of Religion*, Boston: Beacon Press, 1994 [1988], 75–8. On timelessness in childhood, see Cross, *Kid's Stuff*, 86, and childhood as carefree space Antorini, *Brand Community*, 178–9. In the words of Eugen Fink, 'In the autonomy of play action there appears a possibility of human timelessness in time. Time is then experienced, not as precipitate rush of successive moments, but rather as the one full moment that is, so to speak, a glimpse of eternity'; see Eugen Fink, 'The Oasis of Happiness: Toward an Ontology of Play,' *Yale French Studies* 41 (1968), 19–30, here 21; abbreviated version of *Oase des Glücks: Gedanken zu einer Ontologie des Spiels*, Freiburg: Karl Alber, 1957.

59. On the construction of childhood as innocent and sacred, see Colin Heywood, *A History of Childhood* [2001], Cambridge: Polity Press, 2011, 23–31. Gary Cross has many astute observations on how adults frame toys as representing something 'timeless' and how this 'timelessness' is historically determined; see, for instance, Cross, *Kids' Stuff*, 83–92, and on LEGO esp. 220. For an example of specific categories of toys labeled 'timeless,' see John and Elizabeth Newson, *Toys and Playthings in Development and Remediation*, Harmondsworth: Penguin, 1979, 69–96. On Judaic-Christian apocalyptic narrative in astrofuturism, see Thore Bjørnvig, 'Transcendence of Gravity: Arthur C. Clarke and the Apocalypse of Weightlessness,' in Geppert, *Imagining Outer Space*, 127–46; and on religion in relation to pro-space rhetoric and the exploration of space in general Thore Bjørnvig, 'Outer Space Religion and the Ambiguous Nature of *Avatar*'s Pandora,' in Bron Taylor, ed., *Avatar and Nature Spirituality*, Waterloo: Wilfrid Laurier University Press, 2013, 37–58. See also the special issue *Astropolitics* on spaceflight and religion, *Astropolitics* 11.1–2 (2013). On paradise and the heavens, see Eliade, *Dreams*, 99–122. For toys of the 1960s and parents' nostalgia as well as gifts of toys as intergenerational communication, see *Kid's Stuff*, 44–8, 167, 238–9; on toys as carrying cultural encodings, see Fleming, *Powerplay*, 58, 156.

60. Roger D. Launius, 'Perfect Worlds, Perfect Societies: The Persistent Goal of Utopia in Human Spaceflight,' *Journal of the British Interplanetary Society* 56.5 (September/October 2003), 338–49; Michaud, *Reaching*, 293.

61. Ormrod, 'Pro-Space Activism'; on parents trying to fulfill their own fantasies through buying space toys for their children, see Gillam, *Space Toys*, 102.

CHAPTER 8

The Province and Heritage of Humankind: Space Law's Imaginary of Outer Space, 1967–79

Luca Follis

On 4 October 1957 the Soviet Union successfully launched into orbit the world's first artificial satellite. Sputnik seemingly confirmed the notion that earth's inhabitants might soon transcend their terrestrial limits; it stoked ambitions of earthly deliverance even as it reinforced an already tense set of political and military circumstances. According to philosopher Hannah Arendt (1906–75), the launch of Sputnik brought into sharp relief a fundamental tension at the core of the human condition. We are biological beings shaped by our connection to nature and terrestrial existence; this link provides the experiential reservoir we draw from when ascribing meaning to worldly events and making human accomplishments intelligible. At the same time we are increasingly tempted to abandon these anthropological and geological referents in pursuit of the unfolding *telos* of scientific progress. In other words, if human interpretive categories rely on embodied experience, then they are also at odds with the scientific point of view which tends to expose the limits of sensory perception and questions its reliability. The scientist approaches the world through the experimental method: he or she is intent on discovering the 'how' of creation and in harnessing or replicating its forces, not in finding the significance behind these new capacities.[1]

Luca Follis (✉)
Lancaster University, Lancaster, UK
e-mail: l.follis@lancaster.ac.uk

© The Author(s) 2018
Alexander C.T. Geppert (ed.), *Limiting Outer Space*
European Astroculture, vol. 2
https://doi.org/10.1057/978-1-137-36916-1_8

Thus, in Arendt's view, the above approach presents unique 'translation' problems when applied to spaceflight and other technoscientific advances. How does a non-scientist interpret these events given that they defy the basic categories of perception and knowledge we have always relied upon to make the world intelligible? Arendt wondered whether in the face of rapid technological and scientific change humans could remain capable of explicating their achievements outside the paradigms and language of the scientific method. Her worry was that if we failed to do so, we would be alienated from ourselves, trapped on a world that no one can understand or explain in meaningful terms.[2] This is because 'universal science' introduced new modes of representing and encountering the given world, making viable an astounding conceptual leap. It became possible 'to think in terms of the universe while remaining on Earth, and [...] to use cosmic laws as guiding principles for terrestrial action.'[3] The irony of this situation is that we have become so proficient at thinking from what Arendt called the 'Archimedean point'[4] that the secrets of the universe can be discovered without actually leaving earth. Hence Arendt's problem is one of legitimation: science can answer the 'how' of space travel and exploration, but it cannot provide a meaningful account of 'why' we should endeavor to go there in the first place or, more importantly, what it might mean for a human race which is still territorially bounded.

Throughout the 1960s and 1970s technoscientific progress and spatial discovery were presented as the core rationales for space exploration. These terms inform the broad aims that appeared time and again in the agreements and treaties that make up the body of space law. Beginning in October 1967 when the so-called Outer Space Treaty (OST) entered into force, and followed by the Astronaut Agreement (1971), the Liability Convention (1972), the Registration Convention (1975) and the Moon Agreement (1979), nation states embarked upon an intensive set of negotiations designed to shape the kinds of activities and interactions that would be permitted in outer space.[5] Arendt's thoughts anticipated the important ethical and normative questions that animated this codification project. What significance did reaching outer space or the moon hold for countries that were not space capable and might never be? What sort of benefits might developing countries draw from the Space Age and how might these be grounded in terrestrial notions of justice and equity? Ultimately, landing humans on the moon or erecting a settlement there might be empty technological gestures if they were not anchored in a larger normative justification that could give these achievements an independent significance.

This chapter offers a close reading of the negotiations surrounding the Outer Space Treaty of 1967 and the Moon Treaty of 1979 in light of Arendt's concerns. The *corpus juris spatialis* provides legal scaffolding for space exploration, but it also wrestles with a set of normative expectations and commitments that attempt to answer the question of what outer space is and what it can or cannot be for the human race. As expressed in the Outer Space

Treaty of 1967, these expectations largely take the form of a ban; the treaty expresses an international consensus over what outer space should not be or become. It targets the logic of national appropriation, ownership and war making, but it also declares outer space 'the province of mankind.' This lofty phrase speaks to the utopian spirit and idealistic culture that animated the Space Age in the postwar period, even if a lack of consensus over its meaning prefigured the fissures that would develop in the international community during the Moon Treaty negotiations. At that time attempts were made to elaborate upon the positive, material benefits that might flow from this 'province' and what broader principles and procedures would inform their distribution.

This chapter argues that a central stumbling block in negotiation efforts involved divergent interpretations between developed and developing nations over the practical significance and use of space exploration and scientific technology. These disagreements spoke to the broader meaning and purpose of the Space Age but also dovetailed with an international movement that championed notions of distributive justice and equity worldwide. In Arendt's terms, then, rather than a will to flee earth and its terrestrial burdens, the latter reflected a liberation dream grounded in historical experience and oriented toward the transformation of existent global inequalities.

I   Space as a zone of freedom and peaceful exploration: the Outer Space Treaty of 1967

The Outer Space Treaty of 1967 was negotiated in the United Nations (UN) Committee on the Peaceful Uses of Outer Space (COPUOS) and its two subcommittees – the Legal Subcommittee and the Scientific and Technical Subcommittee – through a process of consensus rather than voting.[6] The Soviet Union and the United States set the frame for negotiations by each producing a draft of the treaty. The former was compiled in large part from the 'UN Declaration of Legal Principles Governing the Activities of States in the Exploration and Use of Outer Space' (Resolution XVIII) of 1962 (a resolution adopted by the General Assembly without a vote) while the latter was largely based upon the Antarctica Treaty of 1959.[7] The Russian draft was more extensive and because of its grounding in Resolution XVIII anticipated a number of the topics and formulations of the finished treaty. For example, paragraphs 1 and 2 of the Outer Space Treaty reproduce in almost identical language the first two paragraphs of the 1966 draft proposal of the Soviet Union. The negotiation process involved a line-by-line consideration of the two proposals with the aim of amalgamating their provisions and the suggestions and amendments made by other states. Negotiations took place throughout 1966 during the fifth session of the COPUOS Legal Subcommittee. Delegates debated preferences, comments and substantive proposals in Geneva and New York over the course of 17 meetings between 12 July–4 August and 12–16 September 1966.[8] The final draft was included in the agenda for the UN General Assembly's twenty-first session and adopted on 19 December 1966.

The finished Outer Space Treaty is informed by a commitment to principles of common interest and mutual participation between nation states; a sentiment nowhere expressed as clearly as in the first two paragraphs of Article I:

> The exploration and use of outer space, including the Moon and other celestial bodies, shall be carried out for the benefit and in the interests of all countries, irrespective of their degree of economic or scientific development, and shall be the province of all mankind. Outer space [...] shall be free for exploration and use by all States without discrimination of any kind, on a basis of equality and in accordance with international law, and there shall be free access to all areas of celestial bodies.[9]

The treaty frames space exploration and scientific investigation as strictly peaceful endeavors pursued under conditions of equality and free access. To this end, it forbids nuclear and other weapons of mass destruction in orbit or on celestial bodies, and outlaws military bases, fortifications and weapons testing. It also decouples the link between exploration and future claims of sovereignty or ownership: 'Outer Space, including the Moon and other celestial bodies, is not subject to national appropriation by claim of sovereignty, by means or use of occupation, or by any other means,' Article II reads.[10] The treaty projects the rule of law deep into the future of human civilization and evokes a not-too-distant spacefaring culture where humans will live in bases and stations on other worlds. It posits outer space as a new normative zone emptied of sovereign claims and military jostling, and characterized by transparency, reciprocity, international law and human freedom.[11]

Prior to the treaty, outer space bore the status of *res nullius* (a thing owned by no one) in international law, which meant that the celestial bodies were spaces awaiting claimants and open to possession (that is, *terra nullius*). The Outer Space Treaty seems to merely extend the principles contained in the freedom of the high seas doctrine and the Antarctica Treaty of 1959. Like the high seas, outer space and the celestial bodies cannot be owned or claimed and, much like Antarctica, scientific cooperation under conditions of free access structures relations among different actors. In this sense, the treaty would recategorize outer space and the celestial bodies as *res communis* (that is, a thing held in common).[12]

Although it draws extensively from the Antarctica Treaty of 1959 (for example, in the ban on military use and fortifications) and the Principles adopted by the General Assembly in 1962, there was a clear novelty to the document. Negotiating parties were conscious of the fact that in codifying a set of legal principles for space activity, they were opening a new area of international, extraterrestrial law. But they were also anxious about the length of the negotiation process and apprehensive about the possibility that an agreement might not be reached before the first lunar landing.[13] Some delegations, like Mexico, expressed this view with grandiose language by emphasizing the importance of the historical moment and international law's role as a creative force that could anticipate and shape the direction of world events rather than merely provide them with *ex-post facto* legitimation.[14]

Other nations worried that space law might simply extend and reinforce established structures of inequality and dependency rather than transcend historical patterns. Thus, Mongolia warned against the extension of territorial ambitions outside earth, invoking the ongoing conflict in Vietnam and the history of colonial domination.[15] The Brazilian delegation, on the other hand, emphasized the need to balance the rights and obligations of the 'space powers' with those of the 'non-space powers.' It likened space exploration with emerging developments in atomic energy whereby all nations shared the potential risks of the technology, but only a handful would directly profit from its benefits.[16] Indeed, the positions of Mexico, Mongolia and Brazil neatly summarize the loaded and contradictory expectations different nations brought to the negotiating table. These in turn mapped onto countervailing positions about the key stakes in the agreement and the overall justifications for space law.

The range of postures alluded to above is well captured by probing the significance of the term 'mankind' in the treaty. 'Mankind' appears twice in operative clauses (in conjunction with 'province' and with 'envoys') and is also mentioned twice in the preamble in conjunction with 'benefit.' According to one commentator, the treaty represents the first international formal agreement in which the benefit of 'all mankind' is made the object and concern of an operative clause.[17] The phrase 'province of mankind' appears in the first line of the first article and can be read as the central, legitimating and unifying concept that flows from the agreement.[18] It is only because outer space and the celestial bodies are the province of mankind and must therefore be preserved for future human generations that injunctions on sovereignty, ownership, weapons testing – as well as obligations of reciprocity, aid and rescue – gain logical coherence. Yet, more than one delegate expressed dissatisfaction with the expression and questioned its legal significance. The British delegation, for example, challenged both the meaning of the term and its conjunction with the equal access provisions contained in the second paragraph of Article I. In the UK delegation's view, not only did it seem wide-ranging, but it also 'might place onerous obligations on States. For example, would a State launching a satellite for one country be obliged to provide an unlimited number of launchers and launching opportunities for all other States party to the Treaty if so requested?'[19]

If some thought Article I went too far, other states clearly thought it did not go far enough. The United Arab Republic (UAR) proposed an alternative draft of Article I designed to strengthen the legal force of the phrase 'province of mankind' by giving it concrete meaning rather than leaving it merely as a statement of 'theoretical principle':[20]

The Parties to this Treaty recognize outer space as the province of mankind [...]. States engaged in the exploration of outer space undertake to accord facilities and to provide possibilities to the non-space Powers, to enable them to participate in and to draw benefit from the exploration and the use of outer space for the aim of deriving practical benefits related to their economic and social development.[21]

The UAR draft explicitly linked the benefits of scientific exploration and use of outer space to the economic and social development of undeveloped nations. This was a theme that would also figure prominently in the Moon Treaty negotiations four years later.

Though the UAR draft was not included in the final version of the Outer Space Treaty, the phrase 'province of mankind' on its own proved remarkably contentious in the US Senate. Indeed, in the 1967 hearings before the Committee on Foreign Relations over whether the United States should ratify the treaty, the senators worried that Article I entangled the United States in a new set of international obligations. They questioned Ambassador Arthur J. Goldberg (1908–90), the US representative in the negotiations, over whether the treaty went beyond the free access principle in the high seas doctrine and the stated presidential priorities of Eisenhower and Johnson, as well as the provisions of the National Aeronautics and Space Act of 1958.[22] Section 102(a) of that act states that space exploration will be carried out for the 'benefit of all mankind.' The senators wondered whether 'benefit' was substantively different from 'province' and, if so, in what way? Further complicating matters, Article I stated that the 'use' of outer space (and the moon and other planets) would be carried out to 'benefit and in the interests of' all countries regardless of their 'degree of economic and scientific development.'

Though Ambassador Goldberg attributed the above sentence to Brazil and explained it as a symbolic concession, its appearance in the final text of the treaty illustrates the wider purchase among some delegates of linking techno-scientific achievements with development goals. In this sense, the article was meant to reflect the universal reach of the phrase 'all countries' and the notion that space technology could accelerate the pace of economic progress for undeveloped states.[23] Despite Goldberg's assurances, the senators were not entirely convinced that Article I did not generate new obligations on the United States vis-à-vis the rest of the world. Echoing the comments of the United Kingdom delegate during the treaty negotiations, Senator Albert Gore (1907–98) asked 'to just what are we binding ourselves by this article if we are not binding ourselves, as it provides, to make the use of outer space available to all nations?'[24]

Gore's question reflects the tension between the normative aims and pledges of the treaty, and the practical interpretation of its obligations. In the view of countries like Brazil or the United Arab Republic, the universalist language in Article I spoke to principles of distributive justice. It represented a recognition of the fact that space law could promote greater global equity by tying space exploration, access and exploitation to the interests of the developing as well as the developed world. On the other hand, as the Senate testimony makes clear, the US delegation understood the intent of the article as little more than 'broad goals and objectives' in the spirit of a preamble and 'subject to what can be worked out in specific application.'[25] Indeed, Gore himself characterized the article as the treaty equivalent of a 'political stump speech.' And perhaps more tellingly, when the Senate endorsed the treaty's ratification, it qualified its agreement with the following proviso: 'It is our understanding of the Committee on Foreign Relations that nothing in Article I, paragraph I of the treaty diminishes or alters the right of the United States to determine how it shares the benefits and results of its space activities.'[26]

The above ambiguity and disagreement over the treaty's meaning and obligations seemingly reified a split between 'spacefaring' and 'non-spacefaring,' developed and underdeveloped nations, but it also clarified the distance between the underlying cultural assumptions that informed delegate positions. The Outer Space Treaty prioritized space exploration and scientific discovery as the core rationale for the Space Age, yet throughout human history exploration and discovery have often served as ideological cover for furthering territorial expansion, military strategy or capitalist exploitation. Given that all of these were banned or greatly constrained by the Outer Space Treaty, the question of what underlying set of goals exploration and discovery were meant to promote remained open and undetermined.

For non-spacefaring nations and the developing world an important outcome of the treaty involved its alignment with broad notions of distributive equity, as well as its endorsement of a link between the use and exploration of outer space and its potentially larger role in the development of the human race. Another key consideration shared with European countries flowed from the notion that space exploration and the codification of space law could be channeled as a mechanism to promote world peace and disarmament. Indeed, besides the debate over Article I and its obligations, a central aspect of negotiations involved defining the role of military personnel in space missions and the restrictions against martial involvement.

Though in broad terms all delegates agreed on the importance of peace in outer space, the specifics and practicalities of how this might be accomplished were nuanced and impacted other concerns. Did a ban on militarization also include a ban on military personnel? Did a ban on nuclear and other weapons of mass destruction preclude the deployment of non-aggressive technology with clear military applications? Seen from a strategic perspective many non-contentious issues including whether the treaty applied to all the celestial bodies or the moon, the specifics of state reporting and disclosure requirements of activity and the degree and timing of international access to future bases and facilities, took on a different light.

Thus, though the negotiations did emphasize the utopian promise of a 'province of mankind,' the ideological, cultural and military commitments of the Cold War also weighed on delegates' reading of the treaty's wider implications. For example, the United States prioritized the importance of orbital space for the deployment of spy satellites to assess the location and scope of Soviet capabilities, yet it was also ideologically committed to countering Soviet secrecy (and cloaking its own evolving military use of orbital space) with tropes of openness and international cooperation. The Soviet Union, on the other hand, initially resisted American surveillance efforts but relented once it too developed military applications for its space programs; it countered US calls for openness and cooperation by backing the characterization of outer space as a non-sovereign, 'province' for all mankind.[27]

The Outer Space Treaty was negotiated and finalized in 17 sessions over the course of 1966. Delegates drew inspiration from the International Geophysical Year (1 June 1957–31 December 1958), whose success illustrated the viability of international scientific cooperation, as well as the engineering and technological achievements represented by Sputnik (1957) and Yury Gagarin's Vostok I mission (1961). Additionally, the 1966 negotiations were animated by the expectation of an impending lunar landing, and it is clear that this anticipation gave the talks significant momentum. However, when in 1970 Aldo Armando Cocca (1924–), Argentina's representative to COPUOS, declared that the 'use of the Moon's resources had already begun' and that a new formal instrument was necessary to regulate the exploitation of these materials, he was referring to Apollo 11's return to earth with a score of samples collected on the moon.[28] In the same session, Cocca submitted a 'Draft Agreement on the Principles Governing Activities in the Use of the Natural Resources of the Moon and Other Celestial Bodies' and thus initiated the protracted nine-year negotiations that would ultimately lead to the Moon Agreement.[29] Thus, from the outset and unlike the Outer Space Treaty, the Moon Agreement negotiations took place in a world where not only was human spaceflight beyond earth possible, but ongoing lunar landings were a reality. These developments heightened the perception among delegates that the colonization and industrialization of outer space was, if not imminent, at least foreseeable in the near future. At the same time, they also reified the notion that technoscientific progress might soon outpace law's capacity to serve as ethical and normative compass for the Space Age.[30]

II   The moon as common heritage: the Moon Agreement of 1979

Much like the Outer Space Treaty, the 1979 Moon Agreement's text largely compiled and restated key principles from past UN space agreements.[31] For example, Article III stated that the moon would be used by state parties exclusively for peaceful purposes and prohibited 'any threat or use of force or any other hostile act on the Moon.' It reaffirmed many of the key articles of the Outer Space Treaty banning lunar militarization but extended these to include the moon's orbit. At the same time, the agreement returned to the question of distributive equity and confronted it more explicitly. For example, although Article IV begins by restating Article I of the Outer Space Treaty, it added an important supplement: 'Due regard shall be paid to the interests of present and future generations as well as to the need to promote higher standards of living and conditions of economic and social progress and development in accordance with the Charter of the United Nations.'[32] Thus, the article explicitly linked the general notion of a 'province of all mankind' to both future and current generations and the interests of global development. This shift illustrates the strong interpretive role played by non-spacefaring nations in the drafting of the agreement.

While the Outer Space Treaty negotiations had concentrated on the drafts proposed by the United States and the Soviet Union, it was Argentina that initiated the Moon Treaty deliberations. In 1970 and in response to the Apollo 11 mission, Argentina's representative proposed five supplementary articles to regulate the use and exploitation of the moon's resources. In 1971 the Soviet Union followed with its own draft treaty proposal on the moon for consideration in the General Assembly. And by 1972, the Legal Subcommittee of COPUOS prioritized the drafting of a Moon Agreement and established a working group to consider the rapidly accumulating drafts, proposals and working papers concerning the moon. Between 1972 and 1976, the working group considered and amalgamated the submissions from a wide range of countries including Argentina, Australia, Brazil, Bulgaria, Canada, Chile, Egypt, India, Indonesia, Iran, Italy, Mexico, Mongolia, Nigeria, Romania, Sierra Leone, Sweden, United Kingdom, United States, Soviet Union and Venezuela.[33] Though in 1977 the working group identified only three remaining areas of disagreement – natural resources, reporting requirements and the scope of the projected treaty – many delegates felt that it was the first of these that represented the central impasse. In particular, disagreement flowed between those that backed Argentina's 1970 proposal that the moon's natural resources be designated the 'common heritage of all mankind' (and their benefits be made available to 'all peoples without discrimination of any kind') and those that found the Soviet Union's position more persuasive.[34]

Throughout the nine-year negotiations, from 1970 until 1979, the Soviet Union consistently opposed the notion of 'common heritage' on the grounds that it lacked a precise legal meaning. The delegation argued that while 'heritage' might be likened to civil law terms like 'inheritance' or 'succession' the latter were tied to notions of ownership or possession over a thing and its use. Though space law did recognize a concept of ownership over those space objects and their parts launched by nation states, with respect to the moon or the celestial bodies (which could not be appropriated or owned) there was no possibility of ownership. Accordingly, without a concept of property, terms like 'succession,' 'inheritance' and 'heritage' were meaningless. Drawing upon the Outer Space Treaty and the notion of a 'province of all mankind,' the Soviet Union argued that the celestial bodies 'are available for the undivided and common use of all States [sic] on earth, but are not jointly owned by them.'[35] The Soviet stance was that international law confirms ownership only to the extent that it is already recognized and decreed as valid within the civil law framework of a nation state. In other words, it was impossible to make ownership claims outside of the sovereign scheme because international law does not provide alternative mechanisms for validating such claims.

Argentina's position, on the other hand, explicitly linked the common heritage principle to the 1967 Outer Space Treaty. The delegation argued that this treaty already deviated from established international law doctrine in that it banned ownership claims through conquest or occupation.[36] Yet, rather than reaffirming a ban that applied only to those states capable of reaching space, the Moon Agreement could do more. It would enjoin all countries through the elaboration of a positive corollary: '[a] realization on the part of all states and peoples that they are entitled to the benefits derived from the principles and norms established for outer space and celestial bodies.'[37] These included the equitable sharing of profits from lunar exploitation, supervision of this activity to ensure equal distribution, and the existence of international machinery or institutional authority to regulate and oversee that exploitation. Thus, in this view, the common heritage principle was nothing new but rather was already implied in the Outer Space Treaty and naturally flowed from it:

> The major merit of replacing the vague expression 'province of all mankind' by the more meaningful expression 'common heritage of all mankind' is that in so doing one has specified the commencement of an action, replacing an abstract statement by means of operating, within a specified legal framework.[38]

The link between 'province' and 'common heritage' was also made by the Swedish delegation. They argued that while the former did not necessarily create a property title for mankind in the context of resource exploitation, 'province' clearly implied the related notion of 'common heritage.' And, given that the idea of common heritage had already been found applicable to the seabed and its resources, it made sense to extend that broader usage from a related UN context to the Moon Treaty:

> Once we accept this concept [of heritage], the question of a proper international administration [arises] [...] the idea of such machinery should be recognized. [...] This is, of course, just part of the much larger problem of turning the exploration and exploitation of outer space from its present unilateral or bilateral course into an international undertaking with tangible United Nations involvement.[39]

The Swedish and Argentinian positions were strongly endorsed and a number of countries produced drafts or proposals that designated the moon and the celestial bodies as the 'common heritage of all mankind.' Most delegates stressed that the equitable sharing of benefits and the interests of developing countries should be central considerations in framing exploitation. Egypt, Nigeria and India, for instance, argued that the samples and resources collected from the moon or other planets should be placed squarely under the ownership and control of the United Nations.[40]

Thus, the novel aspect of the Moon Agreement concerned the attempt to establish a framework that would give a more definite legal meaning to the phrase 'province of mankind' by stipulating the terms under which the exploitation and use of lunar resources might be conducted. This framework is contained in Article XI of the treaty. Its first section both references the above phrase and significantly extends it by describing the moon and its natural resources as the 'common heritage of mankind.' The agreement does not explicitly define, contrast or differentiate between 'common heritage' and 'province.' It specifies that the former will find expression in the establishment of an 'international regime, including appropriate procedures, to govern the exploitation of the natural resources of the moon.'[41] It explicitly links the notion of a common heritage to the exploitation of natural resources and outlines the purposes of this new regime in four subsections to Section 7 of Article XI.

The first three subsections involve the orderly and safe development of lunar resources, their rational management and the expansion of opportunities for their use. Though the agreement does not specify how or in what form this regime will function, it nonetheless carries far-reaching implications. When coupled with the injunctions on appropriation and ownership of the moon and its raw materials, these subsections would bring the management, control and exploitation of resources under the jurisdiction of a newly constituted international body or framework whose role would be one of stewardship for the interests of humankind. If the above were not already significantly transformative of existent relations, subsection (d) of Section 7 calls for 'an equitable sharing by all States Parties in the benefits derived from those resources, whereby the interests and needs of the developing countries, as well as the efforts of those countries which have contributed either directly or indirectly to the exploration of the Moon, shall be given special consideration.'

The intent of Article XI is unmistakable: the subsection places the interests of non-spacefaring nations on equitable footing with those of the developed and spacefaring world. It precludes the de facto monopolization of the moon and its resources simply because developing states lack the technological resources and wherewithal to fully partake of their *de jure* rights of open access.[42] At the same time, the above framework remained fully consonant with the principle expressed in 1967 that 'the use of outer space, including the Moon and other celestial bodies, shall be carried out *for the benefit and in the interests of all countries*, irrespective of their degree of economic or scientific development, and shall be the province of all mankind.'[43] The key difference between this formulation and its counterpart in the Moon Treaty is that the latter seeks to transform what had been interpreted by some nations as merely a non-binding aspiration into a legal requirement for any future regime.

Not surprisingly then, the contentious status of 'common heritage' hinged on Article XI and the complex legal ramifications (for example, in the sharing of proprietary technologies or profits) and obligations it might portend for the developed world. In the view of some critics, the customary legal principle of the 'freedom of the high seas' already provided a viable blueprint for the exploitation and use of space resources. Although states may not acquire sovereignty over an area, they enjoy the right to use and exploit the resources as they see fit – as long as they do not impinge upon the rights and activities of others.[44] This doctrine is compatible with the principle of *res communis* that underpinned the Outer Space Treaty, but only to the extent that activities are confined to exploration, research and scientific use. This is because the latter endorses the open and common use of resources, but it also designates those resources as having a common value that should be used for common benefit. Scientific research and use easily fall under the umbrella of common benefit, but any other form of exploitation becomes more problematic because it implies the need for a more formal agreement for their management. Whether this would take the form of common ownership or an international regime, exploitation would be contingent on the approval of all states or some representative body of all states.[45]

Further compounding the interpretive problem, the Moon Treaty intentionally splits 'resource exploitation' from 'exploration, use and scientific research' and places them under different categories. While the latter are contained in provisions very similar to their appearance in the Outer Space Treaty and hence fall under the 'province of mankind' concept, the former falls exclusively under the 'common heritage framework,' which is unique to the Moon Agreement and carries different (and more binding) legal implications. Thus, another way of understanding the legal question generated by the concept of 'common heritage' involves whether the agreement sketches alternative and mutually exclusive frameworks for understanding use and exploitation or whether the two formulations are compatible, as well as historically and conceptually interlinked. And in this sense, the perceived contentiousness of Article XI serves more as a measure of the powerful interests that surround the possibility of exploiting extraterrestrial natural resources than of the divergence between it and the principles contained in the Outer Space Treaty. This is because even under its most expansive understanding, common heritage merely means the establishment of an international regime to manage, monitor and ensure equitable exploitation. It leaves the 'how' or 'what' of this regime and its duties open for future elaboration because the treaty is primarily addressing the question of what makes the moon and the celestial bodies distinctive for humanity. In other words, much of the institutional, organizational and practical specifics surrounding lunar resources and their use remain undetermined in the agreement and open for future elaboration.

III   Justice and equity in outer space

The term 'common heritage' first appeared in 1967 in two UN fora: Argentina's Ambassador Aldo Armando Cocca used it in a COPUOS committee session, and Malta's Ambassador Arvid Pardo (1914–99) applied the phrase to the deep seabed and the ocean floor in his statement initiating negotiations over a new Law of the Sea regime some months later.[46] Its genealogy is linked to the successful independence movements of the 1960s, the ensuing withdrawal of colonial powers and the growing prominence of the new states in the international arena. In particular, the non-aligned movement (an international organization founded in 1961 by Egypt, Ghana, India, Indonesia and other similarly minded nations based on their abstention from alliance with the two superpowers) became a vocal proponent for an ambitious program to restructure the postwar and postcolonial global economic order through, among other things, the redistribution of resources from the North to the South. The actors in this movement were active in a variety of other international fora including the UN and championed calls for a 'New International Economic Order' (NIEO) in the General Assembly.[47]

The 'Declaration on the Establishment of a New International Economic Order' approved in the Sixth Special Session of the UN General Assembly in April 1974 is indicative of the broad range of issues under contention. The document emphasized the right of all nations to exercise sovereignty over their natural resources and economic activities, the right to regulate multinational corporations, called for better terms of trade, as well as greater transfers of science and technology on more beneficial terms.[48] More generally, the document made a compelling case for the catalog of historical injustices (apartheid, colonialism, neo-colonialism, racial discrimination, foreign occupation) upon which the current developed world advantage had been built. Finally, at the forefront of the NIEO was the link between technological progress and social and economic development:

> Technological progress has also been made in all spheres of economic activities in the last three decades, thus providing a solid potential for improving the well-being of all peoples [...]. The benefits of technological progress are not shared equitably by all members of the international community [...]. It has proved impossible to achieve an even and balanced development of the international community under the existing international economic order.[49]

The same coalition of countries and constellation of interests that appeared in the non-aligned movement and that backed the NEIO also figured prominently in the Law of the Sea and Moon Treaty negotiations.[50] One reason for this was that at stake in both treaty negotiations was the future exploitation of potentially transformational resources (the deep seabed or lunar materials). Within the context of a developing world seeking to regain control over the exploitation and extraction of its own domestic natural resources, the two treaties took on greater significance and visibility.[51] Control over these natural resources potentially opened a new realm of economic activity that might help ensure future social and economic development, but it was also meaningless (much like the freedom to exploit and profit from these resources) without the appropriate technological capabilities and scientific expertise to harness them.

The geopolitical context behind the two treaties (or at least the broader power realignment they seemed to envisage) and the fact that they channeled the demands and expectations of the developing world through the adoption of a 'common heritage' framework helped obscure the differences between their actual provisions concerning resource exploitation. Unlike the open-ended characterization of resource exploitation contained in Article XI of the Moon Agreement, the 1982 Law of the Sea Convention actively regulates the exploitation of resources found on the seabed, ocean floor and subsoil beyond national jurisdiction. It sets up an international authority to license and regulate exploitation, a governmental mining firm that will compete with the private corporations it will license, and a decision-making system where each country has one vote. It also details a series of licensing fees and equitable sharing requirements over technology and profits.[52] In contrast, the Moon Treaty merely requires that once it becomes technologically feasible to exploit lunar resources, states pledge to begin negotiations over the shape, content and scope of an international regime for its management.

The conflation of the Moon Treaty's common heritage provisions with the Law of the Sea framework likely doomed its reception by developed countries and private corporations. Unlike other space law documents, the agreement acknowledges existent power asymmetries among nation states and seeks to ensure that space exploration remains sensitive to equity principles. Yet, the emphasis in both the treaty negotiations and legal commentary on what impact the phrase 'common heritage' might have on resource exploitation has displaced the treaty's central normative promise. One might read space exploration as a twentieth-century extension of nineteenth-century colonization processes as some developing states did. In this view, space law with its high-minded rhetoric – province of mankind, common heritage, envoys of mankind – merely provided ideological cover for the construction of new and similarly durable structures of global inequality. The corollary to this interpretation might be that the negotiation process stuttered when developed and newly constituted nations sought to transform these high-minded principles into legally binding commitments. Yet another reading might emphasize that the juridification of outer space was but one aspect of a wider and more extensive legal reaction to the prospect of a future dominated by science, technocracy and the increasing obsolescence of law. In this view, the possibility of outer space as a legal vacuum worked powerfully on the self-identity of the legal profession, generating an elaborate proliferation of legal scholarship on extraterrestrial questions that only abated when the gap between the promised future and existent capabilities was revealed to be a chasm.[53]

Yet, however one interprets the Moon Agreement, it did explicitly seek to transcend the historical status quo through establishing a normative high ground. Since the Outer Space Treaty in 1967, the international community has recognized a new subject of international law: mankind. If this is correct then the Outer Space Treaty and related treaties (in applying to 'mankind') represent the law of the human race as a whole; a concept that is both temporally (previous, current and future generations) and geographically (it comprises all peoples) inclusive. Thus, in their attempts to inflect a set of humanist concerns within the fabric of international law, the space law treaties actively further a new legal entity: humankind.[54] The Moon Treaty should be read in light of this broader project of codification begun with Article I of the Outer Space Treaty and legally elaborated through the concept of common heritage.[55]

In the view of Argentina, Brazil, India, Sweden and the United Arab Republic, the Moon Treaty had to be considered a rejection of the historical primacy of militarization, colonialism and capitalist exploitation in approaching the world and in structuring relationships between states. This would posit space as a novel post-national, post-sovereign, human community whose benefits should be in principle and practice equally shared by all humankind. Indeed, the treaty's chief aim was to place human action in the universe under an international framework that would ensure that whatever occurred there would further, and remain subject to, principles of equitable distribution and mutual engagement. Such an approach would give legal and normative substance to the grand goals of exploration and discovery, parity and mutual benefit enumerated in the previous four space-related treaties; principles that seemingly intimate and prefigure a grander vision of space exploration – visible, for instance, in the formulation of astronauts as 'envoys of mankind' – than they in fact articulate.[56]

However, despite the multilateral nature of negotiations – many countries submitted drafts and comments – and the broad agreement the current form enjoys among all the negotiating parties, the Moon Treaty has to date been ratified by only 15 countries.[57] In 1972 the Apollo program flew its final mission, and both the United States and the Soviet Union shelved prospects for future lunar landings. By the close of the 1970s the expansive spacefaring future that the Apollo missions had helped evoke was refocused on more tangible aspirations. Reusable Space Shuttles and orbital space missions dominated the priorities of the United States and Soviet Union as well as the European Space Agency (ESA), which in 1979 marked its independent launch capability with the successful launch of Ariane 1 on 24 December.

The lukewarm reception of the Moon Agreement (as indicated by its poor ratification record) dealt a powerful blow to the broad imaginary of outer space it outlined. It generated a normative vacuum within which more instrumental understandings of the uses and value of outer space have flourished.[58] An important element in this interpretive shift was initiated by the United States. After supporting the common heritage phrase throughout the long negotiations, it withdrew support at the time of ratification under resistance from special interests and private enterprise in Congress.[59] Much of this was prefigured in the 29–31 June 1980 Congressional testimony on the Moon Treaty of former General Alexander Haig (1924–2010), a former chief of staff at the White House and NATO Supreme Commander, then president of the armaments manufacturer United Technologies Corporation. During the hearing, Haig painted the Moon Treaty as an attack on the very core of the capitalist world system:

> The common heritage concept expressed in the treaty underlies third world efforts directed at a fundamental redistribution of global wealth. [...] Third world countries have indicated they intend to gain control over critical raw materials and to gain access as a matter of right to technology needed to exploit them. [...] Proceeding any sooner with signing and ratification is opposed by United Technologies, because it would doom any private investment directed at space resource exploitation.[60]

According to this broadly shared and revisionist view, the Moon Agreement's poor ratification record should be read as a failure. Negotiators placed too many shackles on free enterprise and private entrepreneurship in an attempt to appease the revolutionary ambitions of the third world. If one could not profit from the investment made either directly through the exploitation of resources or indirectly through the development of technologies that might then have applications on earth, why would one want to go to the moon in the first place?[61] The irony of this perspective is that it fails to register that the treaty was less concerned with the mining and harvesting of resources than it was with the preservation of equity and a shared understanding of space exploration. In this sense, the Moon Agreement should be interpreted as an imperfect attempt to anchor the meaning of space exploration in the human experience not as an inevitable process chained to the juggernaut of the market or scientific progress but as a shared repository for the transcendence of historically imposed and earthly limits. That this position was articulated by a developing world that had long been situated at the wrong end of the negotiating table only gave the treaty an authentically political and transformative character.

IV   The universe as negative space

The perceived failure of the Moon Agreement exposed the limitations of the remaining four treaties. Together they set up a neither/nor framework that advanced an array of injunctions on sovereign appropriation while limiting the traction of a transnational space politics that might form the basis for resolving and transcending them. If outer space cannot be a zone of common property then, even under a modest interpretation of the *corpus juris spatialis*, it cannot be a zone of unrestrained free enterprise either. The capacities of sovereignty that might enable states to validate claims of ownership or lease sections of the moon or the celestial bodies for commercial purposes are similarly interrupted. On the other hand, space law recognizes that states remain the sole actors in outer space. It takes as given that domestic legal frameworks are the core instruments of legal import and relegates the regulation of what will happen in space to bilateral or multiparty agreements or individual emendation by nation states. The ad hoc legal framework suggested above dramatically limits the chances of considering space exploration as anything other than a means to some other end.

What was at stake in the Moon Agreement was more than a disagreement over the future of private exploitation of space resources. It offered the possibility of recovering a universal understanding of what outer space is and how this should guide our actions in the universe. More importantly, it provided a positive corollary and normative supplement to the Outer Space Treaty and the vacuum generated by the latter's powerful bans on sovereignty, appropriation and militarization. Thus, the Moon Agreement sought to preserve the possibility for conceiving and reserving (for future generations) outer space as a place of wonder and contemplation along the lines advocated by Arendt. And it involved the preservation of a platform for human action in the universe and the conditions for a space politics that could temper the empty trajectory of scientific progress or capitalist accumulation. This is what is most striking about the Moon Treaty: instead of focusing on the process by which resource exploitation should unfold, it attempted to articulate why the moon and the celestial bodies are important and valuable to the human race. In other words, the phrase 'common heritage of mankind' in the Moon Agreement serves as a normative and ethical compass as well as a principle of wealth redistribution.

In its absence we are left with the profound irony of Arendt's critique: from the perspective of universal science, knowing the secrets of the universe does not require us to go anywhere. The question then, of why we should go there at all cannot be understood from within this paradigm or that of capitalism, whose sole difference is that it replaces the quest for abstract knowledge with the relentless quest for profit. If space exploration and humanity's role within it will remain untethered from its political and terrestrial premises then the universe will also inevitably be chained to a teleological trajectory whose endpoint is neither articulated nor conceptualized as inherently meaningful.[62]

Notes

1. Hannah Arendt, *The Human Condition*, Chicago: University of Chicago Press, 1958, 261–4.
2. Idem, 'The Conquest of Space and the Stature of Man' [1963], *The New Atlantis* 18 (Fall 2007), 43–55, here 45.
3. Arendt, *Human Condition*, 264.
4. This is a reference to Archimedes's boast that he could move the earth off its axis if given a long enough lever and the ability to stand anywhere in the universe. Arendt adopts the idea to convey the self-defeating standpoint of modern science: the drive to understand nature and the given world objectively from a point outside earth as might be done by her creator; see Arendt, 'Conquest of Space.'
5. The full names of the treaties are: 'Treaty on Principles Governing the Activities of States in the Exploration and Use of Outer Space, including the Moon and Other Celestial Bodies' (1967), 'Agreement on the Rescue of Astronauts, the Return of Astronauts and the Return of Objects Launched into Outer Space' (1971), 'Convention on International Liability for Damage Caused by Space Objects' (1972), 'Convention on the Registration of Objects Launched into Outer Space' (1975) and the 'Agreement Governing the Activities of States on the Moon and Other Celestial Bodies' (1979). The text of these agreements is contained in United Nations, *United Nations Treaties and Principles on Outer Space*, New York: United Nations, 2002; available at http://www.oosa.unvienna.org/pdf/publications/st_space_11rev2E.pdf (accessed 1 October 2017).
6. COPUOS is the United Nations committee for the elaboration, discussion and ratification of space law.
7. UN Committee on the Peaceful Uses of Outer Space, Fifth Session, 'Summary Record of the Fifty-Eighth Meeting,' 20 October 1966, A/AC.105/C.2/SR.58, 2; available at http://www.oosa.unvienna.org/pdf/transcripts/legal/AC105_C2_SR058E.pdf (accessed 1 October 2017).
8. Many of these involved the wording of the inspection and access provisions of bases and activities (Egypt, Italy, Japan), the liability provisions in case of damage (India) and the future binding scope of the agreement (Australia).
9. UN, 'Treaty on Principles Governing the Activities of States in the Exploration and Use of Outer Space, including the Moon and Other Celestial Bodies' [1967], *United Nations Treaties and Principles on Outer Space*, 4.
10. Ibid.
11. See, for example, Article X which discusses the opportunity to observe the launch and flight of space objects; Article XI which lays out requirements for the reporting of space activities, discoveries and results; and Article XII which discusses installations and bases on the moon and other planets as well as the procedure for 'visits'; UN, 'Treaty on Principles,' 6–7. Christy Collis and Phil Graham, 'Political Geographies of Mars: A History of Martian Management,' *Management and Organizational History* 4.3 (August 2009), 247–61, here 251–2; Christy Collis, 'The Geostationary Orbit: A Critical Legal Geography of Space's Most Valuable

Real Estate,' in David Bell and Martin Parker, eds, *Space Travel and Culture: From Apollo to Space Tourism*, Chichester: Wiley-Blackwell, 2009, 47–65, here 53.

12. Christopher C. Joyner, 'Legal Implications of the Concept of the Common Heritage of Mankind,' *International and Comparative Law Quarterly* 35.1 (January 1986), 190–9, here 193–4.

13. See, for example, the opening remarks of the 1966 session by the Chairman of the Committee and the Legal Counsel. UN Committee on the Peaceful Uses of Outer Space, Fifth Session, 'Summary Record of the Fifty-Seventh Meeting,' 20 October 1966, A/AC.105/C.2/SR.57, 2–4.

14. UN Committee on the Peaceful Uses of Outer Space, Fifth Session, 'Summary Record of the Sixty-Second Meeting,' 19 July 1966, A/AC.105/C.2/SR.62, 8.

15. Ibid., 9.

16. UN Committee on the Peaceful Uses of Outer Space, Fifth Session, 'Summary Record of the Seventy-First Meeting,' 21 October 1966, A/AC.105/C.2/SR.71, 17.

17. Carl Q. Christol, *The Modern International Law of Outer Space*, New York: Pergamon Press, 1982, 44.

18. Ibid., 45–8. Having originally suggested that the concept be relegated to the preamble because it lacked a precise legal meaning, the Indian representative noted after significant debate that it should be retained and strengthened because 'a large number of subsequent articles would be based on that introductory statement'; see UN Committee on the Peaceful Uses of Outer Space, Fifth Session, 'Summary Record of the Sixty-Fifth Meeting,' 24 October 1966, A/AC.105/C.2/SR.65, 9.

19. In this context Henry G. Darwin, the UK representative, was discussing the version of Article I that appeared in the USSR draft which is virtually identical to Article I of OST; see UN Committee on the Peaceful Uses of Outer Space, Fifth Session, 'Summary Record of the Sixty-Third Meeting,' 20 October 1966, A/AC.105/C.2/SR. 63, 9.

20. UN Committee on the Peaceful Uses of Outer Space, 'Summary Record of the Fifty-Eighth Meeting,' 20 October 1966, A/AC.105/C.2/SR.65, 8.

21. UN Committee on the Peaceful Uses of Outer Space, 'Report of the Legal Subcommittee on the Work of its Fifth Session (12 July–4 August and 12–16 September 1966),' 16 September 1966, A/AC.105/35, 6.

22. 'The Congress hereby declares that it is the policy of the United States that activities in space should be devoted to peaceful purposes for the benefit of all mankind.' United States Senate, 'Treaty on Outer Space, Hearings before the Committee on Foreign Relations,' Ninetieth Congress, Washington, DC, 7, 13 March and 12 April 1967, 52.

23. Ibid., 53–6.

24. US Senate, 'Treaty on Outer Space,' 12.

25. Ibid., 12–13.

26. Christol, *Modern International Law*, 43.

27. See, for example, Walter A. McDougall, …*The Heavens and the Earth: A Political History of the Space Age*, New York: Basic Books, 1985; Paul B. Stares, *The Militarization of Space: U.S. Policy, 1945–1984*, Ithaca: Cornell University Press, 1985; and William H. Schauer, *The Politics of Space: A Comparison of the Soviet and American Space Programs*, New York: Holmes & Meier, 1976.

28. The six Apollo spaceflight missions (1969–72) returned to earth with 381 kilograms (2,200 separate samples) of lunar rocks, core samples, pebbles and dust from the moon surface; see NASA Lunar Curation; available at http://curator.jsc.nasa.gov/lunar (accessed 1 October 2017).

29. David E. Marko, 'A Kinder Gentler Moon Treaty: A Critical Review of the Current Moon Treaty and a Proposed Alternative,' *Journal of Natural Resources and Environment Law* 8.2 (1992), 293–345.

30. Barton Beebe, 'Law's Empire and the Final Frontier: Legalizing the Future in the Early *Corpus Juris Spatialis*,' *The Yale Law Journal* 108.7 (1999), 1737–73, here 1754–5.

31. As one scholar has noted, of its 21 articles, the treaty contains only six new provisions that are not covered by the previous four treaties; see James R. Wilson, 'Regulation of the Outer Space Environment Through International Accord: The 1979 Moon Treaty,' *Fordham Environmental Law Report* 2.2 (Summer 1991), 173–93.

32. UN, 'Agreement Governing the Activities of States on the Moon and Other Celestial Bodies,' *United Nations Treaties and Principles on Outer Space*, 28.

33. In 1977 the Legal Subcommittee reproduced all the drafts and proposals on the remaining unresolved issues (natural resources, disclosure of missions and treaty scope) between 1972 and 1977; see UN Committee on the Peaceful Uses of Outer Space, 'Report of the Legal Subcommittee on the Work of its Sixteenth Session (14 March–8 April 1977),' 11 April 1977, A/AC.105/196.

34. Article I of the Argentinian proposal designated the natural resources of the moon and the celestial bodies as the 'common heritage of all mankind' and Article IV stipulated that the benefits derived from the use of these natural resources would be made available to 'all peoples without discrimination of any kind.' See UN Committee on the Peaceful Uses of Outer Space, 'Report of the Legal Subcommittee on the Work of the Eleventh Session (10 April–5 May 1972),' A/AC.105/101 May 11, 1972, Annex I, 6–7.

35. The above discussion of common heritage is contained in a working paper from 28 March 1973 entitled 'Question of the Common Heritage of Mankind'; see ibid., 12.

36. Indeed the ban on sovereign appropriation in the Outer Space Treaty is one of the key distinctions between it and the Antarctica Treaty of 1959. At the time the latter was under ratification, seven states (Argentina, Australia, Chile, France, New Zealand, Norway and the United Kingdom) had already made sovereign claims on parts of the territory and the treaty merely 'paused' those claims during the time the treaty was in force; it did not require states to relinquish their sovereign claims. See Antarctica Treaty, 1 December 1959, Article IV. See also Christy Collis, 'Critical Legal Geographies of Possession: Antarctica and the International Geophysical Year 1957–1958,' *GeoJournal* 75.4 (August 2010), 387–95.

37. UN Committee on the Peaceful Uses of Outer Space, 'Report of the Legal Subcommittee on the Work of its Sixteenth Session (14 March–8 April 1977),' 11 April 1977, A/AC.105/196, 14.

38. Ibid., 15.

39. See Eileen Galloway, 'Agreement Governing the Activities of States on the Moon and Other Celestial Bodies,' in *Committee on Commerce, Science and Transportation* [United States Senate], Washington, DC: US Government Printing Office, 1980, 31.

40. UN Committee on the Peaceful Uses of Outer Space, 'Report of the Legal
    Subcommittee on the Work of its Thirteenth Session,' 6–31 May 1974
    A/AC.105/133, 14.
41. See Article XI Sect. 5, The Moon Agreement, *United Nations Treaties and Prin-
    ciples on Outer Space*, 31.
42. Article XI, Sect. 3 of the Moon Agreement, states that neither the moon (includ-
    ing its surface or subsurface) nor any natural resources found there can become
    state property or the property of any 'international intergovernmental or non-
    governmental organization, national organization or non-governmental entity or
    of any natural person.' Although states may erect bases, station personnel and
    equipment, none of these shall form the basis for a claim of ownership. Finally the
    treaty provides only two instances under which the appropriation of lunar resources
    is authorized: the first relates to the events that predate the establishment of the
    international regime sketched in Article II (a clause the United States adamantly
    insisted upon) and the second is described in Article VI (under which reasonable
    appropriation for scientific and peaceful purposes is authorized under the condition
    that those resources can be shared with the international scientific community).
43. UN, 'Treaty on Principles,' 4 [emphasis added].
44. See, for example, Lynn M. Fountain, 'Creating Momentum in Space: Ending the
    Paralysis Produced by the "Common Heritage of Mankind Doctrine,"' *Connecti-
    cut Law Review* 35.4 (2002), 1753–87.
45. David Tan, 'Towards a New Regime for the Protection of Outer Space as the
    "Province of All Mankind,"' *The Yale Journal of International Law* 25.1 (2000),
    145–94, here 161–2; Julie A. Jiru, 'Star Wars and Space Malls: When the Paint
    Chips Off a Treaty's Golden Handcuffs,' *South Texas Law Review* 42.1 (Winter
    2000), 155–82, here 159–60; Marko, 'A Kinder Gentler Moon Treaty,' 309–10.
46. Joyner, 'Legal Implications,' 190–9; Rüdiger Wolfrum, 'The Principle of Com-
    mon Heritage of Mankind,' *Heidelberg Journal of International Law* 43.1 (1983),
    312–37.
47. See, for example, Paul Streeten, 'Approaches to a New International Economic
    Order,' *World Development* 10.1 (January 1982), 1–17; Ajit Singh, 'The Basic
    Needs Approach to Development vs. the New International Economic Order:
    The Significance of Third World Industrialization,' *World Development* 7.6 (June
    1979), 585–606. Philip S. Golub, 'From the New International Economic Order
    to the G20: How the "Global South" is Restructuring World Capitalism from
    Within,' *Third World Quarterly* 34.6 (July 2013), 1000–15.
48. For an extensive discussion of these demands, see James W. Howe, 'Power
    in the Third World,' *Journal of International Affairs* 29.2 (1975), 113–27;
    R.L. Freidheim, 'The "Satisfied" and "Dissatisfied" States Negotiate International
    Law: A Case Study,' *World Politics* 18.1 (October 1965), 20–41.
49. United Nations, 'Declaration on the Establishment of a New International Eco-
    nomic Order,' Sixth Special Session, General Assembly, 1 May 1974 A/RES/
    S-6/3201, 1.
50. Also prominent within these various fora were Latin American states which,
    though independent since the nineteenth century continued to wrestle with Amer-
    ican imperialism and domestic interference. Theotonio Dos Santos, 'The Structure
    of Dependence,' *American Economic* Review 60.2 (1970), 231–6; Robert W. Cox,

'Ideologies and the New International Economic Order: Reflections on Some Recent Literature,' *International Organization* 33.2 (1979), 257–302.

51. Stephen D. Mau, 'Equity, the Third World and the Moon Treaty,' *Suffolk Transnational Law Review* 8.2 (1984), 221–58, here 235–6.

52. UN Convention on the Law of the Sea, Final Act of the Third UN Conference on the Law of the Sea, opened for signature 10 December 1982; available at: http://www.un.org/depts/los/convention_agreements/convention_overview_convention.htm (accessed 1 October 2017).

53. Beebe, 'Law's Empire and the Final Frontier,' 1748–54.

54. At least this was the view of legal scholars like Aldo Armando Cocca. See idem, 'The Advance in International Law Through the Law of Outer Space,' *Journal of Space Law* 9.1 (1981), 13–20, here 14; and idem, 'Prospective Space Law,' *Journal of Space Law* 26.1 (1998), 51–6.

55. Idem, 'Advance in International Law,' 16.

56. The term 'envoys of mankind' is a case in point. It seemingly grants astronauts a new supranational status transcendent of nationhood but makes little sense within the current, mostly national (ESA represents an important exception) exploration. Gabriella Catalano Sgrosso, 'Legal Status, Rights and Obligations of the Crew in Space,' *Journal of Space Law* 26.2 (1998), 163–86, here 164.

57. Australia, Austria, Belgium, Chile, Kazakhstan, Lebanon, Mexico, Morocco, the Netherlands, Pakistan, Peru, the Philippines, Saudi Arabia, Turkey and Uruguay have ratified the agreement; France, Guatemala, India and Romania have signed it. Ratification essentially means that a nation agrees to be legally bound by the terms of a treaty. Signature indicates a willingness to consider a treaty's ratification at some future date but is not legally binding.

58. One can note this shift in the legal literature. Recent articles discuss the Moon Treaty fairly cursorily and only as a means for establishing the importance of unshackling private ownership or entrepreneurship to develop resources. See, for example, Fabio Tronchetti, 'The Moon Agreement in the Twenty-First Century: Addressing its Potential Role in the Era of Commercial Exploitation of the Natural Resources of the Moon and Other Celestial Bodies,' *Journal of Space Law* 36.2 (2010), 489–524; Lawrence L. Risely, 'An Examination of the Need to Amend Space Law to Protect the Private Explorer in Outer Space,' *Western State University Law Review* 26.1 (1999), 47–70; Brandon C. Gruner, 'A New Hope for International Space Law: Incorporating Nineteenth Century First Possession Principles into the 1967 Space Treaty for the Colonization of Outer Space in the Twenty-First Century,' *Seton Hall Law Review* 35.1 (2005), 299–357; Allan Duane Webber, 'Extraterrestrial Law and the Final Frontier: A Regime to Govern the Development of Celestial Body Resources,' *Georgetown Law Journal* 71.5 (1983), 1427–33; Fountain, 'Creating Momentum in Space'; Art Dula, 'Free Enterprise and the Proposed Moon Treaty,' *Houston Journal of International Law* 2.1 (1979), 3–33.

59. Among the groups most vocal against the Moon Treaty were the L5 Society, the National Association of Manufacturers and the Aerospace Industries Association as well as a number of private companies like United Technologies. Additionally, because a variant of the common heritage framework had been an element of the Laws of the Sea Treaty III negotiations (which were ongoing during the

Moon Treaty discussions), groups like the National Ocean Industries Association and the Laws of the Sea Committee of the American branch of the International Law Association also opposed the United States' ratification of the treaty. See Nancy L. Griffin, 'Americans and the Moon Treaty,' *Journal of Air Law and Commerce* 46.1 (1981), 729–63, here 749; and Peter Westwick's contribution, Chapter 12 in this volume.

60. Quoted in Marko, 'A Kinder, Gentler Moon Treaty,' 311–12.
61. Such a position of course ignores a number of compromises embedded in the document and designed to appease the concerns of developed countries. In particular, the common heritage regime depicted in Sect. 7 of Article XI gives special consideration to the interests of developed countries (that is, 'countries which have contributed directly or indirectly') and nowhere in the document does one find a provision which imposes a moratorium on resource exploitation before the common heritage regime takes effect. The Moon Agreement, *United Nations Treaties and Principles on Outer Space*, 31.
62. Arendt, *Human Condition*, 307.

PART III

Grounding Utopias

Transnational Utopias, Space Exploration and the Association of Space Explorers, 1972–85

Andrew Jenks

Participants and observers often interpreted the Space Age in two ways. One vision – dubbed 'cosmopolitics' by a space policy analyst – viewed space exploration as an extension of national power on earth, just as 'the great maritime powers of the past used specific means and instruments, such as navy and naval bases, to achieve and maintain their power position.'[1] According to this narrative, forays into the cosmos served a national and military purpose, extending military prerogatives – and national economic growth – into outer space, just as European powers in the nineteenth and early twentieth centuries had expanded their political, military and economic reach into Asia, Africa and the open seas. This interpretive framework emphasized the dictates of Cold War politics and various national programs for economic growth and strategic nuclear capability. In this interpretation the pursuit of scientific goals through space exploration or utopian projects for unifying people across ideological and national barriers were secondary to military and strategic imperatives. As Foy D. Kohler (1908–90), US ambassador to the Soviet Union during the Kennedy Administration, noted in 1974: 'Whatever else may be said of America's motivations in embarking on its great space undertaking, certainly a compelling reason was to deny to the Soviet Union a monopoly of the benefits – political, strategic and psychological – that went with the forward surge of science-technology in the conquest of space.'[2]

Andrew Jenks (✉)
California State University, Long Beach, CA, USA
e-mail: Andrew.Jenks@csulb.edu

© The Author(s) 2018
Alexander C.T. Geppert (ed.), *Limiting Outer Space*
European Astroculture, vol. 2
https://doi.org/10.1057/978-1-137-36916-1_9

Alternatively, others viewed space exploration as something potentially opposed to traditional power politics. According to this interpretation, the Cold War and geopolitical considerations should play a subordinate role in the history of spaceflight. Instead, these observers focused on the scientific and utopian potential of Space Age discovery. The colonization of outer space, in this view, signaled the final stage of history leading to a more perfect global community. Ecological awareness occupied center stage in the new global consciousness of the Space Age, emerging in response to the cosmic view of earth as a unified yet fragile ecological system. The iconic 1968 'Earthrise' and 1972 'Blue Marble' images of earth, like bookends around the first Earth Day in 1970, would inspire people to be far better stewards of the home planet and to unify against the militarization of outer space and nuclear war. Those images and the dramatic stories of the first space travelers linked environmental science and an emerging ecological consciousness to space travel and to visions of new transnational communities. Manned spaceflight was thus 'a foundation for building a galactic and eventually a universal civilization [...] beyond national and ideological barriers.'[3] By the 1970s, ecological perspectives, utopian ideas about global integration and traditions of astrofuturism began to migrate across Cold War borders, stirring visions of global peace and of a Europe liberated from Cold War divisions. Images of the whole earth from outer space constantly pushed people's imaginations beyond the boundaries on political maps. The cosmonaut Oleg Makarov (1933–2003) remarked: 'Unconsciously, you look for the lines that are usual on such maps, the parallels and meridians; it is strange not to see the markings on the living map.'[4] The East German cosmonaut Sigmund Jähn (1937–) said that 'when it takes 90 minutes to go around the world, there is no need for borders.'[5] That statement reflected a common view that space travel would encourage sympathy for the interests of all people and the health of the world in which they lived, regardless of national affiliation. Space technology was thus midwife to the era of transnationalism – a popular idea in historical studies but one often lacking a concrete chronological focus.

Focusing on manned spaceflight in Europe in the long decade of the 1970s – from the Ostpolitik and détente of the 1970s to its demise in the first half of the 1980s – this chapter explores spaceflight as a window into tensions between cosmopolitics and utopian visions of a new global community. In the words of the Dutch astronomer H.C. van de Hulst, recipient of the 1990 Planetary Award: '[W]hen a system without frontiers, such as science, meets a system with frontiers, such as politics [...] a dialogue between the political and scientific systems must take place in order to minimize the turbulence which could be encountered.' Sandwiched between the two superpowers and under the constant threat of nuclear catastrophe, Europe was at the center of this turbulence zone; it became a staging ground for both Cold War military and economic competition and for bold new experiments in cosmic and terrestrial collaboration. The first part of this chapter discusses early ideas about global forms of consciousness and their connection to space travel and ecological thinking. After briefly examining the first collaborative manned ventures in space, the Apollo-Soyuz Test Project (1975) and the manned Soviet Interkosmos missions (1978–88), I then turn to a group called the Association of Space Explorers (ASE), formed in Paris in 1985 and comprising individuals who had flown into earth's orbit, regardless of their nationality. The ASE's creators viewed themselves as the vanguard of the future, a transnational community to which all people, they believed, would soon belong.

I   The cosmic perspective

Historian Robert Poole has commented on the iconic image snapped on Christmas Eve 1968 from Apollo 8. 'Looking back,' he writes, 'it is possible to see that Earthrise marked the tipping point, the moment when the sense of the Space Age flipped from what it meant for space to what it meant for the Earth.'[6] Yet, even before the Earthrise moment of the late 1960s, competing urges to escape into outer space – as a way of transcending human limitations – and to recognize the need to confront challenges on earth had a long pedigree. Like the astrofuturist ideal pursued in the West by science-fiction writers and engineers such as Arthur C. Clarke (1917–2008) and Wernher von Braun (1912–77), the Russian cosmists at the beginning of the twentieth century believed the survival of the species required the colonization of other planets. Space colonization, in their view, was both the solution to terrestrial problems and the culminating point and rationale for human development. Russian science-fiction writers – *fantastika* in Russian – began representing aviation and spaceflight 'as the catalyst for abolishing all concepts of the national, linguistic, ethno-racial and religious identity.' Combining ideas about social revolution and aerospace technology, they imagined that 'democratic internationalism [...] would pervade the world when everyone had free access to the sky,' a belief later integrated into Soviet representations of the Space Age.[7]

Borrowing from the ideas of the Frenchman Pierre Teilhard de Chardin (1881–1955), the cosmist Vladimir Vernadsky (1863–1945) began contemplating the earth as an integrated planetary system. His breakthrough in the 1930s involved a radical shift in perspective that anticipated later satellite imagery: imagining the earth from the vantage point of outer space. Integrating the increasingly fractured disciplines of geology, biology, chemistry and physics, Vernadsky believed that a description of an organism – including mankind and the earth itself as a 'biosphere' – was not possible without an analysis of the environment within which the organism lived. Toward the end of his life he developed the idea of a 'noosphere.' The noosphere was the stage of cosmic development that followed the biosphere. In this final stage of cosmic evolution, the accumulation of scientific knowledge about the physical environment had allowed people to manipulate the natural world – as well as to produce technologies that could destroy the planet. Vernadsky, however, believed a greater sense of collective moral responsibility would result from recognizing the potential tragedies caused by the growing ability to dominate nature. He anticipated the dark implications of nuclear physics, dying seven months before the dropping of atomic bombs on Hiroshima and Nagasaki. Yet, somehow, amidst the ruins of Nazi devastation in Russia, he remained an optimist. Echoing the ecological consciousness later associated with Rachel Carson's 1962 *Silent Spring*, Vernadsky wrote during the Second World War:

> The planet's face – the biosphere – is consciously and mainly uncon-sciously being chemically and physically changed by Man. As the result of the growth of human culture in the twentieth century, the coastal seas and parts of the ocean are changing more and more dramatically. Man must take more and more steps now to preserve the riches of the sea for future generations.[8]

Vernadsky's cosmic perspective evoked a heightened sense of the plan-et's vulnerability as well as demands for global collaboration to prevent its destruction.

The International Geophysical Year (IGY) of 1957 – when scientists from around the globe pooled resources to study the earth as a unified system – marked a key moment in the conception of space exploration as an interna-tional and collaborative venture, drawing on Enlightenment ideas about science as an inherently international and collaborative endeavor. One scientific participant in the IGY noted that the scientific investigation of outer space, 'has no palpable appeal to nationalism or power politics. [...] Copernican astronomy not only achieved a revolution in science but also changed man's concepts of man and of earth. The onset of the Space Age affords the possi-bility of a comparable impact.'[9] But it was the ability to actually see the earth from outer space – and reproduce that image for mass consumption – that provided the most powerful stimulus to the imagination. When Yury Gaga-rin (1934–68) became the first human in space on 12 April 1961, his gaze shifted almost immediately from outer space to the home planet. He com-mented on the fragility of the earth and on the need 'to preserve and expand this beauty and not destroy it.' That statement, set against the backdrop of an angelic image of Gagarin's helmeted head with his charismatic smile, was reproduced in an iconic propaganda poster that dominated Soviet public space as thoroughly as the Beatles in the capitalist West (Figure 9.1). Ulti-mately, Gagarin's flight, as a Rand political scientist put it in 1963, had 'an emotional appeal for people at large, regardless of their political attitudes. Here is a Soviet success that can be presented as a success for all mankind.'[10] The irony was lost on most observers that Major Gagarin, a fighter pilot

Figure 9.1 The iconic 1961 poster that reproduced, in Gagarin's handwriting, his famous quote: 'Circling the earth in my spaceship, I saw how beautiful our planet is. People, we must preserve and expand this beauty and not destroy it!'
Source: Author's personal archive.

whose job it was to test the viability of ICBM technology, represented a project managed by strategic nuclear rocket forces. Said another Rand analyst: 'Space exploration, to adapt a Maoist formulation, grew out of the nose cone of the ICBM.'[11]

Gagarin's simultaneous status as both a representative of a military industrial complex and a messenger of global peace captures a key paradox of spaceflight. In the realm of politics and popular culture, space technology had the potential to both unite and to separate, to appear as a catalyst for world peace and harmony or, alternatively, as an instrument of global self-destruction. As Robert Poole notes, this was true both of space technology and air travel in general: 'In a swords-into-plowshares movement, long-range air travel with its potential for peaceful interchange was the product of the bomber technology of the war, just as space travel was to be the product of the intercontinental ballistic missile technology of the Cold War.'[12] New satellite images and communications technologies constantly pushed people's imaginations beyond the boundaries on political maps, just as earlier technological innovations had given rise to the idea of the nation state. Benedict Anderson famously pointed to the vital role of print capitalism and newspapers in helping people to imagine themselves as members of the modern nation state.[13] The nation was a community of millions of people who physically could never see each other but who nonetheless imagined themselves – through newspapers, textbooks and images on maps – as having a common national identity. Mary Louise Pratt has suggested that new European technologies for conquering space, starting in the eighteenth century, combined with new ways of representing distant corners of the globe and ultimately produced a planetary consciousness, a 'picture of the planet appropriated and redeployed from a unified, European perspective.'[14] Space exploration and satellite imagery of the earth marked another phase in the development of this planetary consciousness, stimulating ever larger visions of a global human community.

While the Earthrise and Blue Marble images entered popular culture in Western Europe and the United States in the early 1970s, artistic and photographic representations of the earth from outer space – along with messages of universal peace, brotherhood and environmental protection – dominated propaganda poster images behind Europe's Iron Curtain. One of the Soviet Union's major artistic producers of those images was the cosmonaut Aleksey Leonov (1934–). During the 1970s his paintings of earth from space were produced by the millions on propaganda posters, postage stamps and in art galleries. The transnational subject, whose identity emerged from a cosmic perspective, was thus a byproduct of the Space Age. And just as newly emerging national identities in the nineteenth century had made the modern nation state possible, so, too, had the possibility of new transnational identities, facilitated by the view of earth from outer space, marked the beginning of a new phase of global and European integration.

II   Cosmism in policy and practice

Even as the 1968 invasion of Czechoslovakia seemed to solidify Europe's division into two armed camps, Soviets looked toward spaceflight as a way of realizing very different ideas about universal community, albeit with a socialist rather than capitalist emphasis. The Soviets created Interkosmos in 1967, a counter to the European Space Research Organization (ESRO) established in 1964. Soviet commentators waxed philosophical about the potential for space exploration to transform Europe into a no-war zone – a belief forged in the bloody experience of the Second World War in which the Soviets lost circa 30 million lives. The signing in 1967 by the Soviet Union and the United States of a United Nations treaty banning nuclear weapons from outer space was a watershed moment, prompting paeans 'to cooperation and mutual understanding' from Soviet Foreign Minister Andrey Gromyko and President Lyndon Johnson.[15]

Reaching across the Cold War divide was the civilian cosmonaut Vitaly Sevastyanov (1935–2010), an engineer rather than fighter pilot who hosted a popular television science show in the Soviet Union. He became a regular contributor in the English language space-industry press during the 1970s, introducing English-speaking audiences to the utopian conceptions that had partially driven the Soviet space program. Space technology, in his view, was a panacea that would lead humanity down the path toward a more perfect global community. Space exploration, Sevastyanov wrote in 1972, 'has been exerting an ever greater influence on social being and consciousness, it has been changing our traditional ideas about the world and the relationship between society and nature.'[16] Such ideas – along with a desire to view the American system of technological management and production – had inspired the Soviets to work with the United States on the first joint manned space mission, the 1975 Apollo-Soyuz Test Project (ASTP). The project grew out of Nixon's determination to use a collaborative space project to launch his new policy of détente. His point man was Frank Borman (1928–), who first broached the topic in the first ever visit of an American astronaut to Moscow in the summer of 1969. Sending Borman off, Nixon invoked the romantic spirit of collaboration and eschewed the cosmopolitical stance of the earlier Space Race. The United States, he said, was prepared 'to work together with all peoples on this earth in the high adventure of exploring the new areas of space.' NASA Director Thomas O. Paine (1921–92) was under Nixon's orders to collaborate with both Western Europeans and the Soviets.[17]

The result of these overtures, which bore unexpected fruit for many NASA officials, was an agreement signed in Moscow between President Nixon and Soviet Premier Aleksey Kosygin in May 1972. In addition to a broad-ranging program of space-science collaboration, they agreed to develop a docking system between the Apollo and Soyuz capsules. Sevastyanov presented a cosmist interpretation of the project for the readers of *Space World*, an independent magazine devoted to the space industry published in the United States from 1960 to 1988. The high costs of space exploration, he noted, had made collaboration across ideological and national borders imperative. But even more pressing was the need to prevent the devastation of the environment and to find the resources that might solve the Malthusian dilemma, elaborated by the Club of Rome in 1972, that seemed to promise a future of resource deprivation. 'We must go beyond the confines of this planet,' wrote Sevastyanov, 'because of the earth's limited surface and natural population growth and search for other worlds with favorable conditions for existence and development [...] we of the twentieth century must work out a global program for developing space vehicles which would ensure the emergence of human civilization outside the solar system.'[18] Meanwhile, Nixon's desire to use space collaboration as a foundation for détente put the ASTP program on a fast track of development from the agreement in 1972 to the actual docking in July 1975. During that period cosmonauts and astronauts traveled to each other's countries, acquainting themselves with the culture of their erstwhile enemies and developing new friendships. Similar to the fraternity of fliers from the early days of aviation – when military pilots from opposing countries nonetheless developed a deep appreciation of each other's skill and courage – the astronauts and cosmonauts developed a public image and bond that did not fit easily into Cold War conceptions of Russians and Americans as implacable enemies. Those relationships, in microcosm, symbolized the new possibilities for human interactions back on earth made possible by space collaboration (Figure 9.2). Soviet commentators floated various utopian ideas about an international space platform – with broad US and Western and Eastern European collaboration – as a basis for building a new global order based on peaceful technological and scientific exchanges.[19] Far from marking an end to utopian conceptions of spaceflight, the 1970s, at least in the Soviet Union, stimulated the utopian imagination – thanks in large part to ASTP.

NASA, however, was entering a perilous period in its space program in the gap between the last Apollo missions and the first Space Shuttle launch in 1981. American plans to collaborate in outer space were also stymied by concerns that the nation would be giving away its competitive edge.[20] Meanwhile, the Soviets were eager to build on ASTP and promote joint ventures in outer space with Eastern Europeans, if not their Western counterparts. They initiated joint manned missions through the late 1970s, underwriting the costs of the launches – in stark contrast to NASA's insistence that international partners share expenses – and promoting joint manned missions as milestones toward the creation of a new global order of peace. Similar to Antarctic exploration, space science for the Soviets was a kind of neutral space that both advanced knowledge of the cosmos and promoted peace and brotherhood. As one study of transnational ideas in Soviet politics has argued, Soviet leaders in the 1970s and 1980s were highly receptive to using transnational networks of scientists to promote disarmament and the demilitarization of outer space.[21] Its Interkosmos program had emerged in part as an attempt to forge peaceful political ties through programs of scientific collaboration with other nations.

Figure 9.2 The Soviet cosmonaut and head of cosmonaut training, Vladimir Shatalov (1927–), greeting Shoshone tribal leaders in Wyoming in September 1974. Shatalov visited the United States as part of the numerous cosmonaut-astronaut exchanges associated with ASTP, forging new friendships with former enemies and gaining a new appreciation of American culture.
Source: Courtesy of NASA Historical Reference Collection, NASA History Division, NASA Headquarters, Washington, DC.

Eastern Europeans involved in Interkosmos became the first Europeans other than Russians to leave earth's atmosphere, beginning with Vladimír Remek (1948–) from Czechoslovakia in March 1978. The selection of a Czechoslovakian was a symbolic gesture designed in part to patch over ill feelings from the Soviet invasion ten years previously. Remek was followed by a Pole and then the East German Jähn.[22] Thanks to the Soviets, the first Western European, Jean-Loup Chrétien (1938–), flew into space on 24 June 1982.[23] Chrétien's flight was the culmination of a long collaboration across the Cold War divide between France and the Soviet Union. Charles de Gaulle had promoted France as a third way between the Soviet Union and the Anglo–American alliance. Those policies, in part, led de Gaulle to a visit to Moscow in June 1966 to sign a bilateral agreement on space exploration in a number of areas, with a particular focus on joint scientific missions. The French participated in the Soviet lunar landings of the early 1970s. A 1973 Soviet Mars probe carried French instruments. French biological experiments were included in the Soviet space station Salyut 6 in 1978, followed by tests for satellite transmissions between the Soviet Union and France in 1981.[24] As in European politics more generally during the 1960s and 1970s and parallel to the policy of Ostpo-litik in West Germany, France played a key role in imagining a Europe beyond the Cold War divide; and its overtures to the Soviet Union emphasized the international and collaborative traditions of joint scientific missions. Building on Enlightenment ideas about the transnational nature of the scientific commu-nity as a Republic of Letters, those missions were an essential antidote to Cold War tensions – and to the Cold War model of a nationalized and increasingly classified scientific community.

III   Europe at the margins

It was nonetheless difficult to escape the reality that Western and Eastern Europeans were sandwiched politically and geographically between two super-powers armed to the teeth with nuclear weapons. 'Throughout the 1960s and 1970s Western Europe seemed condemned to the status of an also-ran in the headline-grabbing Space Race between Washington and Moscow.'[25] Tensions between the Soviets and their European partners constantly surfaced, despite rhetoric about international brotherhood and collaboration. For example, the original Soviet commander for the Franco-Soviet mission on which Chrétien flew transferred out of the mission due to personal problems with his French comrades. The United States also made it clear that it was a first among equals when it came to its Western European colleagues.[26] A NASA official in 1986 reflected the often superior attitude that irked so many Europeans: 'Partners accepted NASA control because […] they perceived themselves to be junior partners in fact and especially because they did not have ready alternatives to working with NASA if they wished to work at the cutting edge of the space frontier.'[27] Depending on whose rockets they used, Europeans were thus cosmonauts or astronauts, even though the French briefly named themselves 'espationautes,' just as the Chinese created the term 'taikonauts,' thereby hop-ing to nationalize their own path to the stars.[28] In the tradition of cosmopoli-tics, the neutrality of outer space – or its transformation into a staging ground for new types of human community – was constantly challenged by the Cold War tensions between the United States and the Soviet Union.

As détente collapsed in the late 1970s, a group of retired astronauts and active cosmonauts, led by Apollo mission veterans Russell 'Rusty' Schweickart (1935–) and Edgar Mitchell (1930–2016) and Aleksey Leonov on the Soviet side, came up with the idea of an Association of Space Explorers. Edgar Mitchell first suggested the idea in October 1973 in a letter to the White House. He proposed a committee of retired astronauts and cosmonauts 'to function as goodwill ambassadors for peace and unity on a global scale.' Nixon, for his part, was intrigued by the idea, which was later taken up by the Esalen Institute.[29] It was another suggestion of the utopian notions that Nixon had associated with spaceflight since his first inaugural address, which highlighted the Earthrise image in his call for a new international order. Despite the collapse of Nixon's presidency, retired Apollo astronauts, along with active Soviet cosmonauts who had met during various exchanges of the 1970s and on ASTP, kept the idea alive. The group's founders wanted the group to be open to space travelers from all countries, though the initial focus was on Soviet and American space travelers, with no apparent attempt to define just what would constitute traveling into 'space.' Leonov, in particular, was inspired by ASTP's memorable 'handshake in space' and by visions of a fragile planet under the twin threats of nuclear holocaust and environmental destruction. The handshake was the moment when the Soviet cosmonauts and American astronauts opened the docking mechanism between their capsules and shook hands, thereby initiating, it was hoped, the launch of a new era of peace and brotherhood.[30]

Political circumstances, however, quickly shifted the ASE's focus from the United States and the Soviet Union to France. Sure of the superiority of American space technology and seeing outer space as an arena for national competition rather than collaboration, American opponents of collaboration gained official support during the first Reagan Administration.[31] Those views were reflected in NASA's reluctance to work with Western Europeans on joint missions and payloads (whose origins went back to Nixon's initiatives) and to a refusal to renew in 1982 the agreement for widespread collaboration with the Soviet Union in outer space, first signed in 1972 in Moscow by Nixon and Kosygin.[32] Reagan's Star Wars speech in March 1983 and then the downing of a South Korean civilian jetliner in Soviet airspace in September 1983 further heightened Cold War tensions. Meanwhile, many within the astronaut corps considered the ASE a betrayal of US national security imperatives in the wake of the Strategic Defense Initiative (SDI) and plans to upgrade missiles in Western Europe. The astronaut Vance Brand (1931–) noted that the Soviet cosmonauts were not 'free agents,' supposedly unlike the American participants. The Soviet side would invariably try to 'reap any propaganda or political advantages' from the ASE, while putting the astronauts into the uncomfortable position of opposing their own government by participating in the effort.[33] Those sentiments were not uncommon, as Russell Schweickart learned when he polled Apollo veterans about the ASE in order to gage potential support among American space-traveling veterans.

Based on his polling data, it became apparent to Schweickart that it would be difficult to achieve his original goal, as expressed in a letter to the Soviet cosmonaut Aleksey Yeliseyev (1934–): 'We want to proceed in such a way so as not to criticize the views and actions of political leaders and institutions. Rather we will clarify or elucidate problems of importance for humanity, contributing to the unification of efforts toward their solution.'[34] The problem was that any declaration of unified efforts to solve global problems, in the context of the Cold War, was often interpreted as a direct attack on the views and actions of political leaders and institutions in the Reagan Administration and, to a lesser extent, the Soviet Union. The US refusal in 1982 to renew the program of space collaboration with the Soviets was one reflection of this ideological animosity toward joint space missions as a stepping stone to a new international order; another indication was the Reagan Administration's resistance to celebrations of the tenth anniversary of ASTP in 1985 – which stood in stark contrast to prominent Soviet celebrations urging the demilitarization of space.[35] Thus, despite the impressive and growing record of international collaboration in the 1970s, beginning with ASTP and French and Soviet joint efforts, the perceived national and strategic imperatives of the Reagan Administration had stopped the momentum of international space-cooperation initiatives.

IV   France as a neutral zone

The NASA correspondence in the Reagan Presidential Library Archives, especially as compared to NASA records from the 1970s, reveals an obsession with injecting ideological meaning back into the American space program on two fronts: emphasizing the military and national importance of space exploration, and a dogmatic insistence on privatization as the only possible future direction for civilian space initiatives. To the extent that the Reagan Administration was prepared to encourage collaboration of any sort, it emphasized cooperative efforts between the United States and the private sector – and not between NASA and space administrations of other nations. In the words of one Republican legislator, in a letter to the White House in September 1983, the Republican Party needed to 'use space to break out of the psychological "Limits to Growth" mentality, and thereby move away from a Liberal Welfare State toward a Conservative Opportunity Society,' and that meant shifting the emphasis toward government financing of military space applications and leaving civilian applications to the American free-market system.[36]

The renewed emphasis on the military implications of space exploration also meant ever-tightening controls over technology transfer, driven by fears that even collaborative efforts with friendly Europeans might ultimately – through their contacts with the Soviets – give away supposedly decisive American technological advantages. Such sentiments had a chilling effect on the ASE's attempts to build on the precedent of ASTP and build bridges across the Cold War divide. To draw attention away from the politically explosive image of American astronauts making overtures to Soviets without official sanction, the ASE decided to go to Europe with their project and include people who had flown into outer space from both Eastern and Western Europe. Doing so, its founders hoped, would save the ASE from the McCarthyite sentiments of former and active astronauts, many of whom were clearly spooked by the Reagan Administration's aggressive Cold War rhetoric and renewed emphasis on the military and national significance of spaceflight. European participation, it was thought, would also prevent the structuring of the event as a bipolar Cold War project and 'avoid the balancing charade of one-of-theirs, one-of-ours.'[37] The astronaut Jim Irwin (1930–91) met in Europe 'with all of the Euroastros and got a positive response from them in the "let's all get together" theme.' Schweickart also consulted with the popular French ocean-ographer Jacques-Yves Cousteau (1919–97), who recommended that Euro-pean astronauts be brought into the effort.[38] When the Soviets backed out of a planned astronaut-cosmonaut meeting for May 1984, because of the Soviet boycott of the Los Angeles Olympics and because of the US refusal to remove its Pershing II and Cruise missiles from Europe, the organizers had con-cluded that only Europe could provide the neutral path they had been seek-ing. 'Joining the Euroastros [...] may indeed be the best way,' said Mitchell, in an August 1984 letter to Schweickart. To answer charges that the group would be used by the Soviet side against the US government, organizers made it clear that non-Soviet participants represented only their own personal views – and not those of the American government or other non-participating astronauts.[39]

At first, the organizers contemplated Iceland or the Canary Islands as a meeting place, but later decided to shift to France when they found a private sponsor with a château outside of Paris to host the inaugural 'Congress of Planetary Explorers,' held 2–6 October 1985 (Figure 9.3). The conference was attended by 25 space travelers, including eight Soviets, five Americans, a Saudi Arabian, Frenchman, Czechoslovakian, Mongolian, Cuban, Romanian and East German. In addition to lining up the private American corporate sponsor PepsiCo, the ASE secured sponsorship from the French engineer-ing firm Constructions Mécaniques de Normandie, whose president owned the château where the space travelers planned to meet. Equally important, France was the homeland of Western Europe's first space traveler, Chrétien, who would play a key role as the French host. Chrétien was brought into the effort, as Schweickart wrote to him in August 1984, 'to minimize the ten-dency toward political speculation and embroilment in international politi-cal issues' and to help function as a 'bridge' between the Soviets and the Americans, thus adding a 'multi-lateral flavor to the meeting.' Removing the effort from US or Soviet soil was especially critical 'given the timing of this meeting vis-à-vis the US elections and the potential space/strategic weapons talks.' Schweickart also reached out to Ulf Merbold (1941–), the first West German in outer space, and to Wubbo Ockels (1946–2014), the Dutchman who was then an astronaut in training.[40] Both 'Euroastros' played a poten-tially useful role as intermediaries who might lessen the politically challeng-ing aspects of the ASE at a particularly tense moment in US-Soviet relations. Merbold and Ockels were the first Europeans brought into the Space Shut-tle program. They were payload specialists responsible for the European scientific experiments on ESA's Spacelab – given to NASA in exchange for allowing European astronauts to fly on the Shuttle. With a background in science rather than the military, and as representatives of the European Space Agency (ESA), they occupied a seemingly neutral and disinterested

Figure 9.3   The official poster for the Association of Space Explorers' First Planetary Congress in 1985. With writing in English and Russian on either of the earth's poles, the poster suggests a bipolar political frame for the emerging global consciousness of the Space Age.
Source: Courtesy of Association of Space Explorers, Houston, TX.

non-political space as scientists. Whenever challenged by the military and national imperatives of spaceflight, advocates of collaboration relied on the Enlightenment notion of science as a supposedly universal and politically neutral endeavor, a role reserved in this instance for the European payload specialists and their scientific experiments.

V   New agers and spaceflight in the 1970s

The spirit of détente and the counter culture of the 1970s, like the aroma of burning incense, hung over the entire enterprise. The Esalen Institute in Big Sur, California, the ASE's primary sponsor and a driving force in the human potential movement, said it wanted to launch CBMs – Confidence Building Measures – rather than ICBMs.[41] It had initiated the first cosmonaut-astronaut exchanges in the early 1980s through 'hot tub diplomacy.' Since the early 1960s, Esalen had searched for ways to link modern life and science to a mystical sense of unity with the cosmos. The spiritual transformations of many astronauts and cosmonauts, including Russell Schweickart and Edgar Mitchell, who had conducted extrasensory perception (ESP) experiments during his Apollo mission, had convinced Esalen in the 1970s that space exploration just might be the key to the kinds of mental transformations needed to save the world from environmental and nuclear destruction.[42] 'We thought we were astronauts of inner space,' said an Esalen Institute executive, 'about to break through into new realms of consciousness; we wanted to put man on the psychic moon.'[43] At one point, when negotiations between cosmonauts and astronauts to form the new group seemed to reach an impasse, Mitchell went to the blackboard and drew a mushroom cloud below a horizontal line, 'representing the politics of fear and destruction. On the upper half, he drew an image of the whole Earth, symbolizing the perspective from space of a world without boundaries.'[44] By invoking the image of nuclear holocaust, Mitchell highlighted the potential significance of the ASE's deliberations. He wanted to clarify – in response to some criticisms that it was unclear if the ASE wanted to be a 'social club or a think tank,' an 'Elk's Club' for astronauts – that the group was more than just a social gathering of professional space travelers reminiscing about old times and exchanging stories, though there was plenty of that, too. At stake in their actions was nothing less than the fate of the earth.[45] Indeed, ASE members had a sense of being instruments of historical destiny in which space travel marked an end to the conflicts that arose from national competition. In their own minds, they had become models of the future global citizens, ecologically aware and peace-loving. As ASE founding member Sevastyanov put it in a popular Russian-language book translated into English, the space travelers had 'features' that were 'typical of humanity as a whole in later stages of the Space Age.' They would 'increasingly differ from the terrestrial average in their individual, moral and even physical qualities,' thus giving them a special responsibility as civilization's saviors.[46] Those comments echoed a draft of the ASE's mission statement, written by Russell Schweickart and Edgar Mitchell, which claimed that, 'none [...] but we can speak and act from direct experience of being outside of and looking back on the single home of all the life we know in the universe.'[47]

The location for their first meeting had symbolic meaning equal to the grand tasks the founding members had set before themselves. The group thus met in October 1985 'near Paris where Crusaders once passed on their way to Jerusalem.' The image of a crusade captured perfectly the group's sense of world-historical mission. The grounds of the private château seemed to blend European tradition and modernity, seamlessly integrating bucolic landscapes with visions of new worlds to sow and reap. Two artists among the group – the cosmonaut Aleksey Leonov and the astronaut Alan L. Bean (1932–) – brought their works for display at La Géode, a geodesic dome at the Cité des Sciences in Paris. The art was meant to depoliticize the effort, something 'cooperative and non-controversial,' said Alan Bean. As if to tone down the harsh rhetoric of the Cold War, 'meals were conducted in a leisurely European style,' and not in the hurried manner of the Americans. 'There was ample time for fishing and boating on the lake,' and plenty of vodka for the Soviets.[48] Those unstructured exchanges over food and drink were more than rest and relaxation before the main work of the ASE. As with the informal citizen exchanges sponsored by the Esalen Institute more generally, the ASE believed that camaraderie among the elite group of space travelers, the fraternity of fliers, was the first but critical step toward building alliances across political divides and ultimately to forging new types of consciousness. Emotional satisfaction in human exchanges would point the way to higher, spiritual truths. As one Esalen scholar has noted: 'People at Esalen advocated spirituality, emotional disclosure, and men's connection to other men as pathways to a new, better kind of masculinity,' one that would be more in tune with the need for peace rather than the demands of various military-industrial complexes.[49]

While Europe seemed to represent a more relaxed and neutral space than either Moscow or Washington, environmental science provided the ASE with a seemingly non-political thematic focus. Focusing on environmental topics, the ASE convinced Cousteau to give the keynote speech.[50] The intrepid showman and explorer had inspired 'Cousteau Societies' throughout Europe and the world, including the Soviet Union and the United States. His invention of the aqualung allowed people to explore new realms, just as chief engineer Sergey Korolyov's rockets had allowed humans to go into space (Figure 9.4). In line with ASE's hopes for space exploration, Cousteau's undersea activities helped to forge a transnational environmental consciousness, opening new worlds yet highlighting the fragility of the only one people inhabited. He was terrified that space exploration 'is now almost monopolized by the military.' Hunger for knowledge and exploration had inspired the pioneers of outer space, Cousteau claimed. He was struck by the sincerity and friendliness of the sentiments within the fraternity of space travelers. 'They all emphasize that

Figure 9.4 French naval explorer Jacques-Yves Cousteau (1910–97) and Apollo veteran Russell Schweickart (1935–) aboard the legendary research vessel *Calypso*, September 1975. Schweickart would later call on Cousteau to help launch his Association of Space Explorers, invoking similar environmental themes.
Source: Courtesy of NASA Historical Reference Collection, NASA History Division, NASA Headquarters, Washington, DC.

our planet is one, that borderlines are artificial, that humankind is one single community onboard spaceship Earth. They all insist that this fragile gem is at our mercy and that we must all endeavor to protect it.' The exploration of space, in his view, marked the beginning of a new transnational moment in human history, 'the birth of a global consciousness that will help build a peaceful future for humankind.' Citing Teilhard de Chardin, the patron saint of eco-consciousness and inspiration for cosmic thinkers on both sides of the Cold War divide, Cousteau said that space prompted people to contemplate three infinite realms, 'the infinitely big, the infinitely small, and the infinitely complex. And from all the beauty they discover while crossing perpetually receding frontiers, they develop for nature and for humankind an infinite love.'[51] For the ASE, Cousteau provided a vital link to the past – to the heroic European age of exploration from which Cousteau emerged – but also to the new environmental and global consciousness of the Space Age. His ability to connect science and environmental protection to utopian visions of new kinds of human community made Cousteau an ideal spokesperson for the ASE, just as it had also drawn the famous science popularizer Carl Sagan (1934–96) into the ASE. Sagan was a television personality with global name recognition, and one whom the ASE founders actively cultivated as an ally.[52] Like Cousteau, he was the epitome of the cosmopolitan scientist, devoting his services to all mankind, rather than to his nation's secret military-industrial complex, and linking his professional identity to a moral agenda of disarmament and environmental protection. Even more, his status as a scientist suggested a moral position that seemed to stand in the rarified space above politics and ideology.

VI   Limits

If the ASE aimed to transform global consciousness, the Paris meeting was nonetheless a humble beginning. Staging the first meeting in Western Europe, while avoiding the fraught political atmosphere and controversy that would have resulted from a meeting in either the United States or the USSR, necessarily limited the public relations impact of former Cold Warriors embracing across ideological and national barriers. The organization also faced a dilemma arising from the contradictory nature of its mission. Organizers scrupulously claimed to be apolitical, but they aimed to create a new politics opposed to the Cold War policies of the superpowers. Schweickart expressed the contradictory apolitical politics of the group: 'Clearly it is controversial [...] but our intent is not political, except in the largest and highest sense of that word.' Yet, any program advocating the demilitarization of outer space, in the face of national policies such as Star Wars based on its militarization, had to be political and, in the view of one hostile astronaut, 'anti-American.' As one newspaper reporter put it, '[the Association] sticks strictly to the personal and philosophical, carefully avoiding politics, even though it is world politics that association members would like to see change.'[53] The inability to confront the political nature of the entire enterprise was perhaps the greatest limitation encountered by the group.[54] One proposal was to focus on purely technical matters, such as diverting an asteroid from hitting earth or developing a space rescue capability, yet this assumed that technology and politics were separate realms, a questionable claim. Another approach was to seek out scientists and the idea of scientific investigation as a supposedly neutral territory beyond national interests and political ideology, recruiting European members with scientific expertise as payload specialists and enlisting the aid of highly visible scientists and global popular-culture figures such as Jacques Cousteau and Carl Sagan. But the result of such efforts was as much to politicize science as to depoliticize space exploration. The Soviet participants – especially in the wake of Reagan's promotion of SDI – seemed to understand far better than their American and European counterparts that pretending things were not fraught with political meaning was political precisely because it promoted the political status quo. During negotiations in 1984 in Cernay to create the group, the Soviet side agreed that 'we must preserve the image of oneness for future generations, but we also cannot lose contact with the reality of political struggle.'[55] Space technology clearly performed military and political work, enhancing military and strategic objectives even while being presented as part of a peaceful scientific program of unlocking the secrets of the cosmos.[56] Patrick Baudry (1946–), the second Frenchman in orbit who flew aboard a NASA Shuttle in 1985, encountered the political and military nature of American Space Shuttle technology. His flight provided a first test of Star Wars systems, reflecting the dirty little secret of the supposedly civilian NASA Shuttle missions that many carried classified military experiments.[57] Perhaps that was why Baudry, who was recruited to attend the first ASE meeting in Paris in October 1985, ultimately declined the invitation, since NASA had made clear that active astronauts should not participate in the ASE if they wanted to fly again on a Space Shuttle mission. That threat put the lie to the idea held by many astronauts that they were free agents, supposedly unlike their Soviet cosmonaut counterparts.[58]

Political differences were not the only barrier space travelers confronted in orbit. Another limitation was conveying the emotional response that many on both sides of the Cold War had experienced in outer space. During the first ever space walk in 1965, Leonov described the experience as a 'rebirth' that made him acutely aware that 'earthlings are members of one family.' In the early 1970s some American astronauts claimed that the trip into outer space 'sharpened their religious awareness.' A guide to the Houston Space Center in 1973 called this the 'astro-effect.' Like Mitchell, astronaut James Irwin became interested in finding connections between science and spirituality, setting up an organization called 'High Flight Inc.,' headquartered in Colorado Springs, to 'share a message of scientific exploration and religion.'[59] The officially atheist Soviets, of course, did not profess faith in Jesus Christ; but they did view their experience through a spiritual lens – preferring the Soviet ideological emphasis on universal brotherhood and a greater awareness of the need to be better stewards of the earth. Oleg Makarov (1933–2003), a psychologist and cosmonaut in the Soviet space program, noted that cosmonauts were often men of few words. But when he listened back to transcripts of conversations between cosmonauts and mission control, 'within seconds of attaining Earth orbit, every cosmonaut, without exception, be they a dry, reserved flight engineer or a more emotional pilot, uttered the same sort of confused expression of delight and wonder.'[60] Schweickart recalled that 'by going up into space and back down, we were able to leapfrog the normal barriers which exist to touch and experience each other directly. This magic must be shared.'[61]

But how? The main way was through ASE's 1988 publication *The Home Planet*, which its publisher, Addison-Wesley, said 'is a book we really expect to put us on the trade publishing map.' The book received 200,000 orders in advance of its publication. ASE members had made its publication a priority since their first meeting in Paris in 1985. They hoped that the high-quality photographs and accompanying comments from space travelers testifying to the consciousness-transforming experience of spaceflight – 150 space photographs and 200 comments – would make believers of its readers, especially given the relatively limited impact of the mere news that the ASE had formed and former enemies were now embracing and feasting together. The book was republished simultaneously in 1989 across Europe – in Spain, Italy, Sweden, England, France and Holland, with quotes in 16 different languages representing space travelers from 18 different countries.[62]

The ASE believed the book project would create a vital link between the space enterprise and the international public, thus re-energizing public support for space exploration and rekindling utopian hopes for consciousness raising that the space travelers had experienced by flying into outer space. It was the same thought that had inspired Frank White's *The Overview Effect*, published only a year earlier, whose broad impact in both the scholarly community and wider public has been investigated by Danish scholar Thore Bjørnvig. Bjørnvig argues that White's thesis – like the tradition of cosmism that predated it by decades – provided an overall philosophy, indeed a religious outlook, that has connected the space venture to transformations in consciousness and world view – perhaps mirroring the way that the modern nation state as an imagined community in the late nineteenth and early twentieth centuries became an object of sometimes fanatical worship and faith.[63] As with White, the ASE's aim was to make the spiritual and aesthetic qualities of space exploration contagious, and thus politically potent, so that those who had not directly experienced it as space travelers could somehow appreciate it vicariously – and perhaps also be more apt to support funding for space programs.

The Home Planet is a feast for the eyes: the stunning blueness and clouds of the earth against the backdrop of utter blackness, three cosmonauts at sunrise marching toward a launch at the Baikonur cosmodrome, the Shuttle Atlantis lifting off, wispy thin clouds over southern Africa, a massive tropical storm over the Pacific Ocean, brilliant panoramas of deep blue ocean, pictures of astronauts and cosmonauts on space walks. Reviewers were thrilled by the dramatic images, calling the book 'an entrancing taste of the space experience' and focusing on the emotional and aesthetic, rather than scientific, aspects of spaceflight.[64] Given the widespread press *The Home Planet* received – and the substantial sales, along with White's book at nearly the same time – the project was certainly a public relations success for the idea that space exploration had enormous implications for the future of human civilization. The more artistically inclined cosmonauts and astronauts, such as Aleksey Leonov and Russell Schweickart, who went on a tour of the US and USSR to promote the book, believed that paeans to the beauty of the earth could launch people to a place in inner space – the grey matter between their ears – that was beyond their ideological beliefs or national allegiances.[65]

They echoed a point that White had made as well: 'Just as people who had never seen a slave could become adherents of the abolitionist cause, so can those who have never been in space support a vigorous space exploration program.' White called these people 'Terranauts' and while he realized that 'it is not possible to fully replicate the Overview Effect without going into space, similar experiences are available to us all. They can be used as foundations for personal growth and transformation [...] in support of a more peaceful, self-aware, and ecologically careful species.'[66] White's view was a common sentiment among the founders of the ASE, not to mention figures in the counter culture ethos of the 1970s that had inspired its creation. It embodied the utopian hope, also a feature of Russian cosmism, that radical estrangement from familiar space would cause a shift in thinking that would make new forms of cooperation and planetary consciousness possible.

But even among the space travelers, few became Terranauts; they stubbornly remained Frenchmen, Russians, Germans, Bulgarians and Poles. The trick, as the ASE put it, was 'respecting national interests and international relationships, while rising above their limits.'[67] Jean-Loup Chrétien envisioned a unified Europe from outer space but one that still seemed to privilege its Western and Mediterranean parts. 'Around 6 p.m. we were flying north of the Mediterranean, almost right above Marseilles,' he wrote, echoing the tendency among space travelers to remark first upon the view of their childhood home. 'I know this region well, since I lived there more than twenty years.' But then Chrétien began to expand his view, taking in all of 'France, Corsica, Sardinia, Italy, part of Spain.' He also noticed the South – but not North – of England and 'part of Germany.' He then realized that he could see a good bit of the globe, 'all the while distinguishing without difficulty the little details of the place where I was wandering on foot some weeks earlier.' The Bulgarian cosmonaut Georgi Ivanov (1940–) also sought out the comfort of his ancestral homeland when he flew into space. 'I remembered my childhood, my hometown of Lovech, the mountains, my relatives, and of course my mother.'[68] Not even the experience of space travel was powerful enough to replace national and regional identities with something more universal. Yet the experience of radical estrangement from homeland did add another layer of understanding – one that broadened the space traveler's perspective and allowed for an awareness of one's place, not only in a national and local context, but also in the greater cosmos. Ultimately, global and national identities were not mutually exclusive, just as nation states have often successfully integrated regional and local identities into the broader nation.

VII   Imaginative leaps through space

Depending on one's outlook, new technological capabilities could lead to utopia – a 'fantasy about the limits of the possible,' according to Jay Winter – or to some horrific dystopia ending in nuclear and ecological holocaust.[69] But, paradoxically, those two visions were also connected, like two sides of the same coin. The intensity with which members of the ASE pursued the idea of world peace fed off the existential threat of nuclear war and the cosmopolitical calculations that drove some of the interpreters and architects of space exploration. That threat had reached a fever pitch during the first Reagan Administration, manifested in Western and Eastern Europe over US plans to upgrade its missiles on the continent and to use space technology much more explicitly for purposes of enhancing national military power and strategic objectives. So while the end of the Cold War in the late 1980s seemed to promise new possibilities for collaboration, especially with regard to the future of the European Union (which ESA was partly designed to consolidate), it also removed one of the justifications for collaboration in outer space, that is the Cold War itself. The very things that seemed to limit human development in that era – a Europe divided by ideology, military bases, concrete walls and nuclear-tipped rockets – had stimulated utopian hopes for a renewed commitment to world peace and international cooperation, and perhaps even new kinds of global identities, through space exploration. The head of ESA in 1985 captured the spirit of the long decade of the 1970s: 'Space is open and space has no known boundaries,' Reimar Lüst (1923–) wrote. 'It is therefore one of the most ideal domains in which to cooperate across national borders, especially across national borders in Europe. [...] ESA demonstrates that we can work together very successfully in Europe, and in this way ESA is contributing to building a united Europe.' Yet, once the European Union expanded to include nearly all of Europe save Russia, space exploration, especially of the manned variety, was no longer useful, or promotable, as a tool for unification. Or rather, it had outlived its usefulness in that regard.[70]

At the same time, the late 1970s and 1980s marked a transition from the heroic and romantic phase of space exploration to a more technical and commercial one – perhaps embodied best by the collapse of communism and the passing of the distinctly non-commercial spirit of the early Space Race.[71] The Soviet newspaper *Komsomol'skaia pravda*, on 23 December 1988, may have observed the beginning of the new era with an article called 'Money in

Space.' The article, written by a veteran Soviet space journalist, lamented that it was time for the Soviets to stop giving free rides into space and charge for their services, just like the Americans. A quarter century after Gagarin's flight, 'the flurry of excitement and romance surrounding everything having to do with the cosmos has declined and today public excitement about this once loudly glorified area of our life has become quite calm. Some even angrily call cosmonautics a "Potemkin Village".'[72] The days of Interkosmos were over, when the Soviet Union gave 'selfless aid to other peoples in exploration and use of outer space for man's benefit.'[73]

Yet, in other ways the 1970s were the beginning of a new way of imagining human community beyond the nation state or even a united Europe. Ideas about using outer space to transcend traditional political divisions marked a critical juncture in transnational history – and they have not given way completely to more traditional conceptions of space as an arena for national competition. Just as maps, educational systems and newspapers in an earlier era allowed people to imagine themselves as national subjects, images of the globe from outer space permitted people to view themselves as transnational subjects – similar to how cyberspace has also promoted new virtual forms of human association. Those images, as they propagate and multiply, will be a constant stimulant to the imagination, as they were for the millions of readers of both *The Home Planet* and White's *The Overview Effect*. At the 1991 Congress of the Association of Space Explorers, held in a newly unified Berlin, the theme was 'Space Has No Boundaries.' The Congress proclaimed: 'Space exploration has given humanity a new perspective of the artificial borders that separate nations and the role that international cooperation in space plays in transcending these boundaries.'[74]

That is perhaps one of the most enduring aspects of the 1970s and its echo in the ASE, the International Space Station (ISS) and other collaborative space ventures – the imaginative leap through spaceflight that has infused the exploration of outer space with broader political significance; ideas about the need to transcend national differences and become better stewards of the planet. The broader importance of ISS as a vehicle for both exploring space and rearranging affairs back on earth was made clear in light of recent tensions between the United States and Russia in the spring of 2014. The station remained perhaps the major remaining pillar of collaboration between the two. Despite Russian and American threats to end space cooperation because of tensions over Ukraine, NASA reiterated that ISS had grown out of a tradition of cooperation initiated during the Cold War – and would continue to provide a foundation for joint efforts to gain a greater understanding of the home planet. Through the crisis, the fraternity of fliers – European, Russian and American – continued to celebrate triumphal conclusions to international missions on the steppes of Kazakhstan, the Eurasian terminus of international space collaboration and a staging ground for experiments in new forms of human association.

Notes

1. Stephan Freiherr von Welck, 'Outer Space and Cosmopolitics,' *Space Policy* 2.3 (August 1986), 200–5, here 202.
2. Dodd L. Harvey and Linda C. Ciccoritti, *U.S.-Soviet Cooperation in Space*, Miami: Center for Advanced International Studies, University of Miami, 1974, x.
3. Hoover Institution Archives, Association of Space Explorers (ASE), folder 3, 10 September 1984, Paris press release, 6 September 1984 draft of press release; Frank White, *The Overview Effect: Space Exploration and Human Evolution*, Boston: Houghton Mifflin, 1987, 100, 108.
4. Oleg Makarov, 'Preface,' in Kevin W. Kelley, ed., *The Home Planet*, London: Queen Anne Press, 1988, n.p.
5. 'Club's Entry Requirement: Orbit Earth,' *Houston Chronicle* (27 January 1986), 4.
6. Robert Poole, *Earthrise: How Man First Saw the Earth*, New Haven: Yale University Press, 2008, 8; see also his contribution, Chapter 5 in this volume.
7. Anindita Banerjee, *We Modern People: Science Fiction and the Making of Russian Modernity*, Middletown: Wesleyan University Press, 2012, 48–9.
8. Vladimir Vernadsky, *Geochemistry and the Biosphere*, New Mexico: Synergetic Press, 2007, 416.
9. Hugh Odishaw, 'International Cooperation in Space Science,' in Lincoln P. Bloomfield, ed., *Outer Space: Prospects for Man and Society*, Englewood Cliffs: Prentice-Hall, 1962, 105–23, here 106–7, 121.
10. The quote from the Gagarin poster is from a poster in my personal collection; Paul Kecskemeti, 'Outer Space and World Peace,' in Joseph M. Goldsen, ed., *Outer Space in World Politics*, New York: Praeger, 1963, 28–37, here 29.
11. Arnold L. Horelick, 'The Soviet Union and the Political Uses of Outer Space,' in ibid., 43–71, here 44.
12. Poole, *Earthrise*, 41.
13. Benedict Anderson, *Imagined Communities: Reflections on the Origin and Spread of Nationalism*, New York: Verso, 1991.
14. Mary Louise Pratt, *Imperial Eyes: Travel Writing and Transculturation*, New York: Routledge, 1992, 36.
15. Harvey, *U.S.-Soviet Cooperation in Space*, 172–3. See also Luca Follis's contribution, Chapter 8 in this volume.
16. Vitaly Sevastyanov and Arkady Ursul, 'Space Age: New Relationship Between Society and Nature,' *Space World* I-1-97 (January 1972), 31–9, here 31.
17. *Weekly Compilation of Presidential Documents, Week Ending Friday, January 31, 1969*, 190 (located in the NASA Historical Reference Collection, NASA History Division, NASA Headquarters, Washington, DC, 1/3/1 'Borman, Frank NASA Post-Apollo 8'); 'Armstrong Tells Russian Scientists U.S. and Soviet Should Cooperate in Space Projects,' *New York Times* (4 June 1970), 27C. On American overtures to the Soviets, see Thomas O. Paine Papers, Library of Congress Manuscript Collection, Washington, DC, box 23, folder 2. On doubts within the US State Department, CIA and DoD that collaboration with the Soviets was possible, see a transcribed interview with Robert Gilruth from 25 March 1975, in NASA Historical Reference Collection, 1/6/2, 'Gilruth, Robert R. (Bio.).'

18. 'Russian Report,' *Space World* L-2-134 (February 1975), 24–7, here 26.
19. See 'Vperedi – novye mezhdunarodnye polety (Forward to New International Flights),' and 'Na blago chelovechestva (For Mankind's Benefit),' *Krasnaia zvezda* 20 (23 July 1975), 1; 'Vstrecha nad zemloi (Meeting above the Planet),' *Literaturnaia gazeta* (23 July 1975), 1.
20. On military concerns about space collaboration at NASA, see Nixon Presidential Library and Archives, National Security Decision Memorandums, box H-326, NSDM-187 [1 of 2]; 'Détente Draining U.S. Technology,' *The Washington Post* (6 May 1975), B13.
21. Matthew Evangelista, *Unarmed Forces: The Transnational Movement to End the Cold War*, Ithaca: Cornell University Press, 1999.
22. V.I. Kozyrev and S.A. Nikitin, *Polety po programme 'Interkosmos,'* Moscow: Znanie, 1980, 7, 12–13, 64.
23. Guy Collins, *Europe in Space*, Basingstoke: Macmillan, 1990, 123.
24. Ibid., 124–6; 'De Gaulle Invites Soviet Science Tie,' *New York Times* (23 June 1966), 1.
25. Collins, *Europe in Space*, 1.
26. Niklas Reinke, *The History of German Space Policy: Ideas, Influences, and Interdependence 1923–2002*, Paris: Beauchesne, 2007, 472; 'ESA Aims at 2001,' *Space World* V-5-257 (May 1985), 26–9, here 26.
27. Kenneth S. Pedersen, 'The Changing Face of International Cooperation: One View of NASA,' *Space Policy* 2.2 (May 1986), 120–38, here 129.
28. Collins, *Europe in Space*, 126–7.
29. NASA Historical Reference Collection, 1/12/6 'Mitchell, Edgar D. (1966–1976).'
30. Russell Schweickart letter to Walter Cunningham, 24 May 1983, in Hoover, ASE, folder 6.
31. 'U.S. Bursts Technology Bubble for Red Scientists,' *New York News* (22 February 1980), 4.
32. 'The Last Frontier for Trust, NASA Shuns Contact with Cosmonauts,' *San Francisco Chronicle* (27 March 1988), 6. On NASA frustrations with Reagan Administration hostility toward foreign participation in the Space Shuttle program, see a Hans Mark memorandum from February 1983: NASA Historical Reference Collection, OSX/A/2 'Astronauts, Foreign'; 'First European Astronaut Criticizes Shuttle Manning,' *The Washington Post* (20 December 1983), A4.
33. Questionnaire response from Vance Brand, Joseph Kerwin to Russell Schweickart, in Hoover, ASE, folder 3.
34. Schweickart letter to Aleksey Yeliseyev, 19 September 1983, in Hoover, ASE, folder 5.
35. 'U.S. May Boycott Space-Linkup Event,' *The Washington Post* (16 May 1985), A21; 'Cosmonaut Calls for Weapons-Free Space,' *Foreign Broadcast Information Service*, USSR III.143 (25 July 1985), U1. The U.S. Office of Technology Assessment in October 1984 presented a draft argument against the value of collaborative programs with the Soviets: 'The U.S.-U.S.S.R. Intergovernmental Agreement on Cooperative Space Activities: Should it be Renewed?,' NASA Historical Reference Collection, 10/14/3, folder 15592. There was widespread opposition within the US government to the Reagan Administration's refusal to develop joint missions. See *Congressional Record – Senate*, 9 February 1984, 1276–82 for statements in favor of a Congressional Resolution entitled 'East–West Cooperation in Space as an Alternative to a Space Arms Race.'

36. Letter from James Muney to the Reagan White House, 21 September 1983, Ronald Reagan Presidential Library and Archive, Outer Space, 170000–179999.

37. Minutes of 3 March 1984 meeting at Pepsico headquarters, Purchase, NY, in Hoover, ASE, folder 4.

38. Schweickart letter to Captain Jacques-Yves Cousteau, 27 February 1984, in Hoover, ASE, folder 6.

39. Mitchell letter to Schweickart, 23 August 1984, in Hoover, ASE, folder 4; folder 6, Jack Matlock letter to Schweickart, 26 April 1984; Schweickart letter to Robert McFarlane, 11 April 1984.

40. Schweickart letter to Chrétien, 29 August 1984; Schweickart letter to Edgar Mitchell, 4 September 1984; Schweickart letter to Mitchell, 27 July 1984; Mitchell letter to Schweickart 23 August 1984; Schweickart letter to 'Astros,' 15 August 1984, all in Hoover, ASE, folder 5; folder 6, Schweickart letter to Ulf Merbold, 17 March 1984; folder 1, May 1985 Progress Report.

41. John Edwin Mroz letter to James Hickman, 13 February 1984, in Hoover, ASE, folder 6.

42. Walter Truett Anderson, *The Upstart Spring: Esalen and the American Awakening*, Reading: Addison-Wesley, 1983, 310. See also Anthony Enns, 'Satellites and Psychics: The Militarization of Outer and Inner Space, 1960–95,' in Alexander C.T. Geppert, Daniel Brandau and Tilmann Siebeneichner, eds, *Militarizing Outer Space: Astroculture, Dystopia and the Cold War*, London: Palgrave Macmillan, forthcoming (= *European Astroculture*, vol. 3).

43. 'Still on Edge, Esalen Grows Up,' *Chicago Tribune* (23 August 1987), 7.

44. Copy of brochure, 'First Planetary Congress, Association of Space Explorers, October 1985, Cernay, France,' in Hoover, ASE, folder 1.

45. Vance Brand letter to Russell Schweickart, March 1984, in Hoover, ASE, folder 3.

46. V. Sevastyanov, A. Ursal and Yu. Shkolenko, *The Universe and Civilization*, Moscow: Progress Publishers, 1981 [1979], 163, 235.

47. Draft of mission statement, in Hoover, ASE, folder 3.

48. Copy of brochure, 'First Planetary Congress, Association of Space Explorers, October 1985, Cernay, France,' in Hoover, ASE, folder 1; folder 3, questionnaire response from Alan Bean to Russell Schweickart; Sevastyanov, Ursal and Shkolenko, *Universe and Civilization*, 163, 235.

49. Marion Goldman, *The American Soul Rush: Esalen and the Rise of Spiritual Privilege*, New York: New York University Press, 5.

50. Schweickart letter to Cousteau, 10 January 1984, in Hoover, ASE, folder 6.

51. Jacques-Yves Cousteau, 'Foreword,' in Kelley, *Home Planet*, n.p.

52. Schweickart letter to Chrétien, 29 August 1984, in Hoover, ASE, folder 4.

53. Questionnaire response from Harrison Schmitt to Russell Schweickart in Hoover, ASE, folder 3; folder 4, Schweickart letter to Mitchell, 21 August 1984; folder 4, Schweickart letter to astronauts, 30 January 1984; 'Astronauts Compile Book to Support Space Exploration,' *Orange County Register* (30 May 1988), b8.

54. Schweickart letter to Mitchell, 20 August 1984, in Hoover, ASE, folder 4.

55. 'May 1985 Progress Report,' in Hoover, ASE, folder 1.

56. Questionnaire response from Gerald Carr, Joseph Kerwin, Walter Cunningham to Russell Schweickart, in Hoover, ASE, folder 3.

57. Collins, *Europe in Space*, 131–2.

58. Questionnaire response from Gordon Fullerton to Russell Schweickart, in Hoover, ASE, folder 3.
59. 'Rebirth Conveyed by "The Man Who Walked in Space,"' *Evening Bulletin* (17 May 1965), 1; 'Astronauts Find God in Space,' *San Diego Union* (19 May 1973), 9.
60. Makarov, 'Preface.'
61. Schweickart letter to James Hickman, 6 June 1983, in Hoover, ASE, folder 6.
62. Kelley, *Home Planet*. 'Book Notes,' *New York Times* (3 June 1988), C28.
63. Thore Bjørnvig, 'Outer Space Religion and the Overview Effect: A Critical Inquiry into a Classic of the Pro-Space Movement,' *Astropolitics* 11.1–2 (2013), 4–24. See also his contribution to the present volume, Chapter 7.
64. 'The View from the Outside,' *New Scientist* (8 October 1988), 57; 'Lofty Vision From Outer Space,' *Los Angeles Times* (28 October 1988), 14.
65. 'Space Mission on Earth,' *USA Today* (21 October 1988), 2.
66. Draft statement of purpose, 6 September 1984 in Hoover, ASE, folder 3; White, *Overview Effect*, 68, 70, 72–3.
67. Schweickart letter to astronauts, 30 January 1984, in Hoover, ASE, folder 6.
68. Collins, *Europe in Space*, 138; Kelley, *Home Planet*, 81, 89.
69. Jay Winter, *Dreams of Peace and Freedom: Utopian Moments in the Twentieth Century*, New Haven: Yale University Press, 2006, 3.
70. Reimar Lüst, 'Europe's Future in Space,' *ESA Bulletin* 5.44 (November 1985), 8–15, here 8.
71. For the 'romantic era of spaceflight,' see Alex Roland, 'Barnstorming in Space: The Rise and Fall of the Romantic Era of Spaceflight, 1957–1986,' in Radford Byerly Jr., ed., *Space Policy Reconsidered*, Boulder: Westview Press, 1989, 33–52.
72. S. Leskov, 'Den'gi na kosmos,' *Komsomol'skaia pravda* (23 December 1988), 2.
73. Sevastyanov and Ursul, 'Space Age,' here 37.
74. Available at http://www.space-explorers.org/congress/posters/poster7.html (accessed 1 May 2013). The web page with the ASE meeting proceedings, however, is no longer active and has been replaced by images of the posters made for each of the ASE meetings; available at http://www.space-explorers.org/collectibles.html (accessed 1 October 2017).

CHAPTER 10

Architectural Experiments in Space: Orbital Stations, Simulators and Speculative Design, 1968–82

Regina Peldszus

After the 'get up, get around and get down' objective of the Mercury and Vostok programs (1959–63), astronauts and cosmonauts had graduated from supervising automation to manually executing critical operations in Gemini (1961–66) and Voskhod (1964–66), and finally accomplishing highly complex tasks such as docking maneuvers, extravehicular activities and system recoveries in the Apollo program (1963–72). In parallel with mission complexity, spacecraft design also evolved. Initially, the main driver had been 'life and death': engineers had to ensure that there would be no asphyxiation, no depressurization and that adequate nutrition and hydration was provided to the operator in otherwise hostile natural surroundings. From a technical perspective of human systems integration, the emerging concept of habitability in space engineering and the space life sciences was considered synonymous with minimal acceptability of a vehicle environment.[1]

Yet, in the advent of long-duration missions, new mission goals and constraints in orbit meant that a number of practical and conceptual challenges came to the fore. As a function of mission elapsed time, the intricacies of humans in outer space as professional and private individuals with capabilities and limitations, and the more complex aspects of habitability were understood. Beyond traditional, quantifiable, ergonomics-related requirements, such as those of atmosphere, temperature and vibration established in early

Regina Peldszus (✉)
German Aerospace Center (DLR), Bonn, Germany
e-mail: regina.peldszus@dlr.de

© The Author(s) 2018
Alexander C.T. Geppert (ed.), *Limiting Outer Space*
European Astroculture, vol. 2
https://doi.org/10.1057/978-1-137-36916-1_10

handbooks, the more intangible psychological aspects of long-duration missions, with their behavioral and social implications, surfaced as concerns.[2] In response to these emerging requirements, the American and Soviet space programs sought the expertise of disciplines beyond engineering, including architecture, sociology, psychology and industrial design, to work on orbital space stations and experimental ground-based research simulators to mitigate the stressors of prolonged exposure to isolation and confinement. As crammed space capsules gave way to products of a more human-centered design approach, these space habitats represented not only a new generation of architecture, but were also human-made artifacts of the overall sociotechnical system of spaceflight.[3]

Synchronous to a growing interest in foresight during the 1970s on both sides of the Iron Curtain, experts agreed that the discourse on future space systems would need to involve a triad of policy-makers, technologists and the visionary repertoire of the cultural industries who looked 'beyond the technical community' at 'human concerns and desires.'[4] In the cultural and societal context, a third kind of habitat featured prominently in the post-Apollo decade alongside space stations and simulators: the conceptual equivalents of space habitats as part of sets in film productions, whose directors often tapped into authentic subject-matter expertise of aerospace-industry professionals, or consciously adopted an external vantage point. Embedded in the narrative of a fictional mission, these habitats can be retrospectively understood as products of speculative design that constitute an alternative body of materialized thought. These physical models of potential futures or alternative realities in outer space emerged from an often collaborative process that deliberately attempted to (re)configure technological systems and explore their social and political implications, conditions and consequences.[5]

This chapter charts the bandwidth of these three types of space habitats between function and foresight in the 1970s that probed the human-technology relationship by exposing the practical feasibility and conceptual desirability of spaceflight.[6] Adopting the notion of the 'long' 1970s, two demarcations are chosen to bracket the decade: 1968 as the year of the public release of *2001: A Space Odyssey*, the feature-length record of Stanley Kubrick's unprecedentedly detailed design study addressing the scenario of a fictional deep-space mission; and 1982, when the Soviet space station Salyut 7 attained the record of the longest human mission at the time, of 211 days. Between those dates, the American space station Skylab was launched and inhabited; Western Europe's Spacelab module was designed to be flown by the US Space Shuttle first in November 1983; research simulators and extreme environment analogues on both sides of the Atlantic (Tektite II and Bios-3) were operated; and habitat designs created for films produced in or by Europe, such as *Solaris*, *Eolomea* and *Alien*.[7]

As tangible backdrops within which a narrative or mission unfolded, the real, simulated or speculative habitats offered a complex reflection of the human condition in outer space, as much as the challenges posed by a hostile and remote setting. This engagement ultimately built up a body of evidence to support future crewed exploration, yet challenged the very rationale underpinning its agenda at the same time.

I   Space stations: recreating earth in orbit

If during the Apollo era the relationship between humans and technology had been addressed through the functional design of interfaces for short, highly critical operations, missions throughout the following decade were characterized by extended duration and routine.[8] Both the American and the Soviet space programs thus sought to address the more latent challenges of human experience in orbit. Space stations for both programs reflected the effort of creating not just habitable machines, but an internal environment that provided a home base during remote duty.

Formally initialized in 1965, the US Apollo Applications Program aimed at making continued use of Apollo technology in order to retain its highly specialized staff and to understand next-generation hardware needs.[9] This strategy included a crewed orbital platform, Skylab, which was visited by three crews between 1973 and 1975 (Figure 10.1). Concerned about ensuring the usability and comfort of the design, NASA commissioned a range of habitability studies, and eventually the New York-based industrial design office Loewy/Snaith Inc. was tasked to work on Skylab interiors. Loewy's team conducted a range of architectural programming and ergonomic design exercises on habitation systems that addressed longer stays in orbit of between three and twelve weeks, including clothing, storage, crew compartments with dedicated sleep restraints and a galley unit.[10] Experienced in designing for transport and defense contexts, Loewy's group concentrated on creating a sense of fundamental order in layout and configuration to enhance the interplay of human and environment. As part of this process, however, the designers' efforts also demonstrated how private activities from hygiene to food preparation and leisure would be organized and performed by the astronauts as part of housekeeping and self-management. For other leisure aspects, a number of tokens of home were provided through structural and interior features, including a window, a shower and a selection of music. While designers and engineers had provided the broader framework for basic environmental acceptability, even comfort, the crew took ownership of the habitat. They addressed their own cultural needs by observing seasonal holidays and created often humorous objects from onboard scrap materials to be used as props during televised conversations with mission control and the public; they also explored the unique in situ conditions of their temporary home by making the most of the generous interior volume of Skylab and novel forms of physical movement afforded by microgravity (Figure 10.2).[11]

Figure 10.1 Close-up view of Skylab during a fly-around, 28 July–25 September 1973. Against a backdrop of the black sky, the photograph epitomizes the isolation of the habitable platform in a hostile environment.
Source: Courtesy of NASA.

Connections to earth were forged more literally in Soviet habitat interiors during prolonged missions. As early as 1963, the Soviet space program's main design bureau Energiya, based in the Moscow area, recruited an in-house designer to work on interiors of different space habitats they were developing. The architect Galina Balashova (1931–), who was already designing warehouse and factory buildings for the organization, began to develop design concepts for spacecraft, conceptual lunar bases and, from 1976 onwards, space-station interiors for the Salyut program. Balashova's designs aimed to infuse the technical fabric of the habitat with the qualities of an earthly home through familiar furnishings and fixtures, aquarelle images of natural scenes and a pastel color scheme reminiscent of earthly landscapes. Her efforts tied in with a psychological support program implemented by medical personnel

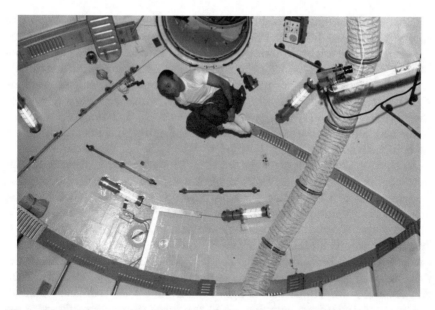

Figure 10.2 Astronaut Alan L. Bean (1932–), commander of Skylab 3, gives a play-fully physical display of the merits of microgravity by doing acrobatics in the dome area of the Orbital Workshop (30 August 1973). Space stations represented a paradig-matic change from the confined interiors of previous generations of spacecraft, yet also brought new challenges through extended inhabitation.
Source: Courtesy of NASA.

to counteract the stressors of a long-duration mission, and provided the cosmonauts with movies and resupplies of fresh fruit. Even the small plant-growth facilities included in the stations for science purposes became a wel-come pastime for the cosmonauts.[12]

Nevertheless, increased mission duration significantly affected the crews through the emotional strain, sense of risk, unusual and uncomfortable envi-ronment and the responsibility and pressure for the successful accomplish-ment of their missions. The final iteration of the Soviet space station program, Salyut 7, which was operational from May 1982 to June 1986, topped the records of the post-Apollo decade with a 211-day mission in 1982. The onboard diary that cosmonaut Valentin Lebedev (1942–) kept during this stay exemplified the technical and behavioral challenges the crews of the pre-vious decade had faced. Interpersonal issues were set off or exacerbated by conflict with mission control, a tight work schedule, equipment malfunction, cumbersome housekeeping and occasionally severe physical ailment. Beyond the many positive episodes and despite the often bizarre beauty of the space environment described by astronauts and cosmonauts alike, monotony and melancholy emerged as a key feature of orbital isolation and confinement. In fact, Lebedev detailed the issue of monotony, even boredom, in such great length that the publishers of his account later omitted several passages that regularly reiterated his emotional state. Yet, exactly these passages illustrate the shifting emphasis from purely physiological towards social elements. Isolated as a two-man crew from the rest of their peers and humanity in general, with limited resupply or rescue, apparently mundane events like watching a transmission of a high-profile football game, with celebrating teams and cheering crowds, could trigger a bout of nostalgia and homesickness that dampened, almost questioned, the overall exhilarating and rewarding experiences of a mission.[13]

In contrast, human spaceflight activities in Eastern and Western Europe were concerned with much shorter mission experiences, as they evolved in tandem with their respective partners. From 1977 onwards, the Interkosmos program allowed nationals of the Warsaw Pact and other states, starting with the Czechoslovak Vladimír Remek (1948–), the first European in outer space, to take part in missions to Salyut. Soviet findings on space research were exchanged annually also with West European parties, such as the European Space Agency (ESA) and, prior to its formation in 1975, its precursor entities. NASA also sought cooperation with European multilateral space organizations, culminating in the first major cooperative venture, the Spacelab program.[14] In planning since mid-1972, Spacelab was a reusable scientific laboratory to be housed in the payload bay of the US Space Shuttle. The habitat module's German prime contractor, Entwicklungsring Nord (ERNO), maintained a dialogue with US contractors and benefited from the experience of astronauts, including former Skylab crewmembers, who spent time at ERNO's site in Bremen or took part in week-long simulations stateside.[15] By 1976, the consortium had constructed a range of low-fidelity mock-ups to troubleshoot human-machine interface issues, and full-scale, functional engineering modules to evaluate layout, verify subsystems, train the crew and showcase developments to European industry and public.[16] Although Spacelab predominantly served as a platform for experiments and mainly catered for the work-related needs of the crew, habitability and psychosocial aspects were considered important. By galvanizing different stakeholders from life science and habitat technology in Europe, the seed for a continued interest in longer-duration habitability issues was laid out, and Spacelab later became the blueprint for the European Columbus module of the International Space Station (ISS).[17]

Essentially, the acute remoteness experienced by Apollo astronauts would not be replicated in distance during the 1970s and early 1980s, but the duration of this experience had been increased by orders of magnitude. Although the Apollo forays to the surface and the orbital far side of the moon had been much more distant, lower-earth orbit, if inhabited for weeks or months on end, became a remote place despite its proximity to earth. Distance became temporal, with the ensuing monotony and routine punctured by occasional high risks. With greater remoteness came autonomy, even emancipation from the ground, an acute sense of isolation and the realization of dependency on the lifeboat of a habitat. Physically tied to 'home' in orbit, space station designs had by necessity a functional priority and, arguably more by intention, an aesthetic emphasis on home comforts. As complex machines they had to provide basic earthly conditions in a radically unusual environment; however, in their often literal export of earthly design paradigms and an interpretation of home that was bordering on the pastoral more than the radical, they turned attention inward, back towards earth. The designers succeeded in reconfiguring the logic of the engineered vehicle environment to make it usable in incremental steps to allow acclimatization to a new environment, but did not engage in further experimentation by abandoning references to earth and zooming into the idiosyncrasies of orbit and the space environment, at which all astronauts and cosmonauts marveled. Space habitats essentially remained within the pull not only of physical gravity instead of venturing out further, but also firmly in the grip of the routine banalities of living and working in lower earth orbit.

II   Recreating space on earth: simulators and extreme environments

The character of remoteness was explored in more depth in human research laboratories and simulators, to understand the behavioral and technical dimensions of life in isolated and confined environments. For Skylab training and ground-based in-flight troubleshooting, NASA devised high-fidelity emulators of physical space conditions, culminating by the end of the 1970s in a growth in life-science research activities and operational simulation capabilities in space agency, academic and industrial contexts in the United States. Much of the emphasis was, however, placed on aspects of human performance in anticipation of shorter future Shuttle missions.[18] Meanwhile, the Soviets concentrated on the longer behavioral narratives of extended isolation. In pursuit of a 'vision of a biological future,' they developed extensive expertise in the development and operation of closed-loop life support systems with a range of designated research complexes, as life sciences and biomedical engineering became important fields within the human spaceflight domain. The isolation facility at the Institute of Biomedical Problems (IBMP) at the Academy of Medical Sciences in Moscow had already hosted studies in the range of 366 days with small crews in a spacecraft model consisting of chambers mimicking accommodation and greenhouse models, where three test subjects lived on water and oxygen regenerated from human waste and dehydrated food. In view of further future autonomy, by 1972, the closed-loop environment simulator Bios-3 at the Institute of Biophysics in Krasnoyarsk, the Siberian wing of IBMP, was constructed.[19]

Bios-3 was internationally hailed as the most advanced closed-loop environment simulator of its time. It resembled a small space station featuring compartments for algae and plant production, quarters for a crew of three, a toilet, galley, airlock and a control room. In 1972 and 1977, two studies of 180 days each were undertaken in the facility, where three participants interacted with the outside through video, phone, voice-loop and windows. Although physically successful in achieving a closed and quasi-closed loop of recycling water, air and waste, the range of provisions and activities onboard, including a television, newspapers, food preparation and correspondence, did not prevent the daily routine being perceived as monotonous.[20] Again, designers were brought in: at IBMP, a small team of architects and industrial designers worked with psychologists to translate scientific recommendations on countermeasures for monotony into design strategies to be implemented by engineers. Their prototypes included virtual reality devices showing natural themes to be used during tedious exercise sessions, and light, sound and imagery recreating seasonal and daily rhythms to be included in the simulator environment.[21]

Through organizational and architectural design features, research and training simulators recreated the situational conditions of spaceflight to elicit an authentic effect on the human, rather than the actual environment itself.[22] The facilities were usually located in (sub)urban settings and could not replicate the full extent of the orbital environment: the participants experienced isolation and confinement but, with the exception of hyperbaric chambers, were not exposed to an existential risk of life. Other facilities were deliberately placed in actual hostile settings analogous to outer space in order to expose the participants to a real operational environment and to explore habitability needs in isolated research contexts. As IBMP grew in the 1970s to a staff of about 2,000 scientists and technical personnel, it expanded from space research to supporting human missions, and translated the expertise from its aerospace portfolio toward extreme environments in the polar, high-altitude and undersea areas for industrial and exploration-related endeavors.[23]

Also NASA's Human Behavior Program, besides studying submarines, contributed to the installation of stationary habitats – Tektite I and II designed by General Electric – on the seabed off the coast of Florida for behavioral research, with crews of four on missions ranging up to 20 days in 1969 and 1970.[24] Using closed-circuit television, video and audio recording, two observers colocated with the mission control team on shore monitored crew activities around the clock, examining conversations, meal behavior, hygiene and sleep patterns. Their records of one of the ten two-week Tektite II missions read like a taxonomy of the banality of life in a remote duty post. The crew's antics of bantering and their gossip about the onshore support personnel sharply contrasted with the conveyed ordinariness of housekeeping, discussing choices of music or cleaning diving equipment. While several aspects of the habitat were in later evaluations found to be of poor usability, the men dubbed their abode 'Tektite Hilton' for its 'air-conditioned comfort,' good service from top-side and pleasant amenities, feeling 'guilty' about the relative luxury of their logistics and musing on the psychosocial challenges of astronauts.[25] The self-conscious discussion of the design and operational concept of their own simulator was a reflection on the firsthand experience of a possible future exploration context. Subsea habitats, which proliferated globally during the 1970s, were in fact seen as worthy and valid outer space analogues by most strategists, psychologists and astronauts themselves and, indeed, offered a real and potentially dangerous environment. Nevertheless, the inhabitants, with a sense of both humility and disappointment, never believed they matched the sense of risk of the space environment, nor the ardor of operating in it. Only having the high-octane Apollo missions as a recent reference, the aquanauts thought their experience was inferior – while, actually, their social dynamics reflected or even preempted the interpersonal issues observed on Salyut, Skylab and later stations. This sentiment also resonated with their Russian colleagues: simulation participants there wrote vignettes and stories about life in the isolation chamber, one of them lucidly pointing out to his readers that he was 'no superman but an ordinary person like yourself.'[26]

Other exploration settings were even more extreme, such as the Antarctic, whose research bases functioned in much the same way as isolation chambers. Teeming with civilian research teams and military personnel stationed there on temporary or permanent bases established in the 1950s and 1960s, Antarctica combined remoteness, isolation and the ensuing autonomy with an acutely extreme setting and limited accessibility for resupply or rescue. There were already harrowing accounts of psychosocial grievances of US soldiers deployed for overwinter stays of 14 months and more, which NASA recorded during a high-profile fact-finding trip, although interest was initially placed on logistical constraints and technical requirements. A Russian surgeon had even conducted an emergency self-appendectomy at the extremely isolated research base Vostok that was also used for behavioral research to inform the Salyut program.[27] Antarctica became a key analogue for the psychological research of the major space programs, and later also for European psychologists and designers who could draw on the experience of the various European bases established between the 1950s and late 1970s, including the French Dumont d'Urville Station (1956) and the Henryk Arctowski Polish Antarctic Station (1977). An entire landscape functioned as a laboratory for the psychological challenges of spaceflight and was systematically mined for evidence.[28] It was necessary to simulate outer space on earth in order to understand the different technical requirements and behavioral issues connected to hostile surroundings, confinement of personnel, limited resupply or rescue during the long, dark winter period and a complete dependency on technology for shelter and science operations. In the course of this process, however, extreme environments still remained areas of exploration in their own right, rather than serving as mere simulation media. As lunar ambitions on both sides of the Atlantic atrophied, territories such as the deep sea or the Antarctic high plateau entailed real, acute logistical challenges that were arguably as exotic as outer space and worthy of tackling, but did not involve the extent of investment necessary for orbital outposts. They fore-shadowed that a future of exploration would be difficult or even impossible to scale up beyond an aggregation of seasonal or permanent remote bases, and reinforced that comparably extreme experiences could be obtained closer to the ground.

III   Speculative design: space habitats on film

Still further extreme experiences could be thought-experimented on within the framework of another kind of 'simulator' architecture – film sets. If the real space stations of the 1970s featured 'depressing' experiences, the picture was also bleak in film.[29] Rather than reconciling real needs and constraints through the design of stations or testing future technology and behavioral narratives through simulation, the speculative habitation designs in films of the post-Apollo era explored the implications and values of these technologies. Until the late 1950s, space-related feature films including documentary fiction had predominantly depicted affirmative futures through their architectural designs. Fastidiously catering to a textbook range of habitability requirements, they allowed their heroic and wholesome crews to dwell in clean, functional quarters as they undertook international joint ventures of interplanetary exploration and science.[30] After the Apollo program, however, a subtle demystification of fiction set in, with astronauts operating within a realistic context that largely mirrored actual constraints of spaceflight.[31] This was foreshadowed in Kubrick's 1968 film *2001: A Space Odyssey*, both from a conceptual perspective, but also due to its topicality, as it considered a possible post-Apollo agenda of interplanetary flight. Kubrick's effort essentially consisted of directing a team of former and present aerospace professionals through a detailed foresight exercise, which positioned its architectural, interior and industrial design as much in the canon of space systems as in that of the cultural industries. The production stressed particularly the future relationship of human and technology: The film's protagonists onboard the *Discovery* spacecraft integrate seamlessly with their habitation system; their composure and sense of duty mirror its clean interiors. Both human and technology are reliable elements of the overall mission, until the perfect interplay disintegrates to reveal the limitations of both.[32]

Not all productions had the means or inclination to achieve the same degree of technical rigor and finesse as Kubrick's machinery was able to deliver. Still, particularly Eastern Europe produced a limited number of films that offered quite idiosyncratic proposals of space systems and human activity. Just a few years after *2001* and before Skylab was operational, *Solaris* (1972) pushed the limits of the human experience off towards grim anticipated troubles in extended space missions. The film solemnly uses progressively more disorderly habitats as markers of the deteriorating mental state of the crew: The space station *Solaris* that orbits the eponymous planet is in a state of neglect and chaos; corridors are littered with waste and accommodation is blocked with makeshift barricades.[33] A reiteration of this process occurs with the protagonist, a visiting psychologist and his personal quarters, which at first appear serene with a bed wrapped in a protective transparent cover, but are then destroyed in parallel with the breakdown that the protagonist himself undergoes. Other crew members have fastened small paper strips to an air vent that emit sounds mimicking those of leaves rustling in the wind; an onboard, wood-paneled library is filled with almost self-consciously selected artifacts of terrestrial high culture. These are sophisticated quarters that fulfill aesthetic, intellectual and practical wants of high-profile scientists, who are, in the course of their mission, reduced to a forlorn group of disturbed individualists, swerving between rationalizing the unexplained events they experience on the one hand, and treasuring the almost pathetic reminders of home on the other: the space habitat is no substitute.

Like *Solaris*, the joint East German, Soviet and Bulgarian 1972 production *Eolomea* commences its plot on earth. In its beginning sequence, the protagonist, a skilled captain, now recalcitrant freighter pilot, moves along a deserted beach. His career has reached a nadir of senselessness, and he voices his frustration about poor working conditions on remote duty, but also about a lack of motivation and ambition in view of a complete alienation from higher rationales for going into outer space in the first place: 'Never cosmos again, never!' His vow functions as the placeholder for the outlook of an entire new generation that only reluctantly still regard themselves as part of a collective. They explore outer space not in highly motivated multicultural teams of cosmonauts, but as groups of individuals with their own personal needs and desires, who nevertheless contribute to a complex institutional machine in the grander programmatic framework of a mission.[34] This is particularly evident when the protagonist talks to an older colleague who still holds dear the previous ideals of scientific exploration and technology, represented by the remote asteroid base that the two inhabit – a small geodesic capsule that appears rugged but ramshackle. Deployed there as a consequence of threefold disciplinary action, the cosmonauts continue to violate regulations, ignoring a no-fly rule and consuming smuggled alcohol by a Christmas tree made from instrument scraps. When their witticisms tip over into serious conversation, they vent their underlying discontent, even cynicism about the organizational bureaucracy of spaceflight, its poor planning, shortages in supply and the lack of programmatic rationale of the scientific council that administrates the program.[35] As portraits of possible alternative versions of the present rather than distant futures, the two films, *Solaris* and *Eolomea*, feature habitats whose inhabitants are prisoners of earthly constraints. Outer space has become not a solution or destination, but yet another setting for the depths of the human psyche to unravel within a literal and programmatic void, and, at that, one where human beings appear distinctly out of place. Space may be extreme, but has long been betrayed by distinctly down-to-earth issues that stubbornly follow the protagonists to even the farthest depths of the solar system: rules, regulations and bureaucracy, coupled with a lack of home comforts and meaningful reward.

However, while the dystopian markers of *Eolomea* and *Solaris* were still concealed within a once utopian quest of progress, exploration or science, the outlook concluding the decade of the 1970s implies a future not only sub-verted by individual members of a spacefaring organization, but by organiza-tions themselves. In *Alien* (1979), the corporate owner of the mining ship *Nostromo* holds sinister motives that go beyond the pragmatism of the crew's initial reluctance to react to a call for help, a decision that is only reverted since contractual terms dictate it. Filmed in Shepperton Studios in the United Kingdom, the production featured an interlocking full-scale set, much like a simulator. The habitat design is aging but functional and neutral; the crew is clad in a sartorial hybrid of informality and corporate service. These are no explorers or heroes with ideals, but cargo-haulers who do not encounter outer space with awe, inquisitiveness or even doubt. Immune to its former allure, but acutely aware of its dangers, they consider the setting a wasted ter-rain to be traversed on long journeys – a nuisance usually spent in resource-saving stasis. In its depiction of a corporate setting, stripped of all but the most banal personal markers, *Alien* was highly topical at the time, as the mining of the lunar surface, prospecting of asteroid resources, orbital man-ufacturing and processing capabilities were recognized as urgent technology projection needs. While *2001* still ventured out into space, now space evolves from being a place of interest, longing and adventure to a work environ-ment and a threat, mitigated and tolerated at best, but mostly loathed. Space exploration evolved into space exploitation, as corporations had entered deep space, even more so than the distinct groups of scientific disciplines or national cultures.[36]

Unlike most of their predecessors, the designs of the 1970s provided the backdrop to scenarios that surpassed clear-cut conflicts, emergencies or achievements. Instead, life in outer space was no longer posited as an adven-ture, a collaborative achievement or a personal accomplishment, but as a state to be endured, circumvented or, if possible, abandoned entirely. These themes resonated in the 1970s, whether as basic human conditions indi-vidualism or skepticism about organizations in view of industrial deregula-tion, declining faith in government programs and large-scale public efforts, and the emergence of contractors.[37] The possible motives for interplanetary travel that some stakeholders had put forward – wonder, beauty, romance, novelty and adventure – had been replaced with largely materialistic driv-ers: prestige, defense and the development of new industries.[38] An arc of motives was manifested by post-Apollo speculative design: of the attraction of outer space where no efforts were spared to make spaceflight happen even if it involved uncertainty; to a hiatus of reluctance, where humans went not necessarily for science, but to pay for a simple life on earth; and to finally succumbing to only going into outer space and braving its dangers if that yielded a significant return. After being a place of interest, where epic narratives of the human condition unfolded, outer space ceased to be a place only; it became mere territory and resource. This reflected the process occurring in other environments on earth, of the formerly blank spots on maps that were explored or 'conquered' for political or athletic competition, which then became the subject of large-scale science and international interest, to finally be transformed into commercial or geopolitical resources. However, even if it involved staying alive in a hostile terrain just to prospect its riches, this thematic development also did away with much of the conquering rhetoric in favor of an idea of caution. Outer space was no more a wonderful or lush place where humans cooperated peacefully, and literally and figuratively expanded humankind's horizons. Instead, it was lethal, dark, cold and distant, where mistakes could cost a life and everything was calculated and systematized, where the cost of supporting – or the value of – an individual had to be precisely measured down to the single breaths of the correct oxygen mixture, and human presence was justified in terms of investment.

IV   The post-Apollo period as design catalyst

The post-Apollo period was a crucial time for Europe, as important capabilities and relationships developed around the founding of ESA in 1975, and through the participation in Interkosmos in the Eastern Bloc. Yet, with no own operational experience beyond the growing partnerships with the two dominant space players of the United States and the Soviet Union, European space efforts had no reason to consider longer duration mission needs, beyond the technical and organizational requirements for shorter stays in the Spacelab module or visits to Salyut. The fledgling development of crewed capabilities was framed with space futures that painted a much more critical picture of outer space. Intriguingly, European stakeholders were actively partaking in spaceflight, but their speculative engagement also questioned this effort at the same time.

While it was a kick-starter for Europe, the post-Apollo period appeared as a slow burner for the established space powers. Fueled by the Space Race and spurred by an exploration agenda, the Apollo program had achieved tremendous, disruptive successes in a short time. Now the teething problems of extended missions had to be tackled incrementally. In that sense, the 1970s represented a crucial incubating period for designing and using a new paradigm of habitable platforms in new mission scenarios that pushed the Apollo experience further, not in terms of more distant or spectacular explorations, but dramatically increased duration. Spaceflight was consolidated as a hostile setting and a complex technological infrastructure. As in other frontiers of the 1970s, such as the nuclear industry where accidents and incidents occurred on both sides of the Iron Curtain, the potential problems of interacting with highly complex systems came to the fore, alongside the conceptual reframing of such infrastructures as sociotechnical systems, by recognizing the entwined nature of human, technological and organizational elements.

Through a string of experimental platforms, post-Apollo operations, research and speculation leveled the optimistic expectations and momentum of the 1960s. A more realistic view on the human condition in outer space emerged unglamorously, shrewdly and piecemeal, and diagnosed the uncomfortable, mundane, humbling and perilous qualities of spaceflight. The Russian and US space agencies responded to the practical challenges by integrating other non-technical disciplines and galvanizing interdisciplinary teams to chart and formalize the field of habitability. This served as lead-time for important subsequent developments. As strands of systems engineering, life sciences and design converged in the course of the decade, the human was understood to be a critical element of the system with unique capabilities and limitations. However, the process of painstakingly stepping closer to a previously proclaimed future of a sustained presence in outer space revealed how demanding the feasibility on the relatively small scale of orbital habitation for research purposes, and how ambivalent its desirability, would be. Concerning utopia, it was a caveat of a decade: less optimistic, more cautious, but certainly twofold in that it offered both a pragmatic promise of a possible future in outer space – more precisely, in lower earth orbit – and a debunking of the greater rhetoric of dominion over the solar system.[39]

As representatives of a discipline, industrial designers and architects themselves had to integrate into engineering-dominated organizations whose constituents had not fully internalized the design ramifications of focusing on the very element they sought to sustain for extended periods – the human operator. Even though the designers' involvement was sought as early as the beginning of the 1960s and championed at the top level, on the workshop floor and in planning meetings, both Balashova's and Loewy's teams faced similar obstacles. Their struggle for acceptance was anchored in the traditional engineering paradigm of emphasizing purely technical functionality; if there was a focus on the human among the hardware and software of a space system, it was in the sense of being 'wetware' to be kept physically alive in order to perform tasks set by principal investigators, technologists and policy-makers. In direct opposition to their engineering colleagues, the industrial designers saw machines – and Salyut, Skylab and Spacelab were essentially complex habitable machines – as an extension of, and structured around, the human. Besides asserting their alternative view on systems, the designers had to justify their involvement, and in reflecting on their experience, gave an insightful and somewhat scathing account of communication barriers between the different professional cultures.[40]

The three habitat types discussed in this chapter echoed similar themes of the human condition, but approached them from different perspectives. Space stations were explicitly designed at the time, or analyzed in hindsight, as 'experiments in design,' 'prototype steps,' 'links with the future' or 'analogues' of extreme settings to develop operational capabilities for extended space missions. In that sense they were somewhat affirmative of a future in outer space by clinging to a presence in orbit as their systems reconciled technical and behavioral constraints of orbit with the desire to operate within this environment.[41] In contrast to serving practical foresight and feasibility purposes, speculative designs flipped this around. They projected *in medias res* of a space future or alternative present, taking it as a premise, only to then chip away at it by introspecting and uncovering 'bottom-up' issues which, finally, questioned the rationale of outer space as a desirable future. In lieu of assuming its conventional remit of problem-solving as in the traditional engineering or industrial-design context, speculative design as manifested in film elicited reflection on values related to sociotechnical systems, provided impulse for discussions where answers were less important than questions, and functioned as an anchor for conjecture. This bifurcation from previously optimistic stances into the affirmative and non-affirmative challenged the raison d'être of the three types of architectural artifacts: to what extent were real space stations required if their experience could be replicated in a simulator or envisioned in speculative scenarios? Or, vice versa, did hypothetically inhabited habitats at the periphery of conceptual development become obsolete in the course of real space operations and developments in the space sector?[42]

If understood as complementary mechanisms, both remained critical for a comprehensive debate on space futures throughout the post-Apollo decade. The rubrics of fact and speculation were not contrary but interlocking, even ingrained within each other on a continuum of design that yielded powerful heuristic potential.[43] Just as the actually inhabited space stations became more refined and complex and a source of operational evidence, speculation retained its relevance through its evolution into critical commentary. On the one hand, speculation could exhaust the range of behavioral extremes and potential dangers to be observed or induced in stations or simulators, since they were unbound by the technical constraints of operations and the research ethics of simulators. On the other hand, the more complex operations became and the more diverse the possible options about a future in outer space, the more varied foresight methodologies were required to trial these options. Theoretically, this may have rendered contemporary thinking about space futures more robust: the divergence into pragmatic vision of the future through complex operational feats, and into undermining this future at the same time, were both necessary to figure out feasibility and strategy in parallel; for there could not be an applied effort without a larger agenda, nor an agenda or vision without a tentative impression of fundamental technical and behavioral practicalities. Whether this mutual reciprocity was really rooted in practice remains to be understood; but in reading the three habitat types on par, this chapter contributes toward the symmetry in the debate of technology artifacts in outer space by articulating their relevance and logic as architectural foresight tools.

Ultimately, as many of the original design practitioners were absorbed back into sectors outside spacecraft design or space industry, the fate of the physical instantiations of their designs discussed in this chapter is of interest. Much like their simulator and set counterparts, many designs developed for actual space programs never made it into orbit due to a variety of scheduling, organizational and financial constraints, facing the dead-end of program closure or routine rejection during the design iteration process. Whether intricate and technically functional or made to full scale from cardboard in an assembly hall of an industrial prime contractor or production studio, they served their purpose as 'architectural experiments.' They disintegrated in factory backyards or storage in daily view of their designers; they were disassembled or recycled, or replicas and records archived in museums and corporate archives. In the case of real operational hardware, much of it was recycled or burned up during atmospheric re-entry. In a sense, if viewed through the lens of a possible future beyond earth orbit, they all remained speculations, and their demise was their final common denominator.

Yet, the narratives they hosted, and which were captured in technical reports, scientific papers and feature films, did prevail, and with them, expertise in, and views on, human systems integration. While as artifacts, the habitat designs were products of processes of both problem solving and critique, also as an overall discipline, design straddled a twofold methodological remit of being the executive agent of, and the interpreter between, the technical disciplines and the behavioral sciences. The concrete design efforts undertaken in the post-Apollo period were a marker of the increasing sensitivity to multi-faceted human requirements that transcended the seemingly incontrovertible priority of mere physical survival, and in the process catalyzed the literal unearthing of empirical and experimental evidence for routine operations in orbit that was extensively reviewed during the 1980s to inform future space systems. The resulting systematic advocacy for the human in a technical system was crucial not only for space hardware in the 1970s, but also for its legacy. As the behavioral sciences and some disciplines of design, particularly of space architecture, began to emerge more formally within the aerospace domain, their expertise propagated prolifically, if sometimes inconspicuously, far into the systems of the Russian MIR and the International Space Station, and to next-generation simulation and analogous facilities.[44] Although the vision of an interplanetary future was curtailed in the post-Apollo period, or at least remained practically elusive, in its candor the experience of designing and operating habitable platforms in real and speculative space signified a reality-check that, for the first time, offered an invaluable baseline of hard evidence and critical impulse in parallel, and provided opportunity to comprehensively map out, understand and discuss futures in outer space.

Notes

1. The colloquial summary of Mercury objectives was articulated by David M. Jones, 'Near Future Applications,' in Lawrence L. Kavanau, ed., *Proceedings of the National Meeting of the American Astronautical Society, 21–23 February 1966, San Diego, California*, Washington, DC: American Astronautical Society, 1966, 121–4, here 122. Alongside aspects such as survivability and training, habitability emerged as one of the key concepts of the wider field of Human Systems Integration in aerospace; see Harold R. Booher, ed., *Handbook of Human Systems Integration*, New York: Wiley, 2003 and NASA, *System Engineering Handbook (Revision 1)*, Washington, DC: NASA, 2007, here 45; Joseph P. Loftus Jr., 'An Historical Review of NASA Manned Spacecraft Crew Stations,' in Gerald P. Carr and Melvin D. Montemerlo, eds, *Aerospace Crew Station Design: Proceedings of the Course Given at the International Center for Transportation Studies (ICTS), Amalfi, Italy, 19–22 October 1983*, Amsterdam: Elsevier, 1984, 3–22.

2. NASA, *Habitability Data Handbook*, vol. 2: *Architecture and Environment*, Houston: NASA Johnson Space Center, 1972; Thomas M. Fraser, *The Intangibles of Habitability During Long Duration Space Missions*, Washington, DC: NASA, 1969.

3. Thomas P. Hughes, 'The Evolution of Large Technological Systems,' in Wiebe E. Bijker, Thomas P. Hughes and Trevor Pinch, eds, *The Social Construction of Technological Systems: New Directions in the Sociology and History of Technology*, Cambridge, MA: MIT Press, 2012, 45–76, here 45.

4. Matthew Connelly speaks of a general and institutional preoccupation with the future; see idem, 'Future Shock: The End of the World as They Knew It,' in Niall Ferguson, Charles S. Maier, Erez Manela and Daniel J. Sargent, eds, *The Shock of the Global: The 1970s in Perspective*, Cambridge, MA: Harvard University Press, 2010, 337–50, here 338. The role of science fiction as conducive to technology development was acknowledged by Lee L. Farnham, 'Overview of Space Systems Technology,' in Charles J. Donlan, ed., *A Collection of Technical Papers: Proceedings of the AIAA/NASA Conference on Advanced Technology and Future Space Systems, 8–10 May 1979, Hampton, VA, 1979*, Washington, DC: AIAA, 1979, 1–4, here 1. For applied examples, see, for instance, Arthur C. Clarke, *Report on Planet Three and Other Speculations*, London: Pan Books, 1972, and Wernher von Braun, *Das Marsprojekt: Studie einer interplanetarischen Expedition*, Frankfurt am Main: Umschau, 1952 (*Project Mars: A Technical Tale*, Burlington: Apogee Books, 1971). Clarke widely articulated the merits of cultural practices such as speculation in the context of actual space efforts, including those related to a post-Apollo agenda, for instance in 'Epilogue: Beyond Apollo,' in Neil Armstrong, Michael Collins and Edwin E. Aldrin Jr., eds, *First on the Moon: A Voyage*, London: Michael Joseph, 1970, 371–419, here 386, and Arthur C. Clarke, *The Promise of Space*, London: Hodder & Stoughton, 1968, xix.

5. The notion of speculative design emerged as a critical perspective on industrial design in the context of corporate future studies in the late 1990s to address 'notopia [...] driven by real human needs' through mock-ups or 'design models' that simulate an operational artifact. See Anthony Dunne, *Hertzian Tales: Electronic Products, Aesthetic Experience, and Critical Design*, 2nd edn, Cambridge, MA: MIT Press, 2005, 89; and idem and Fiona Raby, *Design Noir: The Secret Life of Electronic Objects*, Basel: Birkhäuser, 2002, 6. For a

recent discussion of speculative design in contrast to specification-driven engineering or industrial design in technology contexts, see Carl DiSalvo and Jonathan Lukens, 'Towards a Critical Technological Fluency: The Confluence of Speculative Design and Community Technology Programs,' in *Proceedings of the 8th Digital Arts and Culture Conference: After Media – Embodiment and Context, December 12–15, 2009, Irvine, CA*, Irvine: University of California, 2009, 1–5, here 1–2; David A. Kirby, 'The Future is Now: Hollywood Science Consultants, Diegetic Prototypes and the Role of Cinematic Narratives in Generating Real-World Technological Development,' *Social Studies of Science* 40.1 (February 2010), 41–70; and Ian Miles, 'Stranger than Fiction: How Important is Science Fiction for Futures Studies?,' *Futures* 25.3 (April 1993), 315–21. For an examination of the cultural practice of using manifested models to support arguments in spaceflight, see William R. Macauley, 'Crafting the Future: Envisioning Space Exploration in Post-War Britain,' *History and Technology* 28.3 (September 2012), 281–309.

6. In 1971 ultimate space visions still encompassed 'the exploration of the planets, to expand man's capability to live and work in space,' in view of 'extend[ing] man's domain throughout the solar system'; see Wernher von Braun, 'NASA's Plans for the 70s,' in Michael Cutler, ed., *International Cooperation in Space Operations and Exploration: Proceedings of the American Astronautical Society's 9th Goddard Memorial Symposium, 10–11 March 1971, Washington, DC*, San Diego: Univelt, 1971, 7–20, here 7–8.

7. *2001: A Space Odyssey*, directed by Stanley Kubrick, USA 1968 (Metro-Goldwyn-Mayer); *Solaris*, directed by Andrey Tarkovsky, USSR 1972 (Creative Unit of Writers & Cinema Workers/Mosfilm/Unit Four); *Eolomea*, directed by Herrmann Zschoche, DDR/USSR/BUL 1972 (VEB DEFA-Studio für Spielfilme/Künstlerische Arbeitsgruppe Berlin Mosfilm/Boyana); *Alien*, directed by Ridley Scott, UK/USA 1979 (Brandywine/Twentieth Century Fox).

8. For an in-depth examination of human-machine interfaces of the Apollo onboard computer, see David A. Mindell, *Digital Apollo: Human and Machine in Spaceflight*, Cambridge, MA: MIT Press, 2008.

9. Summer C. Berger, *Interview with G.E. Mueller*, transcript, NASA Oral History Program, Washington, DC: NASA, 1998, 16; Courtney G. Brooks, Roland W. Newkirk and Ivan D. Ertel, *Skylab Chronology: The Story of the Planning, Development, and Implementation of America's First Manned Space Station*, Washington, DC: NASA, 1977.

10. Raymond Loewy and William Snaith, *Habitability Study, Earth Orbital Space Stations: Final Report*, Washington, DC: NASA, 1972; Caldwell C. Johnson, *Skylab Experiment M487 Habitability and Crew Quarters*, Houston: NASA Johnson Space Center, 1975.

11. For instance, in order to celebrate Christmas, the crew of Skylab 4 fabricated a Christmas tree made from leftover food cans onboard, which they filmed for a holiday transmission. The notion of Skylab as a 'home' has been explored widely, including by Henry S.F. Cooper Jr., *A House in Space: The First True Account of the Skylab Experience*, London: Granada, 1977; David Hitt, Owen Garriott and Joe Kerwin, eds, *Homesteading Space: The Skylab Story*, Lincoln: University of Nebraska Press, 2011; and Ben Evans, *At Home in Space: The Late Seventies into the Eighties*, New York: Springer, 2012.

12. Balashova's work on habitats and spacecraft received significantly less attention than her US design colleagues at the time. A recent oral history and comment on her archive is provided by Andrey Kaftanov, *Architecture in Orbit: First Realizations, Galina Balashova Works 1963–1990*, Geneva: EMA Foundation and Union of Moscow Architects, 2000; and Maryina Demydovets, 'Auftrag mit Seltenheitswert: Galina Balashova gestaltet Räume für das Leben in einer anderen Welt,' in Philipp Meuser, ed., *Architektur für die russische Raumfahrt: Vom Konstruktivismus zur Kosmonautik. Pläne, Projekte und Bauten*, Berlin: DOM, 2013, 132–59; Grujica S. Ivanovich, *Salyut – The First Space Station: Triumph and Tragedy*, New York: Springer, 2008; Dennis Newkirk, *Almanac of Soviet Manned Space Flight*, Houston: Gulf Publishing, 1990, 111.

13. Valentin Lebedev, *Diary of a Cosmonaut: 211 Days in Space*, New York: Bantam, 1983 (2nd edn 1990). Lebedev's account – particularly his responses to boredom through humor – is discussed by Debbora Battaglia, 'Coming in at an Unusual Angle: Exo-Surprise and the Fieldworking Cosmonaut,' *Anthropological Quarterly* 85.4 (Fall 2012), 1089–106; the experience of monotony and different coping strategies, including practical jokes of Lebedev, Bean and others, are examined in a comparative review of onboard astronaut diaries by Regina Peldszus, Hilary Dalke, Steven Pretlove and Chris Welch in 'The Perfect Boring Situation – Addressing the Experience of Monotony during Crewed Deep Space Missions through Habitability Design,' *Acta Astronautica* 94.1 (January 2014), 262–76.

14. Roger M. Bonnett and Vittorio Manno, *International Cooperation in Space: The Example of the European Space Agency*, Cambridge, MA: Harvard University Press, 1994, 82. See also Tilmann Siebeneichner's contribution, Chapter 11 in this volume.

15. Eleanor H. Ritchie, *Astronautics and Aeronautics, 1976: A Chronology*, Washington, DC: NASA, 1984.

16. Henning Tolle, 'European Participation in the Post-Apollo-Program,' in Lewis Larmore and Robert L. Gervais, eds, *Proceedings of the Space Station, Safety and International Sessions of the American Astronautical Society 16th Annual Meeting, 8–10 June 1970, Anaheim, CA*, Washington, DC: American Astronautical Society, 1970, 385–400, here 391–2; Bernard Deloffre, 'Spacelab Program – Status Review,' in William J. Bursnall, George W. Morgenthaler and Gerald E. Simonson, eds, *Proceedings of the American Astronautical Society 21st Meeting, 26–28 August 1976, Denver, Colorado*, San Diego: Univelt, 1976, 33–54, here 40; Hans E.W. Hoffmann, 'Spacelab Program – Technical Status and System Capabilities,' in ibid., 55–79, here 63, 70.

17. Günther Seibert, *Spacelab and its Utilization for Biomedical Experiments*, Paris: European Space Agency, 1975.

18. Joan Vernikos, 'Life Sciences in Space,' in John M. Logsdon, ed., *Exploring the Unknown: Selected Documents in the History of the U.S. Civil Space Program*, vol. 6: *Space and Earth Sciences*, Washington, DC: NASA, 2004, 267–440, here 281. Sheryl L. Bishop, 'From Earth Analogs to Space: Getting There from Here,' in Douglas A. Vakoch, ed., *Psychology of Space Exploration: Contemporary Research in Historical Perspective*, Washington, DC: NASA, 2011, 47–78, here 67; Saul B. Sells, 'The Taxonomy of Man in Enclosed Space,' in John E. Rasmussen, ed., *Man in Isolation and Confinement*, Chicago: Aldine, 1973, 280–303; Glenn Bugos, *Atmosphere of Freedom: Sixty Years at the NASA Ames Research Center*, Washington, DC: NASA, 2000.

19. John Allen, *Biosphere 2: The Human Experiment*, London: Penguin, 1991, 12–15; Brian Harvey and Olga Zakutnyaya, *Russian Space Probes: Scientific Discoveries and Future Missions*, New York: Springer Praxis, 2011, 327.
20. Frank B. Salisbury, Josef I. Gitelson, Genry M. Lisovsky, 'Bios-3: Siberian Experiments in Bioregenerative Life Support,' *Biological Science* 47.9 (1997), 575–85.
21. Konstantin Savkin, 'The Space of Cosmic Experiences,' *Project Russia* 15.1 (April 2000), 49–51; Andrey Kaftanov, 'From Science to Fiction: Igor Kozlovs Works 1970–1995,' ibid., 25–31.
22. For a synthesis on their operation and scientific value from the life sciences perspective, see Dietrich Manzey and Christophe Lasseur, *Review of European Ground Laboratories and Infrastructures for Sciences and Support Exploration: Ideal Facility for Psychological Research*, Noordwijk: ESA, 2002; and Viktor M. Baranov, *Simulation of Extended Isolation: Advances and Problems*, Moscow: Slovo, 2001.
23. Oleg G. Gazenko, 'Milestones of Space Medicine Development in Russia: Establishment and Evolution of the Institute of Biomedical Problems,' *Journal of Gravitational Physiology* 4.3 (March 1997), 1–4.
24. Matthew J. Ferguson, *Use of the Ben Franklin Submersible as a Space Station Analog*, Huntsville: NASA, 1970. For field research guidelines and a detailed transcript of in situ observations, see, for instance, Nicholas Zill, *Project Tektite Behavior Observer's Handbook*, Washington, DC: NASA, 1968, and David P. Nowlis, *Tektite II Habitability Research Program: Day-To-Day Life in the Habitat*, Huntsville: NASA, 1972; for a comparative overview on habitat design, see James F. Parker and Martin G. Every, *Habitability Issues in Long-Duration Undersea and Space Missions*, Arlington: Office of Naval Research, 1972.
25. Nowlis, *Day-To-Day Life*, here 21, 28, 35.
26. Yury Gagarin and Vladimir Lebedev, *Psychology and Space*, Moscow: MIR Publishers, 1970, here 266.
27. Vladislav Rogozov and Neil Bermel, 'Auto-appendectomy in the Antarctic: Case Report,' *British Medical Journal* 339.12 (2009), 1420–2. A technical visit by high-level NASA representatives is recounted in Ernst Stuhlinger and Fred I. Ordway, *Wernher von Braun: A Biographical Memoir*, Malabar: Krieger, 266–7, and in Wernher von Braun, 'Space Man's Look at Antarctica,' *Popular Science* 190.5 (1967), 114–16, 200.
28. Richard S. Lewis and Philip M. Smith, *Frozen Future: A Prophetic Report from Antarctica*, New York: Quadrangle, 1973; Gro M. Sandal, Gloria R. Leon and Lawrence Palinkas, 'Human Challenges in Polar and Space Environments,' *Reviews in Environmental Sciences and Biotechnology* 5.2–3 (July 2006), 281–96. For comprehensive outlines on habitability, see Albert A. Harrison, Yvonne A. Clearwater and Chris P. McKay, eds, *From Antarctica to Outer Space: Life in Isolation and Confinement*, New York: Springer, 1991; and for a review of evidence, Jack W. Stuster, *Space Station Habitability: Recommendations Based on a Systematic Comparative Analysis of Analogous Conditions*, Mountain View: NASA Ames Research Center, 1986, which was later expanded to *Bold Endeavours: Lessons from Polar and Space Exploration*, Annapolis: Naval Institute Press, 1996 by the same author. For the meta-notion of Antarctica highlighting the limits of exploration, see Stephen J. Pyne, 'The Extraterrestrial Earth: Antarctica as Analogue for Space Exploration,' *Space Policy* 23.3 (August 2007), 147–9.

29. Evans, *At Home in Space*, here 284.
30. *Doroga K Zvezdam (Road to the Stars)*, directed by Pavel Klushantsev, USSR 1957 (Leningrad Popular Science Film Studio); *The Conquest of Space*, directed by Byron Haskin, USA 1955 (Paramount Pictures); and *Der schweigende Stern (First Spaceship on Venus)*, directed by Kurt Maetzig, DDR/PL 1960 (VEB DEFA-Studio für Spielfilme/DEFA Gruppe Roter Kreis/Film Polski/Iluzjon), whose ideological dimension is discussed by Stefan Soldovieri, 'Socialists in Outer Space: East German Film's Venusian Adventure,' *Film History* 10.3 (1998), 382–98.
31. Brian Stableford, 'Space Flight,' in John Clute and Peter Nicholls, eds, *The Encyclopedia of Science Fiction*, London: Orbit, 1993, 1135–6, here 1135; Peter Krämer, *2001: A Space Odyssey*, Basingstoke: Palgrave Macmillan, 2010, 74.
32. Arthur C. Clarke spoke of a 'virtual symbiosis' with the spacecraft; see idem, *2001: A Space Odyssey* [1968], New York: Roc Books, 2000, 189. See also Volker Fischer, 'Designing the Future: On Pragmatic Forecasting,' in *2001: A Space Odyssey – Exhibition Catalogue*, Frankfurt am Main: Deutsches Filmmuseum and Deutsches Architekturmuseum, 2004, 102–19; Herb A. Lightman, 'Filming *2001: A Space Odyssey*,' in Mario Falsetto, ed., *Perspectives on Stanley Kubrick*, New York: Hall, 1996, 149–57, here 149. The involvement of NASA representatives and their visit to the set of *2001* is described by Regina Peldszus, 'Speculative Systems: Kubrick's Interaction with the Space Industry for *2001*,' in Tatjana Ljujic, Peter Krämer and Daniel Richards, eds, *Stanley Kubrick: New Perspectives*, London: Black Dog, 2015, 198–217.
33. The idea of deteriorating quarters to illustrate the disposition of the crew is also employed in *Dark Star*, directed by John Carpenter, USA 1974 (Jack H. Harris Enterprises).
34. Compare the notion of introspection and the 1970s as 'Me Decade'; Niall Ferguson, 'Crisis, What Crisis? The 1970s and the Shock of the Global,' in idem et al., *Shock of the Global*, 1–21, here 2, 13.
35. Note, however, that the protagonist eventually succumbs to the requests of his colleagues to partake in a mission that will outlast his lifetime; for an analysis, see Dolores L. Augustine, *Red Prometheus: Engineering and Dictatorship in East Germany, 1945–1990*, Cambridge, MA: MIT Press, 2007, 228–30.
36. Jerry Grey and Christine Korp, eds, *Space Manufacturing III: Proceedings of the 4th Princeton/ AIAA Conference, 14–17 May 1979*, New York: AIAA, 1979, iii; Anthony J. Calio, 'Opportunities for Space Exploitation to Year 2000,' in Charles J. Donlan, ed., *A Collection of Technical Papers: Proceedings of the AIAA/ NASA Conference on Advanced Technology and Future Space Systems, 8–10 May 1979, Hampton, VA, 1979*, Washington, DC: AIAA, 1979, 581–7, here 581.
37. This occurred across the parties of the Cold War, see, for instance, Thomas Borstelmann, 'The Shock of the New,' in Ferguson et al., *Shock of the Global*, 351–4; Bruce L. Shulman, *The Seventies: The Great Shift in American Culture*, Cambridge, MA: Da Capo Press, 2002, xii, xv; Charles S. Maier, 'Two Sorts of Crisis? The "Long" 1970s in the West and the East,' in Hans Günter Hockerts, ed., *Koordinaten deutscher Geschichte in der Epoche des Ost-West-Konflikts*, Munich: Oldenbourg, 2004, 49–62, here 49.
38. Arthur C. Clarke, 'Man and Space,' in Eric Burgess, ed., *Voyage to the Planets: Proceedings of the American Astronautical Society's 5th Goddard Memorial Symposium, 14–15 March 1967, Washington, DC*, Washington, DC: American Astronautical Society, 1968, 9–22, here 12.

39. Compare the characteristic 'controversial decade' in Hartmut Kaelble, 'The 1970s: What Turning Point?,' *Journal of Modern European History* 9.1 (2011), 18–20.
40. For the experience of Loewy's team, see Fred Toerge and Charles A. O'Donnell, 'Space Station Habitability: Its Form Relationship to Man,' in Lewis Larmore and Robert L. Gervais, eds, *American Astronautical Society, Annual Meeting on Space Stations, 16th, Anaheim, CA, June 8–10*, Washington, DC: American Astronautical Society, 1970, 163–71. The analogy of 'wetware' is successfully employed by behavioral scientist Sheryl Bishop when discussing the role of the human element in space systems with engineering colleagues.
41. B. John Bluth and Martha Helppie, *Soviet Space Stations as Analogs*, Washington, DC: NASA, 1987; Evans, *At Home in Space*, 129; Robert R. Gilruth, 'Manned Laboratories in Space,' in S. Fred Singer, ed., *Proceedings of the 2nd International Orbital Laboratory Symposium at the 19th International Astronautical Congress, 18 October 1968, New York*, Dordrecht: Reidel, 1968, 1–10, here 2, 10; Walter F. Burke, 'U.S. Industrial Views on NASA's Plans for the 70s,' in Michael Cutler, ed., *Proceedings of the American Astronautical Society's 9th Goddard Memorial Symposium, 10–11 March 1971, Washington, D.C.*, Washington, DC: American Astronautical Society, 1971, 37–47; William B. Taylor, 'Applications of Saturn-Apollo Systems,' in Lawrence L. Kavanau, ed., *Proceedings of the National Meeting of the American Astronautical Society, 21–23 February 1966, San Diego, CA*, Washington, DC: American Astronautical Society, 1966, 125–52, here 127.
42. For these two questions, Peter Krämer and Alexander Geppert are thanked.
43. Alexander C.T. Geppert, 'European Astrofuturism, Cosmic Provincialism: Historicizing the Space Age,' in idem, ed., *Imagining Outer Space: European Astroculture in the Twentieth Century*, Basingstoke: Palgrave Macmillan, 2012, 3–24, here 21 (= *European Astroculture*, vol. 1). Brian Stableford, 'Near Future,' in *The Encyclopedia of Science Fiction*, 856–8, here 856; Wade L. Huntley, Joseph G. Bock and Miranda Weingartner, 'Planning the Unplannable: Scenarios on the Future of Space,' *Space Policy* 26.1 (February 2010), 25–38.
44. Katherine J. Dickson, *Space Human Factors Publications: 1980–1990*, Washington, DC: NASA, 1991; Mary M. Connors, Albert A. Harrison and Faren R. Akins, *Living Aloft: Human Requirements for Extended Spaceflight*, Mountain View: NASA Ames Research Center, 1985. For an overview on the nascent discipline of space architecture, see Brent Sherwood, 'What Is Space Architecture?,' in A. Scott Howe and Brent Sherwood, eds, *Out of this World: The New Field of Space Architecture*, Reston: American Institute of Aeronautics and Astronautics, 2009, 3–6. Current simulators and analogues include the underwater habitat NEEMO operated by NASA; the Antarctic research station Concordia, which is used as analogue for the behavioral dimension of long-duration missions by ESA; the Mars500 simulation chamber at IBMP; and many initiatives of other space agencies or learned societies in areas as diverse as the polar regions, volcanic areas, the subsea, and laboratory environments in government research facilities.

CHAPTER 11

Spacelab: Peace, Progress and European Politics in Outer Space, 1973–85

Tilmann Siebeneichner

'Currently the Space Shuttle is emerging in the US and the Europeans are building the Spacelab. Both devices herald the beginning of a new era in manned spaceflight,' Hermann Oberth (1894–1989), Germany's well-known pioneer of rocketry and astronautics, wrote enthusiastically in 1976 about 'Europe's laboratory in space.' Developed by the European Space Agency (ESA) and designed to take the scientific exploration of outer space to a new level, it represented Western Europe's contribution to the American Post-Apollo Program (PAP).[1] The Spacelab was an integral component of the Space Shuttle, together technically termed the Space Transportation System (STS), and Oberth was not the only one who would connect their development to a new era in the exploration of outer space. This era was cast as one in which not only the permanent presence of human beings in the heavens was well within reach, as Oberth went on to assure his readers, but as one that would also see Europe rise up to become a major space power.[2] Its participation in the PAP, having formally begun in 1973, would not only ensure Europe's entry into manned spaceflight but would also close the frequently invoked 'technological gap' that separated the United States and Western Europe and, it was hoped, raise the latter's reputation in the world.[3] Forty years later, the Shuttle – although out of service since 2011 – still enjoys an iconic status among space enthusiasts and scientists alike.[4] Its counterpart, the Spacelab, however, seems almost forgotten.

Tilmann Siebeneichner (✉)
Zentrum für Zeithistorische Forschung, Potsdam, Germany
e-mail: tilmann.siebeneichner@fu-berlin.de

© The Author(s) 2018
Alexander C.T. Geppert (ed.), *Limiting Outer Space*
European Astroculture, vol. 2
https://doi.org/10.1057/978-1-137-36916-1_11

Although being hailed as an important step toward the permanent use of outer space in its time, present-day observers have a more critical view regarding Europe's 'masterpiece in space.' Failing to attract significant scientific and commercial interest and being used far less than projected, this 'outsized thermos bottle' – as it was popularly characterized by *Time* magazine in 1983 – nowadays is regarded first and foremost as a prestige project of little scientific and technological significance. ESA officials ridiculed the Spacelab as 'Europe's most expensive gift to the US since the Statue of Liberty,' but former director Reimar Lüst (1923–) insisted: 'ESA had to pay the price of Spacelab to acquire the basics of manned spaceflight.' However, with Western Europe joining the PAP there was more at stake than mere space politics, as this chapter demonstrates. As German historian Helmuth Trischler has argued, it is 'precisely the politically tinged character of manned spaceflight' that leads into the heart of Spacelab's historical significance.[5]

Upon completion of the Apollo Program, the high hopes of the pioneering days of the Space Age gave way to outer space realpolitik, reflecting a reintensifying Cold War confrontation on earth from the late 1970s onwards.[6] The two superpowers increasingly considered outer space as vital to their national security interests, and when US President Ronald Reagan (1911–2004) announced his Strategic Defense Initiative (SDI) in the early 1980s, outer space became a battlefield in its own right.[7] Facing ongoing criticism for turning the heavens into a dystopian sphere, European participation in the PAP thus had an important symbolic dimension. Of course, financial motives also played a role, with NASA facing shrinking budgets after its successful lunar landings. But from a European perspective, the Space Age truly took off in the 1970s with the appearance of ESA on the scene and its envisioned entry into manned spaceflight, echoing what military historian Alex Roland has called the 'romantic era of spaceflight.'[8] Thus, having the Western European space community onboard was attractive primarily because it offered the US government an opportunity to demonstrate to the world that it was still committed to the utopian aspects of space exploration.

Beginning in 1973 and focusing on the cultural hopes and political goals that accompanied the development and use of Spacelab, it became clear that this program was only moderately successful in countering the 'American challenge' and establishing Europe as a major space power.[9] Echoing the utopian dimension of astroculture, Spacelab primarily and most importantly served as a symbol for peace and progress.[10] Its early success – as a research platform used by scientists from various nations – and its ultimate failure – as an important step toward a massive human presence in outer space – thus transcend mere space politics and offer insight into cultural mentalities and political dynamics of the late Cold War.

I   Peace

'In these troubled times when our newspapers are full of reports of violence and tragedy, it is reassuring to be able to draw attention to human achievements,' Erik Quistgaard (1921–2013), then director of ESA, told the public during the Spacelab Flight Unit Delivery Ceremony in Bremen in November 1981. 'The success of the NASA Space Shuttle flights has been one such achievement and, on a different scale, the European Spacelab programme is another,' Quistgaard emphasized.[11] Developing the Spacelab was indeed no small technological achievement for Europe, but its historical significance can only be fully understood when taking into account those 'troubled times.' Clearly, Quistgaard was referring to the various political, ecological and economic crises that shattered the planet throughout the 1970s, paving the way for rather pessimistic perceptions of humanity's future as illustrated by the influential study *Limits to Growth*.[12] Quistgaard was possibly alluding to the heavens as well. Despite the Apollo-Soyuz Test Project, the first joint US-Soviet spaceflight in July 1975 which had gained much attention as a symbol of the policy of détente, outer space was increasingly perceived as a sphere of permanent crisis and confrontation. Rumors about the development of so-called killer-satellites and anti-satellite weapons (ASATs) made worldwide headlines in the late 1970s, reflecting what Russian observers would first term a 'military-technical revolution' (MTR).[13] Arguing that computers, space surveillance, long-range missiles, communications systems and information technology as well as their integration into conventional forces were creating a new way of waging war, this MTR promoted the 'revolutionary' synthesis of new technologies, military systems and organizational and operational adaption which relied increasingly on satellites and other space-based assets.[14] By the early 1980s, outer space had become a predominantly dystopian sphere, dominated by killer-satellites and laser guns, as the cover of the German weekly news magazine *Der Spiegel* suggested (Figure 11.1). This cover anticipated what Ronald Reagan would affirm some 15 months later with his Strategic Defense Initiative: In depicting existing spacecraft such as the American Space Shuttle and the Soviet Soyuz engaged in outer space combat, with the shuttles intercepting one Russian satellite and destroying another, it gave its readers a vivid expression of how warfare in outer space might play out in a near future.

While the Soviets had indeed armed some of their space stations with a revolver canon in the 1970s, the conception of laser canons used to attack and destroy enemy spacecraft was still far ahead of its time.[15] Nevertheless, mixing fact and fiction to generate powerful impressions of outer space warfare was widespread, as illustrated by the highly successful 1979 James Bond-movie *Moonraker*, in which the Space Shuttle played a prominent role.[16] Albert Broccoli (1909–96), the film's producer, had originally intended to adapt *For Your Eyes Only* but chose *Moonraker* instead, due to the rise of

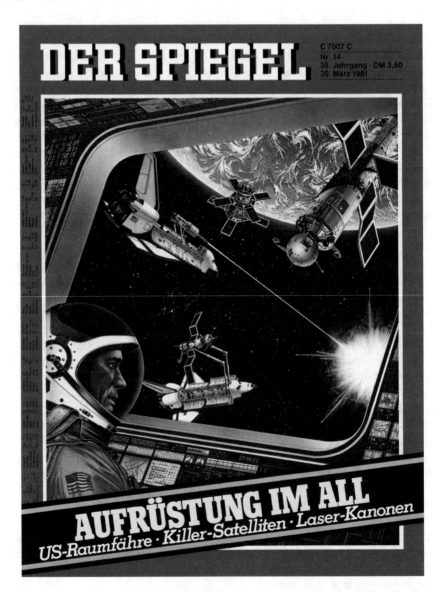

Figure 11.1 'Arms Build-up in Outer Space: US Space Shuttle, Killer Satellites, Laser Guns.' A military confrontation between US and Soviet spacecraft in low-earth orbit as illustrated by the West German weekly *Der Spiegel* in 1981.
Source: *Der Spiegel* 35.14 (30 March 1981).

the science-fiction genre in the wake of the *Star Wars* success. *Moonraker* began 'where all the other Bonds end' – an obvious reference to *Star Trek: The Original Series* (1966), which took its cast to 'where no one has gone before.' In *Moonraker*, Her Majesty's secret agent is confronted with a villain, Hugo Drax, who plots to wipe out the world population and to substitute it with a master race.[17] To this end, Drax creates a giant space station that in the course of the movie is attacked by US spacecraft. Although critics have assessed the movie as 'the series' most outlandish entry,' its spacecraft were clearly modeled on the (then yet to fly) Space Shuttle, leading Broccoli to promote *Moonraker* in 1979 by claiming that it was not about science fiction but about science fact.[18]

The Space Shuttle was the centerpiece of the American PAP, a reusable spacecraft that would be 'sufficiently flexible and versatile to respond to policy changes, take advantage of technological discoveries and generally operate on an ad hoc basis when that becomes desirable,' as an early assessment promised.[19] A 'sufficiently flexible' spacecraft that could 'operate on an ad hoc basis' responding to sudden policy changes did not necessarily imply a future military utilization but neither did it rule out such an option. When US President Richard Nixon (1913–94) announced the PAP in January 1972, he did not comment on the military implications of the Shuttle's design and purpose. However, as space historian John M. Logsdon has recently emphasized, Nixon was not an advocate of continued space exploration as the PAP officially promised. Rather, he was inclined to emphasize the Shuttle's potential value for national security issues. As early as 1969, a joint Department of Defense (DoD)/NASA study stressed its use for 'the interception and inspection of objects in space.' The report noted that 'future unknown satellites could operate for days or weeks, posing a threat ranging from intelligence gathering to delivery of a nuclear weapon' and suggested that 'a national ability to intercept, inspect, and determine the purpose of (as well as destroy, if necessary) unknown satellites is vital.'[20] In stressing the Shuttle's capability to 'intercept, inspect [...] as well as destroy' other spacecraft, this report not only envisioned a scenario quite similar to the one the *Spiegel* would depict eleven years later. It also demonstrates how much military interest would inform and influence the Shuttle's design and development.

Although well before SDI, the overlap of military and civilian ambitions within the PAP received considerable attention throughout the 1970s and made the Soviets believe that the Shuttle was in effect an anti-satellite weapon.[21] Former Gemini astronaut Thomas Stafford (1930–) fueled these fears in a speech he gave to Congress in 1981, stating that 'NASA developed the Shuttle but it is not obvious that they are best equipped to operate it – operation of the Shuttle by the Air Force is certainly a possibility.'[22] Spacelab, on the other hand, did not seem to offer such possibilities. As a reusable and modular laboratory housed in the cargo-bay of the Shuttle, it was officially designed to 'support a wide spectrum of missions for peaceful purposes,' as stated in the Memorandum of Understanding (MOU) that NASA and ESA's predecessor, the European Space Research Organization (ESRO), signed in July 1973. The Americans would continue to stress this point in due course.[23] When then US Vice President George H.W. Bush (1924–) welcomed the first Spacelab to the United States in February 1982, he praised it as a 'tool of peace.' Emphasizing that the European contribution was strictly for science's sake, Bush was quoted declaring 'before the first astronauts left the moon, they left a plaque that said "We came in peace for all mankind." Now man is returning to space once again, this time with all the tools for peace.' While Bush referred to the lunar landings in order to substantiate the peaceful intentions of the PAP, Spacelab Program Director Johannes Ortner (1933–) addressed more recent criticism that the United States faced for the large role that the military played in their space effort. Teaming up with the United States, he assured the audience, would boost the exploration of outer space for the benefit of peace: 'We're proud to say the European Spacelab is designed for peaceful purposes.'[24]

The idea for a Sortie Can or Sortie Module – as the Spacelab was originally called, referring to the modular nature of the laboratory that would allow for a variety of uses depending on the nature of a specific mission – evolved as the funding for a possible space station was pushed into the 1980s. Negotiations over Europe's role in the PAP had been difficult and disappointing. Originally very ambitious, these negotiations had been progressively restricted in scope.[25] Having the Europeans build the Spacelab was a compromise – neither was there any real need for it nor did specific plans exist on how to use it. As a European delegation reported in July 1972, the motives behind the American offer 'were purely political and commercial or technical factors had practically no influence.'[26] Rather, having the Europeans participate was meant to demonstrate the United States' willingness to cooperate internationally, and to promote progress and peace in the use of outer space.

To this end, ESA seemed the perfect partner. In contrast to both the two superpowers and individual member states which made full use of the dual-use-character of space technology, the European space community advocated only the peaceful use of outer space and strongly promoted multilateral projects in space.[27] Since various ESA member states were neither part of NATO nor of the European Economic Community (EEC), the organization represented to some degree a 'third way' in space exploration that sought to avoid classical Cold War borders. The establishment of ESA in fact resulted directly from the negotiations about European participation in the PAP. The European space community had been in tatters when confronted with the US offer to join the program, mainly due to the failure of its 'Europa'-rocket.[28] In two package deals of December 1972 and July 1973, all existing European space-related organizations such as the European Launcher Development Organization (ELDO), ESRO and the Conférence européenne des télécommunications par satellites (CETS), were transformed into a single institution, knitting together the different expert communities of space science, launcher technology and satellite development, while paving the way for a more effective cooperation in the exploration of outer space.[29]

The US offer, however, prompted mixed responses within the newly established space community. Some member states, especially France, opposed investing in a program in which it clearly played a junior role and depended on US support for regular access to outer space. Instead, they advocated the development of independent launch capabilities.[30] Others, with Germany in the lead, felt that the choice Europe faced was either an independent space effort of limited and retrograde character or cooperation with an endeavor that was at that time considered to be the most advanced and which promised to reduce the much discussed technology gap between Europe and the United States.[31] In the end, it was agreed to pursue both the development of independent launch capabilities and participation in the PAP, reflecting a new strategy within the European space community that opted for assigning the technoscientific leadership and ultimate responsibility for a certain project to one country rather than dividing the work equally among all member states.[32] Thus, with the development of independent launch capabilities led by France, Germany became the driving factor behind Europe's participation in the PAP.

The German government was most eager to jump at the opportunity offered by the US government for reasons transcending mere space politics, as a then senior official in the Federal Ministry for Research and Technology (Bundesministerium für Forschung und Technologie, BMFT), Wolfgang Finke, recalled in 1998: 'The German government looked for opportunities to demonstrate her attachment to the Western camp and especially her reliance on the United States without jeopardizing her new policy toward the USSR and her neighbors in Eastern Europe.'[33] Obviously, for the federal government, participation in the PAP was not merely an end in itself but an opportunity to take a stand in contemporary Cold War politics. Central to West German thinking was the need to reassure its Western allies that it was a reliable ally even as it opened up toward the Eastern bloc.[34]

Due to its exposed situation in Cold War Europe, Germany proved an ideal fit as it was known to be most determined in the preservation of peace, both in Europe and on a global scale. The German peace movement was strong, advocating the abolition of nuclear weapons from West Germany and Europe in general, and it would vehemently oppose NATO's Double-Track Decision in 1979.[35] Having the Germans onboard, however, also represented some sort of *Vergangenheitspolitik*, that is, the complicated process of overcoming the Nazi past. As is well known, German engineers had pioneered the development of rockets, resulting in the infamous 'vengeance' weapons.[36] This led German post-Second World War research to distance itself from any military space activities after 1945, promoting peaceful and multilateral projects instead.[37] Nixon, on the other hand, is reported to have said that the prospect of letting a German (or a Japanese, for that matter) fly onboard the Space Shuttle appealed to him since it would symbolize a renewed friendship and signal a joint future.[38] US media were keen to stress 'the unity in exploration' when reporting about the Spacelab, a perspective that would be echoed by the Federal Minister for Research and Technology, Heinz Riesenhuber (1935–): 'Peace results from collective work, not from the prevention of war. This is exactly why spaceflight is a contribution to the preservation of peace.'[39]

II Progress

A stamp devoted to Spacelab indicates how significant the West German government considered this project (Figure 11.2).[40] This stamp was issued as early as 1975 as part of a stamp series dedicated to industrial and technological products. By choosing to dedicate a stamp series to industrial and technological products six years prior to the Space Shuttle's maiden flight and even eight years prior to that of the Spacelab, the German government intended to demonstrate its progressive outlook and intentions. Symbolically reaching for the stars, the Spacelab stamp was not only one of the first stamps issued, but due to its value – 40 Pfennige – it was also one of the most widely used stamps well into the 1980s. Designed by Beat Knoblauch and Paul Beer, more than two billion Spacelab stamps were issued, making it one of the most reproduced stamps of the entire series.

Modeled on an early NASA sketch of the STS (Figure 11.3), the German stamp not only featured the Spacelab but showed the Space Shuttle as well. Remarkable, however, is a small but significant difference: while the NASA sketch featured the national insignia which attributed the different parts of the STS to the institutions that developed them – Spacelab is marked with the caption ESRO while the caption NASA can be spotted on the Shuttle's tailplane. These are missing in the German stamp.

Although cooperation with NASA was intended to strengthen ties to the United States and within Europe, this stamp suggested that Spacelab was an exclusively German product. Of course, the German government had agreed to bear 55 percent of the financial costs incurred and Spacelab was assembled by Bremen-based VFW-Fokker/ERNO, but it was by no means a German product alone. More than 40 companies from all over Europe were involved in a project that was backed by Italy (15.6 percent), France (10.3 percent) and the UK (6.5 percent) as well.[41]

Cultural historians consider stamps as 'cultural business cards,' useful and important in conveying central ideas and orientations of value-based

Figure 11.2　The West German stamp dedicated to the 'Weltraumlabor' (space laboratory), issued in 1975 as part of the stamp series 'Industrie und Technik' (industry and technology).
Source: Courtesy of Deutsche Post/ Bundesministerium der Finanzen.

Figure 11.3 This NASA sketch of the Space Transportation System from the early 1970s depicted its two main components, the Shuttle and the Spacelab, and attributed them to NASA and ESRO, respectively.
Source: Courtesy of NASA.

communities such as the nation.[42] In the German Democratic Republic, 83 stamps devoted to (manned) spaceflight were issued between 1957 and 1990, promoting the progressiveness of the German 'socialist' state.[43] It is certainly possible that the Spacelab stamp was meant to counter such claims, reclaiming 'progress' for the Federal Republic. Along these lines, it not only introduced what Quistgaard would substantiate during the Spacelab Flight Unit Delivery Ceremony in 1981: an exceptional human achievement that promised peace and progress. The stamp also illustrated the federal government's claim for fame, reflecting on the one hand the little regard it obviously had for its European partners also involved in the project and on the other the high esteem and ongoing fascination attributed to human spaceflight.

Interestingly enough, concrete purposes connected to Spacelab were never discussed in detail in Germany. 'The Federal Government is convinced that [...] the future use of manned orbital systems is an important step towards the comprehensive utilization of spaceflight for research, technology and economic objectives,' the Federal Ministry for Research and Technology stated in 1978, and showed itself highly satisfied that the government was part of what was deemed the 'most progressive spaceflight-project of the future.'[44] As the stamp indicates, spaceflight was key to the self-understanding of German governmental elites and their modernist narratives of progress. When addressing Spacelab, German government representatives regularly referred to it as a 'key technology.'[45] Although the term implied that definite fields of application did not exist for Spacelab, it set a strong normative tone.[46] Why discuss specific purposes when its social relevance was perfectly clear, then undersecretary Volker Hauff (1940–) had asked rhetorically in 1977?[47] Quite obviously, German officials promoted spaceflight as the ultimate proof of a nation's progressiveness, and their commitment to the project served to highlight Germany's commitment to a future characterized by 'big technologies' and their groundbreaking ambitions.[48] Rather than discussing possible purposes of Spacelab, the federal government preferred to refer to spaceflight pioneers such as Hermann Oberth or Wernher von Braun (1912–77), echoing a form of romantic enthusiasm for a utopian future that seemed ordained by this new technology and comparing contemporary spaceflight with Christopher Columbus's historic departure into the 'new world.'[49]

Spacelab captured both political and popular attention. Various pop artists responded to the project, though not necessarily sharing the enthusiasm voiced by the federal government. An entire song was devoted to Spacelab on the highly acclaimed album *Mensch-Maschine*, released by the Düsseldorf-based band Kraftwerk in 1978.[50] Taking up issues such as spaceflight, robotics and urbanization, the album was inspired by Fritz Lang's 1927 movie *Metropolis* and is believed to address a future that threatens to transform humankind into servants of complex machines.[51] The same holds true for another Düsseldorf-based band, Fehlfarben, who mentioned Spacelab in their 1980 hit single 'Ein Jahr (Es geht voran).'[52] Fehlfarben took a critical stance toward the progress symbolized by Spacelab, singing 'Spacelabs fall'n auf Inseln, Vergessen macht sich breit – es geht voran' ('Spacelabs crash on islands, oblivion plants itself – things are progressing'). Although the lyrics give the impression that they were actually referring to the dangerous debris plummeting down from the decommissioned American space station Skylab in 1979, causing considerable media attention in Germany (see Figure 1.4 above), the band denied the futuristic promises connected to the exploration of outer space, highlighting instead failure and public indifference.[53]

German media, however, were keen to raise public interest in Spacelab. Echoing the government's position that technology posed the central means of shaping the future, the Westdeutsche Rundfunk (WDR), along with other European broadcast stations, began reporting from Bremen where the laboratory had been manufactured since 1977.[54] TV shows and popular-science books devoted to Spacelab claimed to witness the beginning of a new era. First of all, this era was characterized by relying on reusable spacecraft such as the Space Shuttle or the Spacelab. Such economy promised to make future spaceflight much safer and cheaper, thus allowing for an ever-intensifying use of outer space on an almost everyday basis. Second, flying into outer space on a regular basis would also ensure that 'spaceflight would under no circumstances be the business of only a small group of specially selected people,' popular-science books claimed.[55] Instead, it would soon become accessible to (almost) everyone.

An advertisement by the German Aerospace Center, the Deutsche Forschungs- und Versuchsanstalt für Luft- und Raumfahrt (DFVLR, renamed Deutsches Zentrum für Luft- und Raumfahrt, DLR, in 1997), published in various major German newspapers in 1977, seemed to substantiate this spectacular claim (Figure 11.4). Obviously, the DFVLR was looking for astronauts with solid scientific qualifications but addressing the public through widely read newspapers and asking for only a minimum of qualifications gave the impression that spaceflight was no longer reserved for specially selected people. Spacelab was heralded 'as a milestone in the drive to obtain practical rewards from the exploration of space. [...] In a Spacelab flight which lasts for a week or longer, scientists could come up with results that improve our fundamental knowledge of biological mechanisms and the behavior of materials,' the *New Scientist* claimed in preparation for the Spacelab's maiden flight. 'The researchers could also do the ground work for new industries which capitalize on the special conditions of space, in making substances that may be difficult, or downright impossible, to manufacture on Earth.'[56] Despite such high expectations, ESA authorities were disappointed by the responses from the scientific community.[57] There were various reasons why scientists took little interest in Spacelab, most notably the short duration of the missions and the high costs involved. But they all make clear that its actual use was less important than the symbolic significance of the project. By pointing out how much Spacelab would contribute to the scientific exploration of outer space, the governments involved stressed its significance as a 'tool of peace.' Its significance as a 'tool of progress' derived in part from its scientific purpose but, as the advertisement by the DFLVR indicates, implied a more fundamental prospect.

From the very beginning, the Spacelab project had been promoted as an important step toward a permanent human presence in space, echoing what space journalist Dwayne Day has termed the 'Von Braun Paradigm.' Arguing that it is far more difficult to establish a human presence in outer space than to explore it from an outpost already *in situ*, German spaceflight pioneer Wernher von Braun was instrumental in promoting space stations as the

Figure 11.4   Published by the DFVLR in 1977, this job advertisement conveyed the impression that spaceflight would become accessible to (almost) everyone in the very near future. All that was required was an age no older than 47, a body height measuring at least 153 cm and no more than 190 cm, good health and thorough knowledge of the English language. This copy shows the markings of Ernst Messerschmid, who participated in the German D-1 mission in 1985.

Source: Courtesy of Deutsches Zentrum für Luft- und Raumfahrt (DLR).

first step toward the conquest of outer space.[58] Along these lines, popular-science books about Spacelab published in Germany regularly featured pictures of gigantic space stations and referred to space-colonization pioneers such as Princeton physicist Gerard K. O'Neill (1927–92). O'Neill made headlines in the 1970s advocating the colonization of space on a large scale. His ideas, published in the award-winning *The High Frontier: Human Colonies in Space*, earned him an ever-growing group of devotees who went on to found the L5 Society and promoted massive settlement in outer space.[59] O'Neill gained credit and support from NASA and his works fueled expectations that human emigration into outer space would commence in 2013, with more people living in space than on earth a hundred years later, as German books on the Spacelab enthusiastically reported.[60] 'The realization of massive and permanent space stations is already within reach,' Hermann Oberth assured the readers in his preface to a book on Spacelab. His words echoed what ESRO's General Director Alexander Hocker (1912–96) had announced in 1973: 'Spacelab is the indispensable element to transform the Shuttle into a first generation space station.'[61] Along these lines, the book on Spacelab, for which Oberth had written the preface, featured pictures of a future space station called Alkyone which was to be constructed by combining several Spacelab-modules.

Popular-science books dedicated to Spacelab show that spaceflight was still presented as a great adventure. Compared to older spacecraft, however, the STS seemed a comfortable spaceship with a private bunk for each astronaut, a toilet very similar to those on earth, as West German astronaut Ulf Merbold (1941–) recalled, and plenty of food including 'shrimp cocktail, lobster, turkey and roast beef.'[62] Working in the module was rather like spending a week inside a rail freight container, as astronauts who flew in the laboratory would later comment, since its walls were lined with clinically arranged experiment racks to make maximum use of the limited space available.[63] However, being fully pressurized, Spacelab would allow astronauts to dispense with the clumsy suits that had characterized the early space missions and work in shorts and shirt-sleeves, making for an atmosphere of informality. In his personal account, Merbold wrote in detail not only about the scientific experiments he conducted onboard the Spacelab but also on how much he enjoyed sharing meals with his fellow astronauts or listening to classical music in his bunk, giving the impression that manned spaceflight was a demanding task but also a pleasant experience that would become accessible to more and more people in the near future.[64]

III   Politics

In the end, Spacelab was used only 16 times.[65] The explosion of the Space Shuttle *Challenger* in January 1986 dampened expectations voiced by Ulf Merbold and others that spaceflight would become as normal and as safe as taking the bus.[66] The disaster put the Space Shuttle program on hold for two-and-a-half years, and all previously planned Spacelab missions along with it. But the STS also did not turn out to be an economical means of transport, and, due to the longer-than-envisaged servicing required after each flight, the Space Shuttle was able to make only ten flights a year instead of the 60 originally planned. In total, Spacelab spent 181 days in space and enabled 110 astronauts to work on 720 experiments. Although it was used by scientists from countries such as Canada, Belgium, the Netherlands, Switzerland, Italy and Japan, it also failed to become a long-lasting symbol for peaceful and international cooperation in space.

Contrary to ESA's self-perception of being 'one of the melting pots [...] in which nationalist preoccupations have to give way to wider, more promising visions,' as voiced in a report from 1984, the 'success' of Spacelab stimulated national rivalries and claims for leadership as the German stamp had already demonstrated.[67] None of its flights was devoted entirely to ESA scientific payload, despite the fact that at least two such missions had been planned.[68] Instead, national missions were conducted, two of them, Spacelab-D1 (1985) and Spacelab-D2 (1993), by Germany. 'The Federal Republic of Germany and other European contributors can be especially proud of this achievement,' President Ronald Reagan praised the German effort in a conference call with West German Chancellor Helmut Kohl and crew members of the Space Shuttle *Columbia* during the Spacelab's maiden flight in early December 1983. 'The shuttle is demonstrating that technology can be used to bring people together in a new spirit of enterprise and cooperation to better their lives, ensure the peace of mankind,' Reagan continued.[69] While both the politicians and the astronauts were keen to stress the significance that this mission had for peace and progress on earth, they also addressed national characteristics. The Spacelab's 'top performance' had demonstrated Germany's 'high qualification, a sense of responsibility and a pioneering spirit,' the Federal Minister of Research and Technology, Heinz Riesenhuber, claimed.[70]

Conservative German newspapers like *Die Welt* joined in, celebrating Spacelab for 'surpassing all expectations' and pointing out that 'even prosaic scientists rejoiced over the flood of new, ground-breaking insights' while more liberal-orientated newspapers like *Die Zeit* sensed a 'hoax in space' and titled the whole project as 'useless and expensive.'[71] The federal government promptly awarded Ulf Merbold the *Bundesverdienstkreuz*, the Federal Republic's Order of Merit, and claimed a leading role within the European space community. In due course, German politicians vigorously promoted ambitious projects in outer space such as the development of a small European shuttle, Hermes, and – based on the Spacelab – the construction of a space station that would provide Europe with its own space infrastructure. This did not necessarily fit with ESA's agenda but was intended to illustrate Germany's own newly gained status as a superpower in outer space.[72] 'In the late twentieth century the political status of a nation, its reputation within the international community, will depend essentially on its willingness and its capabilities to explore space,' the Gesellschaft für Auswärtige Politik, an influential German think tank, claimed, causing the German weekly *Der Spiegel* to comment: 'No one is able to cope with this concentrated lobby. Quite obviously, manned spaceflight seems essential no matter how much it costs and no matter if its benefits reasonably relate to the expenditure.'[73] *Die Zeit* was equally skeptical, calling Spacelab an 'undertaking that would barely stand a critical inquiry' and criticizing a 'nationalistic self-adulation.'[74] Instead of forging ties within Europe, the 'success' of Spacelab generated national climates of opinion that separated the European space community into rival factions who argued over priorities in ESA's agenda.[75]

The more prominence US military ambitions in outer space gained in the early 1980s, the more troubled relations became both within the Western European space community and with its transatlantic partner, the United States.[76] Spacelab had not even completed its maiden flight when Reagan announced his plans for a space-based missile defense system in March 1983. His SDI immediately drew heavy criticism from within the United States and from abroad.[77] But it also challenged Europe's commitment to the peaceful use of outer space as governments of larger states such as the United Kingdom and Germany felt tempted to either join in or to consider plans for an autonomous European Defense Initiative.[78] Smaller member states such as Denmark strongly opposed any military involvement and voiced their apprehension that Spacelab might be used for SDI. 'ESA is clearly not involved in any SDI project,' its directorate responded to an inquiry from the Danish Ministry of Research in May 1986. However, it also conceded that 'ESA has no right to control. Spacelab is considered as a component of the Space Shuttle under US jurisdiction and control.'[79]

To the US government, the worries about the use of Spacelab voiced by member states of the Western European space community were merely a question of semantics. For ESA the term 'peaceful purposes' implied opposing any military activity in space. The United States, however, preferred a more subtle approach, differentiating between 'aggressive' and 'non-aggressive purposes' in outer space. Moreover, to legitimate its approach, the United States could refer to the Outer Space Treaty, the basic legal framework of international space law.[80] This treaty, initiated by the United Nations in 1967, backed the US point of view in that it forbade placing weapons of mass destruction in outer space and condemned any 'active' militarization of the heavens. 'Passive' militarization, however – that is, the stationing of non-aggressive military assets such as satellites used for reconnaissance, surveillance and communication purposes – was tolerated by the treaty.[81] Along these lines, a laboratory designed for scientific research obviously served no aggressive purposes and did not conflict with international space law even if it was used for experiments in the context of SDI (as it would be).[82]

These conflicts about the proper use of Spacelab refer to utopian and pragmatic approaches toward outer space. Contrary not only to the United States but also to some of its own member states that showed interest in SDI, ESA promoted an unambiguously pacifist stance. This utopian approach, however, ignored the dual-use capabilities of space technology in general and by conceding the Americans 'full control over the first SL unit,' as the Memorandum of Understanding stated, contributed to a covert militarization of the heavens.[83] Its directorate's answer to the Danish Ministry of Research stressed ESA's utopian stance but also demonstrated the junior role the Europeans had in the PAP. ESA not only lacked any influence on how the transatlantic partner would use 'their' hardware, but the Europeans also had to adapt their own missions to US requirements regardless of whether they fit ESA's policy, as the STS-61-A mission of November 1985 illustrates. In Germany, this mission is known as D1 since it was funded by Germany and, for the first time ever, its payload operations were controlled from the German Space Operations Center in Oberpfaffenhofen and not from the regular NASA control centers. However, while officially designated as a German mission and dedicated to the scientific exploration of outer space, it was also committed to deploying a US spy satellite.[84]

IV Conclusion: covert militarization?

While 'Europe's masterpiece' in outer space is still being discussed with regard to its scientific and technological significance, contextualizing Spacelab within the broader historical and political context of the late Cold War shows these aspects were of secondary importance to the governments involved, at least on the part of the United States. What mattered from the outset was Spacelab's symbolic value as a 'tool of peace,' as US Vice President Bush put it in February 1982. However, while the transatlantic cooperation served to demonstrate the Western camp's commitment to a peaceful and multilateral exploration of outer space, it also contributed to a covert militarization of the heavens. Early evaluations of the usefulness of the Space Shuttle had stressed its military potential. To reassure both friend and foe that the STS was not merely a 'tool of war' the US government opted for an additional space laboratory. Obviously dedicated to the scientific exploration of outer space, ESA was invited to build the Spacelab, thereby gaining access to manned spaceflight.

For the newly constituted Western European space community, the successful lunar landings had set the stage for manned spaceflight to back a nation's prestige in a way no other technological feat could. Bringing more people into space on a permanent basis not only seemed the next logical step in the ongoing exploration of outer space, it also served as proof of each nation's progressiveness. The utopian goals of the early Space Age, however, eventually gave way to more pragmatic approaches toward the use of outer space even within the Western European space community. ESA's commitment to the peaceful exploration of outer space served US interests at a time when Cold War confrontation re-intensified and outer space became a battlefield in its own right, both in fact and in popular fiction. From early on, negotiations about Western Europe's participation in the PAP had given the impression that there was no practical need for a scientific laboratory in outer space. Along these lines, the 'success' of Spacelab was not so much about expanding the scientific knowledge of outer space. Rather, it raised the strategic interest in the heavens among ESA member states as well, which were discussing plans for a European Defense Initiative. Although smaller member states strongly opposed any military activity in outer space, ESA's contribution to the PAP de facto contributed to the militarization of outer space by providing a tool that would also prove useful in the context of SDI. Ignoring the dual-use character of space technology in general, ESA had chosen a utopian stance that increasingly conflicted with outer space realities. More than concealing a militarization, however, Western Europe's participation in the PAP covered a fundamental caesura in the perception and use of the heavens.

Putting the Spacelab to use not only seemed to substantiate the peaceful intentions of both the United States and Europe in the ongoing and – or so it seemed – ever expanding exploration of outer space. It also revitalized fantastic visions about giant space colonies and massive human settlement in outer space at a time when mankind's outlook went 'back and inward rather than forward and outward.'[85] Spacelab was presented as the first step toward a permanent human presence in space that would no longer be limited to specially trained astronauts, allowing an ever-increasing number of ordinary people to travel into outer space. In line with plans about massive human settlement in space promoted by the likes of O'Neill, Spacelab reflected a space enthusiasm that seemed straight out of the heyday of the early Space Age and consequently found support among renowned spaceflight pioneers such as Hermann Oberth.

In the end, plans for a massive human settlement or even an autonomous European infrastructure in outer space proved as far-fetched as Reagan's SDI. After the collapse of the Soviet empire there was no more need for a space-based missile defense system. Spacelab had become politically and technically obsolete. Although it was in use until 1998, political and popular attention had focused increasingly on the space station Freedom, an advanced technological platform that would bind together the 'nations of the free world,' as Reagan announced in 1984. The Western European space community – with Germany again in the lead – was eager to participate in a project that would eventually lead to the International Space Station and build on their Spacelab experience with a research module called Columbus. When the space station eventually got off the ground in 1995, it owed much more to Russian expertise, the Russian Space Agency having joined the program in 1993.[86]

Given the enthusiasm that accompanied Spacelab from its invention to its active use, one might get the impression that the disenchantment often ascribed to American astroculture after the successful lunar landings does not fit for 1970s Europe. In fact, European spaceflight underwent its 'formative experience' only after space enthusiasm had reached its peak in the United States.[87] However, as the criticism voiced by German popular bands and liberal media makes clear, enthusiasm toward Spacelab was politically orchestrated. Despite advertisements such as the one issued by the DFLVR in 1977, it did not generate as much public interest as the human spaceflight programs of the 1960s. Neither did Spacelab contribute to a European space infrastructure in the way that its (less spectacular) sister-project Ariane did.[88] From the very moment ESA joined the PAP, there was more at stake than mere space politics, echoing the 'politically tinged character of manned spaceflight' as well as a new earth-centered approach to outer space. Making the Europeans join the PAP was not so much a technoscientific undertaking in the ongoing exploration of outer space as a political move that reflected global Cold War politics.

Notes

1. Hermann Oberth, 'Preface,' in Werner Büdeler and Stratis Karamonlis, *Spacelab: Europas Labor im Weltraum*, Munich: Goldmann, 1976, 7. On Hermann Oberth, see Hans Barth, *Hermann Oberth: Vater der Raumfahrt*, Esslingen: Bechtle, 1991.
2. Oberth, 'Preface,' 7.
3. John Krige, Angelina Long Callahan and Ashok Maharaj, *NASA in the World: Fifty Years of International Collaboration in Space*, Basingstoke: Palgrave Macmillan, 2013; John Krige, Arturo Russo and Lorenza Sebesta, *A History of the European Space Agency, 1958–1987*, vol. 2: *The Story of ESA, 1973–1987*, Noordwijk: ESA, 2000; for the Post-Apollo Program, see also John M. Logsdon, *After Apollo? Richard Nixon and the American Space Program*, Basingstoke: Palgrave Macmillan, 2015.
4. Literature on the Space Shuttle is extensive; for an overview, see Malinda K. Goodrich, Alice R. Buchalter and Patrick Miller, *Toward a History of the Space Shuttle: An Annotated Bibliography. Part II: 1992–2011*, Washington, DC: NASA, 2012, and Joseph N. Pelton, 'The Space Shuttle: Evaluating an American Icon,' *Space Policy* 26.4 (November 2010), 246–8.
5. See Niklas Reinke, *Geschichte der deutschen Raumfahrtpolitik: Konzepte, Einflussfaktoren und Interdependenzen 1923–2002*, Munich: Oldenbourg, 2004; 'A Giant Workshop in the Sky,' *Time* 121.49 (28 November 1983), 73–4; Reimar Lüst quoted after Krige et al., *NASA in the World*, 121; Krige et al., *Story of ESA*, 604; Helmuth Trischler, 'Contesting Europe in Space,' in idem and Martin Kohlrausch, *Building Europe on Expertise: Innovators, Organizers, Networkers*, Basingstoke: Palgrave Macmillan, 2014, 243–75, here 273.
6. James C. Moltz, *The Politics of Space Security: Strategic Restraint and the Pursuit of National Interests*, Stanford: Stanford University Press, 2008; for the late Cold War, see Leopoldo Nuti, ed., *The Crisis of Détente in Europe: From Helsinki to Gorbachev, 1975–1985*, London: Routledge, 2009. See also Bernd Greiner,

Christian Th. Müller and Dierk Walter, eds, *Krisen im Kalten Krieg*, Hamburg: Hamburger Edition, 2008.

7. See Günter Paul, *Aufmarsch im Weltall: Die Kriege der Zukunft werden im Weltraum entschieden*, Bonn: Keil, 1980; Hans-Günter Brauch, *Angriff aus dem All: Der Rüstungswettlauf im Weltraum*, Berlin: Dietz, 1984. The literature on SDI is vast and ever-increasing, but a comprehensive historical account is still missing. See only Columba Peoples, *Justifying Ballistic Missile Defence: Technology, Security and Culture*, Cambridge: Cambridge University Press, 2010; Frances FitzGerald, *Way Out There in the Blue: Reagan, Star Wars, and the End of the Cold War*, New York: Simon & Schuster, 2001; and, for a European perspective, Bernd W. Kubbig, ed., *Die militärische Eroberung des Weltraums*, 2 vols, Frankfurt am Main: Suhrkamp, 1990.

8. Alex Roland, 'Barnstorming in Space: The Rise and Fall of the Romantic Era of Spaceflight, 1957–1986,' in Radford Byerly Jr., ed., *Space Policy Reconsidered*, Boulder: Westview Press, 1989, 33–52, here 35.

9. Gerhard A. Ritter, Margit Szöllösi-Janze and Helmuth Trischler, eds, *Antworten auf die amerikanische Herausforderung: Forschung in der Bundesrepublik und der DDR in den 'langen' siebziger Jahren*, Frankfurt am Main: Campus, 1999; Claudia Hiepel, ed., *Europe in a Globalising World: Global Challenges and European Responses in the 'Long' 1970s*, Baden-Baden: Nomos, 2014.

10. For the concept of astroculture, see Alexander C.T. Geppert, 'European Astrofuturism, Cosmic Provincialism: Historicizing the Space Age,' in idem, ed., *Imagining Outer Space: European Astroculture in the Twentieth Century*, Basingstoke: Palgrave Macmillan, 2012, 3–24, here 6–9 (= *European Astroculture*, vol. 1); and idem, 'Rethinking the Space Age: Astroculture and Technoscience,' *History and Technology* 28.3 (September 2012), 219–23.

11. Foreword by the Director General of the European Space Agency, held at the Spacelab Flight Unit Delivery Ceremony (23 November 1981), Historical Archives of the European Union/European Space Agency (hereafter HAEU/ESA), 7980.

12. Donella H. Meadows, Dennis L. Meadows, Jørgen Randers, William W. Behrens III and Club of Rome, *The Limits to Growth: A Report for the Club of Rome's Project on the Predicament of Mankind*, New York: Universe Books, 1972. For the ever-increasing literature on the 1970s, see only Martin H. Geyer, 'Auf der Suche nach der Gegenwart: Neue Arbeiten zur Geschichte der 1970er und 1980er Jahre,' *Archiv für Sozialgeschichte* 50 (2010), 643–69, in addition to Alexander Geppert's introduction to this volume.

13. See R. Craig Nation, 'Programming Armageddon: Warsaw Pact War Planning, 1969–1985,' in Nuti, *Crisis of Détente in Europe*, 124–36; Dima P. Adamsky, 'Through the Looking Glass: The Soviet Military-Technical Revolution and the American Revolution in Military Affairs,' *Journal of Strategic Studies* 31.2 (March 2008), 257–94; on rumors about killer-satellites and ASATs see, for example, 'Notfalls rammen,' *Der Spiegel* 31.41 (3 October 1977), 260.

14. Colin S. Gray, 'John Sheldon, Space Power and the Revolution in Military Affairs: A Glass Half-Full?,' *Airpower Journal* 13.3 (Fall 1999), 23–38; see also David M. Walsh, *The Military Balance in the Cold War: US Perceptions and Policy, 1976–1985*, London: Routledge, 2008.

15. See Cathleen Lewis, 'Space Spies in the Open: Military Space Stations and Heroic Cosmonauts in the Post-Apollo Period, 1971–77,' in Alexander C.T. Geppert, Daniel Brandau and Tilmann Siebeneichner, eds, *Militarizing Outer Space: Astroculture, Dystopia and the Cold War*, London: Palgrave Macmillan, forthcoming (= *European Astroculture*, vol. 3); and Asif A. Siddiqi, 'Soviet Space Power During the Cold War,' in Paul G. Gillespie and Grant T. Weller, eds, *Harnessing the Heavens: National Defense Through Space*, Colorado Springs: United States Air Force Academy, 2008, 135–50.

16. *Moonraker* (James Bond), directed by Lewis Gilbert, USA/UK/FR 1979 (United Artists). The movie became the highest grossing film of the series with $210,300,000 worldwide, a record that stood until 1995's *Goldeneye*; see Werner Greve, *James Bond 007: Agent des Zeitgeists*, Göttingen: Vandenhoeck & Ruprecht, 2012.

17. For the famous Star Trek quotation, see Dwayne A. Day, 'Boldly Going: Star Trek and Spaceflight,' *The Space Review* (28 November 2005), available at http://www.thespacereview.com/article/506/1 (accessed 1 October 2017).

18. James Bonds Top 20, IGN (17 November 2006), available at http://uk.ign.com/articles/2006/11/17/james-bonds-top-20?page=2 (accessed 1 October 2017); John Cork and Bruce Scivally, *James Bond: Die Legende von 007*, Bern: Scherzverlag, 2002, 177. On the role of technology in the James Bond series, see Marcus M. Payk, 'Globale Sicherheit und ironische Selbstkontrolle: Die James-Bond-Filme der 1960er Jahre,' *Zeithistorische Forschungen* 7.2 (2010), 314–22.

19. Myron S. Malkin, 'The Space Shuttle,' *American Scientist* 66.6 (November–December 1978), 718–23, here 722.

20. Joint DoD/NASA Study of Space Transportation System (16 June 1969), quoted after Logsdon, *Richard Nixon and the American Space Program*, 165.

21. See Report on DoD Utilization of STS/SL (6 June 1979), HAEU/ESA, 7944.

22. Report to ESA HQ (23 September 1981), HAEU/ESA, 7940.

23. Memorandum of Understanding between the National Aeronautics and Space Administration and the European Space Research Organization for a Cooperative Programme concerning Development, Procurement and Use of a Space Laboratory in Conjunction with the Space Shuttle System (14 August 1973), Paris/Washington, DC, in Douglas R. Lord, *Spacelab: An International Success Story*, Washington, DC: NASA, 1987, 437–59.

24. 'Spacelab Welcomed by Bush,' *Today: Florida's Space Age Newspaper* (6 February 1982), HAEU/ESA, 7980.

25. See Krige et al., *NASA in the World*, chapters 4 and 5, 89–111 and 112–24, respectively; Krige et al., *Story of ESA*, 521–36.

26. Ibid., 525.

27. The term 'dual use' refers to the fact that every piece of space technology can be applied either for civilian or military purposes. See Roger Handberg, *Seeking New World Vistas: The Militarization of Space*, Westport: Praeger, 2000, 55; for European military activity in outer space, see Alasdair McLean, *Western European Military Space Policy*, Aldershot: Dartmouth, 1992. See also the introduction to the forthcoming third volume in this European astroculture trilogy, Geppert et al., *Militarizing Outer Space*.

28. For the early history of the Western European space community, see Krige et al., *History of ESA*, vol. 1; Kevin Madders, *A New Force at a New Frontier: Europe's Development in the Space Field in the Light of its Main Actors, Policies, Law and*

Activities from its Beginnings up to the Present, Cambridge: Kluwer, 1997; Guy Collins, *Europe in Space*, Basingstoke: Macmillan, 1990.

29. See Trischler, 'Contesting Europe in Space,' 266.
30. For French space politics, see Walter A. McDougall, 'Space-Age Europe: Gaullism, Euro-Gaullism, and the American Dilemma,' *Technology and Culture* 26.2 (April 1985), 179–203; and Andreas Hasenkamp, *Raumfahrtpolitik in Westeuropa und die Rolle Frankreichs: Macht – Nutzen – Reformdruck*, Münster: Lit, 1996.
31. See Gottfried Greger, 'Why and How is the Federal Government of Germany Promoting the Utilization of Spacelab?,' *Proceedings of the Royal Society of London: Series A, Mathematical and Physical Sciences* 361.1705 (May 1978), 143–50.
32. Trischler, 'Contesting Europe in Space,' 266.
33. Wolfgang Finke, 'Germany and ESA,' in Robert A. Harris, ed., *The History of the European Space Agency: Proceedings of an International Symposium, London, 11–13 November 1998*, Noordwijk: ESA, 1999, 37–50, here 43.
34. For Germany's Ostpolitik, see Carole Fink and Bernd Schäfer, eds, *Ostpolitik, 1969–1974: European and Global Responses*, Cambridge: Cambridge University Press, 2009.
35. Susanne Schregel, *Der Atomkrieg vor der Wohnungstür: Eine Politikgeschichte der Neuen Friedensbewegung 1970–1985*, Frankfurt am Main: Campus, 2011; Philipp Gassert, *Zweiter Kalter Krieg und Friedensbewegung: Der NATO-Doppelbeschluss in deutsch-deutscher und internationaler Perspektive*, Munich: Oldenbourg, 2011.
36. See Michael J. Neufeld, *The Rocket and the Reich: Peenemünde and the Coming of the Ballistic Missile Era*, New York: Free Press, 1995.
37. See Reinke, *Geschichte der deutschen Raumfahrtpolitik*.
38. Logsdon, *Richard Nixon and the American Space Program*, 162.
39. 'Spacelab Welcomed by Bush,' *Today: Florida's Space Age Newspaper* (6 February 1982), HAEU/ESA, 7980; for Heinz Riesenhuber, see 'Ich schmeiße Geld nie unnötig raus,' *Der Spiegel* 41.32 (3 August 1987), 46–51, here 46.
40. Several countries devoted stamps to the Space Shuttle (others, like Switzerland or Mauretania, devoted stamps to the Spacelab as well) but only after the first successful flight of the Space Shuttle (or the Spacelab, respectively).
41. Other countries involved in the project were Belgium, Spain, the Netherlands, Denmark, Switzerland and Austria, each contributing less than 5 percent to the cost; Collins, *Europe in Space*, 71.
42. See Gottfried Gabriel, 'Ästhetik und politische Ikonographie der Briefmarke,' *Zeitschrift für Ästhetik und Allgemeine Kunstwissenschaft* 54.1 (2009), 1–10; Markus Göldner, *Politische Symbole der europäischen Integration: Fahne, Hymne, Hauptstadt, Pass, Briefmarke, Auszeichnungen*, Frankfurt am Main: Peter Lang, 1988; Alexander Hanisch-Wolfram, *Postalische Identitätskonstruktionen: Briefmarken als Medien totalitärer Propaganda*, Frankfurt am Main: Peter Lang, 2006.
43. Elisabeth Schaber, 'Der rote Weltraum: Die künstlerische Darstellung von Raumfahrt auf Briefmarken der DDR,' *Neue Kunstwissenschaftliche Forschungen* 1 (October 2014), 48–60; see also Jana Scholze, 'Ideologie mit Zackenrand: Briefmarken als politische Symbole,' in Dokumentationszentrum Alltagskultur, ed., *Fortschritt, Norm und Eigensinn: Erkundungen im Alltag der DDR*, Berlin: Links, 1999, 174–91.

44. Greger, 'Why and How,' 143; see also Reinke, *Geschichte der deutschen Raumfahrtpolitik*, 146.

45. 'Key technologies' promise the improvement of new technological fields of application which are deemed essential for overall economic growth; see Susanne Fohler, *Techniktheorien: Der Platz der Dinge in der Welt des Menschen*, Munich: Fink, 2003.

46. See Krige et al., *Story of ESA*, 568.

47. Reinke, *Geschichte der deutschen Raumfahrtpolitik*, 147.

48. See Johannes Weyer, 'Größendiskurse: Die strategische Inszenierung des Wachstums sozio-technischer Systeme,' in Ingo Braun and Bernward Joerges, eds, *Technik ohne Grenzen*, Frankfurt am Main: Suhrkamp, 1994, 347–85.

49. See Büdeler, *Spacelab*, 19; Ulf Merbold, *Flug ins All: Von Spacelab zur D1-Mission: Der persönliche Bericht des ersten Astronauten der Bundesrepublik*, Bergisch-Gladbach: Gustav Lübbe, 1986, 339.

50. 'Spacelab' (5:51), featured on the Kraftwerk album *Mensch-Maschine*, Kling Klang Schallplatten/EMI/Warner Music Group 1978; the album peaked at number 12 in the German charts and at 9 in the UK.

51. See Rüdiger Esch, *Electri_City: Elektronische Musik aus Düsseldorf*, Berlin: Suhrkamp, 2014; for pop recordings as historical source, see Bodo Mrozek, 'Geschichte in Scheiben: Schallplatten als zeithistorische Quelle,' *Zeithistorische Forschungen* 8.2 (2010), 295–304.

52. 'Ein Jahr (Es geht voran)' (2:51), featured on the Fehlfarben album *Monarchie und Alltag*, EMI 1980. The album peaked at number 37 in Germany but has since been considered as one of the most influential German rock albums of all time; in 2001 it received a golden record for having sold more than 250,000 copies.

53. For Skylab, see the contributions by Alexander Geppert and Regina Peldszus, Chapters 1 and 10 in this volume.

54. For media coverage of spaceflight in West Germany, see Bernd Mütter, 'Per Media Ad Astra? Outer Space in West Germany's Media, 1957–1987,' in Geppert, *Imagining Outer Space*, 149–69.

55. Büdeler, *Spacelab*, 93; popular books dedicated to Spacelab frequently referred to NASA flight plans from the 1970s that envisioned up to 60 flights a year; see Hermann-Michael Hahn, ed., *D1: Unser Weg ins All*, Braunschweig: Westermann, 1985, 49.

56. 'Scientists Gain a Foothold in Space,' *New Scientist* (22 September 1983), 75.

57. See Krige et al., *Story of ESA*, 572–4.

58. Dwayne A. Day, 'The Von Braun Paradigm,' *Space Times* 33 (November–December 1994), 12–15; see also Alexander C.T. Geppert, 'Infrastrukturen der Weltraumimagination: Außenstationen im 20. Jahrhundert,' in Kunst- und Ausstellungshalle der Bundesrepublik Deutschland, ed., *Outer Space: Faszination Weltraum*, Bonn: Nicolai, 2014, 124–7.

59. The L5 Society derived its name from the so-called Lagrange points, stable points in the solar system where a spacecraft can maintain its position without expanding energy. See Gerard K. O'Neill, *The High Frontier: Human Colonies in Space*, New York: William Morrow, 1976; for the space colonization movement during the 1970s, see also Ryan J. McMillen, *Space Rapture: Extraterrestrial Millennialism and the Cultural Construction of Space Colonization*, PhD thesis, University of Texas, Austin, 2004. On the L5 Society, see W. Patrick McCray, *The Visioneers: How*

a Group of Elite Scientists Pursued Space Colonies, Nanotechnologies, and a Limited Future, Princeton: Princeton University Press, 2012; and, in particular, Peter Westwick's contribution, Chapter 12 in this volume.

60. See Hahn, Unser Weg ins All, 205.

61. Krige et al., Story of ESA, 571.

62. Merbold, Flug ins All, 140.

63. See Collins, Europe in Space, 72.

64. See Merbold, Flug ins All, 139–46, 150 and 155.

65. For a comprehensive mission list, see Reinke, Geschichte der deutschen Raumfahrt-politik, 330–2.

66. For the Challenger disaster, see Rick Houston, Wheels Stop: The Tragedies and Triumphs of the Space Shuttle Program, 1986–2011, Lincoln: University of Nebraska Press, 2013; Ann Larabee, '"Nothing Ends Here": Managing the Challenger Disaster,' in Steven Biel, ed., American Disasters, New York: New York University Press, 2001, 197–220.

67. Norman Longden and Duc Guyenne, Twenty Years of European Cooperation in Space: An ESA Report, Noordwijk: ESA, 1984, 229.

68. Reinke, Geschichte der deutschen Raumfahrtpolitik, 163.

69. Ronald Reagan, 'Remarks During a Conference Call with Chancellor Helmut Kohl of the Federal Republic of Germany and Crewmembers of the Space Shuttle Columbia (5 December 1983),' available at https://www.reaganlibrary.archives. gov/archives/speeches/1983/120583a.htm (accessed 1 October 2017).

70. 'Schwindel im All,' Die Zeit 37.51 (16 December 1983), 55.

71. See ibid. and 'Nutzlos und teuer,' Die Zeit 38.4 (20 January 1984), 48.

72. See Johannes Weyer, 'Taktische Spiele im All,' Die Zeit 34.17 (22 April 1988), 36–40; idem, 'European Star Wars: The Emergence of Space Technology through the Interaction of Military and Civilian Interest-Groups,' in Everett Mendelsohn, Meritt R. Smith and Peter Weingart, eds, Science, Technology and the Military, Dordrecht: Kluwer, 1988, 243–88.

73. 'Gegen diese Lobby kommt keiner an,' Der Spiegel 41.32 (3 August 1987), 40–3, here 41.

74. 'Nutzlos und teuer,' Die Zeit 38.4 (20 January 1984), 48.

75. Trischler, 'Contesting Europe in Space,' 273.

76. Minute on US Space Policy (8 July 1982), HAEU/ESA, 7944.

77. Karsten Zimmermann, 'Reagans "Star Wars"-Rede,' in Kubbig, Militärische Eroberung des Weltraums, vol. 1, 56–93; Sean N. Kalic, 'Reagan's SDI Announcement and the European Reaction: Diplomacy in the Last Decade of the Cold War,' in Nuti, Crisis of Détente in Europe, 99–110.

78. See Hans Günter Brauch, ed., Star Wars and European Defence: Implications for Europe. Perceptions and Assessments, Basingstoke: Macmillan, 1987; Geoffrey Lee Williams and Alan Lee Williams, The European Defence Initiative: Europe's Bid for Equality, Basingstoke: Macmillan, 1987.

79. See the correspondence between the Danish Ministry of Research and the ESA Directorate (26 May 1986), HAEU/ESA, 9967.

80. Legal Aspects of the Use of Spacelab for the Purposes of SDI (3 October 1985), HAEU/ESA, 9967. See also Sean N. Kalic, US Presidents and the Militarization of Outer Space, 1946–1967, College Station: Texas A&M University Press, 2012, and, above all, Luca Follis's contribution, Chapter 8 in this volume.

81. See 'Treaty on Principles Governing the Activities of States in the Exploration and Use of Outer Space, including the Moon and Other Celestial Bodies'; available at http://disarmament.un.org/treaties/t/outer_space/text (accessed 1 October 2017). The Stockholm International Peace Research Institute (SIPRI) estimated that towards the end of the 1970s at least 60 percent of those satellites stationed in near-earth orbit were used for military purposes; see Reinke, *Geschichte der deutschen Raumfahrtpolitik*, 281.

82. See 'SDI-Tests im Spacelab,' *Der Spiegel* 39.35 (26 August 1985), 14.

83. 'Memorandum of Understanding between the National Aeronautics and Space Administration and the European Space Research Organization for a Cooperative Programme concerning Development, Procurement and Use of a Space laboratory in Conjunction with the Space Shuttle System,' in Lord, *Spacelab*, 454.

84. 'In letzter Minute,' *Der Spiegel* 41.3 (12 January 1987), 83–5; for the US spy satellite GLOMR (Global Low Orbiting Message Relay) see http://nssdc.gsfc.nasa.gov/nmc/spacecraftDisplay.do?id=1985-104B (accessed 1 October 2017).

85. For the 'Post-Apollo paradox,' see Alexander Geppert's introduction, Chapter 1 in this volume.

86. On the ISS, see John Krige, 'Space Collaboration Today: The ISS,' in idem et al., *NASA in the World*, 249–65.

87. Trischler, 'Contesting Europe in Space,' 269.

88. This series of expendable launch vehicles, internally code-named LIIIS (the French abbreviation for third-generation substitution launcher), proved commercially successful. To this day it provides autonomous launch-capabilities for the Western European space community; see Krige et al., *Story of ESA*, 455–520.

CHAPTER 12

From the Club of Rome to Star Wars: The Era of Limits, Space Colonization and the Origins of SDI

Peter J. Westwick

In 1972 the Club of Rome released a report titled *The Limits to Growth*. The report heralded what would be known as the 'era of limits,' when demographic and environmental pressures would force humanity to scale back its ambitions and live with less.[1] A decade later, in 1983, US President Ronald Reagan announced his plan to build a shield in outer space against ballistic missiles. What became known as the Strategic Defense Initiative (SDI) represented a remarkably ambitious vision of literally high technology erasing the threat of nuclear weapons.

The Club of Rome and SDI are familiar historical events, but they seem entirely disconnected. This chapter connects them, showing how the cultural and intellectual context of the 1970s – specifically, the idea of an impending crisis in human society – helped lead to the military program known as SDI. The path ran through space-colonization groups such as the L5 Society and the science-fiction community. The perceived threat to human survival galvanized a group of technological optimists to propose outer space not only as a solution to the Malthusian crisis of overpopulation, energy depletion and environmental damage, but also to the possibility of nuclear Armageddon. This quest for a technological fix led to surprising political alignments, with countercultural types embracing technology amid the backlash against science and technology

Peter J. Westwick (✉)
University of Southern California, Los Angeles, CA, USA
e-mail: westwick@usc.edu

© The Author(s) 2018
Alexander C.T. Geppert (ed.), *Limiting Outer Space*
European Astroculture, vol. 2
https://doi.org/10.1057/978-1-137-36916-1_12

after the 1960s and Vietnam, and also linking up with libertarians and others on the political right pushing missile defense as well as space colonization. The convergence included interesting individual, institutional and intellectual connections, as space-colony enthusiasts mingled with nuclear-weapon designers, and pro-missile defense arguments cohabited with enlightened discussion of biofeedback and the Gaia hypothesis.

These unusual intersections highlight a historiographical need to look off the beaten path for the origins of SDI. The need is especially acute for SDI, since the path has been trodden heavily by actors from the Reagan Administration in their memoirs and interviews. The current historical literature relies a great deal on these post-facto accounts, which impose an overly coherent perspective focused on high policy and almost exclusively on the United States.[2] Looking beyond the United States – and outside high government offices – yields new perspectives. The Soviets, for example, took a different view of SDI, one based on their own history and culture, and one that American scientists and strategists were slow to recognize.[3] Current scholarship on SDI also concentrates, with good reason, on the context of the Cold War, especially the renewed sense of confrontation of the late 1970s and early 1980s after the collapse of détente, and the resultant redoubling of the arms race. Europe, China and Japan, however, viewed SDI not as a Cold War, US-Soviet strategic standoff, but rather as a high-tech economic stimulus in a new era of global industrial competition.[4] This chapter keeps the United States in the foreground, but it broadens the focus further to include developments from earlier in the 1970s and beyond the usual Cold War framework. It thus reveals how the era of limits gave rise to the vast ambitions of SDI.

I   Space colonies and space forts

The Club of Rome's manifesto was just one of several books in the late 1960s and early 1970s, including *The Population Bomb* and *The Coming Dark Age*, to highlight the effects of population growth, environmental pollution and energy and resource scarcity, and so herald an emerging era of limits.[5] Among the solutions proposed in this new era – back-to-the-land, resource conservation and appropriate or alternative technology – was a push for space colonization, led most notably by Princeton physicist Gerard K. O'Neill (1927–92), who in the early 1970s proposed human space settlements based in giant free-floating cylinders (Figure 12.1). In 1974 O'Neill's ideas landed on the front page of the *New York Times*, herald of a widespread movement of technological enthusiasts arguing that humanity could avoid imminent limits by escaping earth.[6]

In 1975, for example, G. Harry Stine (1928–97), an aerospace scientist, model rocket builder and science-fiction author, identified 'a deepening worldwide megacrisis. This megacrisis has many facets. It's an energy crisis. It's a population explosion.' But Stine had good news: 'A far-reaching and final revolution is going to take place in our lifetimes. This revolution has

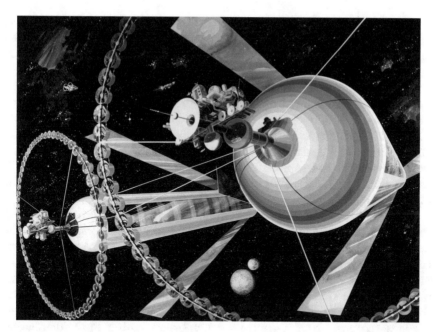

Figure 12.1   Exterior view of a proposed space colony, from studies held at NASA's Ames Research Center in the 1970s to follow up Gerard K. O'Neill's ideas. Artwork by Rick Guidice.
Source: Courtesy of NASA.

already started. [...] It is the Third Industrial Revolution, the final industrial revolution, the exploitation of the Solar System.' Stine's revolution would be driven by the entrepreneur and the capitalist, a stance that resonated especially with libertarians and helped earn the book an introduction by Barry Goldwater (1909–98), a figurehead of conservative politics. Stine, Goldwater enthused, had provided 'a happy combination of hope and optimism, which are vital ingredients of America's greatness. [...] It provides a real challenge to the gloom and doom crowd.'[7]

In 1975, Carolyn (1946–) and Keith Henson (1942–), two devotees of O'Neill's space-colony ideas, formed the L5 Society in Tucson, Arizona. They named it after one of the gravitational equilibrium points in outer space where a space colony could be built. The L5 Society eagerly propagated ideas for space colonies along with literally far-out schemes for lunar and asteroid mining, solar-power satellite systems, solar sails and ion drives. It also provided an early forum for discussion of space-based missile defense. The L5 Society never counted more than 10,000 members, and it might seem like a fringe group, a bunch of *Star Trek* fans off in a corner babbling about ion drives. But it wielded outsized political influence. L5 members, for instance, lobbied to kill the Moon Treaty in 1979, for fear that it would outlaw property rights and hence stall space colonization.[8]

The L5 Society's founders originally responded to the threat to human survival posed by demographic explosion and environmental depletion. Soon, however, the group contemplated that other threat to human survival, the nuclear arms race. The 1970s witnessed an alarming new phase of the arms race, as the Soviet Union matched and then, by 1978, exceeded the earlier build-up of the American arsenal; by that time, each side boasted 25,000 nuclear weapons, including thousands of MIRVed warheads on strategic missiles. Harry Stine's book on the prospects of outer space noted that the current Malthusian crisis was exacerbated by the presence of nuclear weapons, and quoted Krafft Ehricke's warning that 'if we slip back into the Dark Ages this time, we do so with our fingers on the nuclear trigger.'[9]

In May 1976, Keith Henson launched a debate with an article observing that solar-power satellites could run powerful space-based lasers for missile defense. This article is one of the earliest public discussions of such space lasers. Henson at the time was ambivalent about the prospect, questioning the technological, moral and political problems of space-based, beam-weapon missile defenses and noting, 'It is with considerable reluctance that L-5 News opens discussion of this subject.'[10] The article indeed provoked a response from space-colony supporters fearing the militarization of outer space; the editors observed that 'military aspects of space seem to be the Society's most consistent source of controversy.'[11] The L5 News editors seemed to relish the debate, presenting a series of articles on military space, including excerpts from books downplaying fears of space war and defending military space activity. When one reader suggested that the L5 News avoid discussing military uses of outer space, for fear of appearing to support them, the editors replied, 'We'll stop when our under-sand respirators arrive.'[12]

L5's apparent embrace of military space soon came to center on space-based lasers for missile defense and their potential to solve the threat of nuclear war. By June 1979 Keith Henson had abandoned his earlier ambivalence. In an article titled 'Space Forts: Or "Where Are You Obi Wan Kenobi?",' Henson presented a hopeful survey of space-based defenses against the nuclear 'Sword of Damocles' and mocked a recent, skeptical Scientific American article on space weapons as having 'the flavor of a 1900 essay on the impossibility of heavier than air flight.' Henson urged military funding of mass drivers and solar sails, which might serve as celestial tugboats to deliver asteroids into earth orbit. Such asteroids could then serve as 'space forts' bristling with lasers or particle beams. A few months later Henson proposed a variant, 'Space Foxholes: Or Beetle Bailey in Orbit,' using material mined from the moon or from Mars's moon Deimos, transported by solar-sail tugs, to protect military satellites.[13] His wife Carolyn wrote a subsequent article titled 'Lase the Nukes!' that similarly backed space-based lasers for missile defense as a way to 'end our fear of nuclear terror.'[14]

Keith Henson's articles suggested, however, that he was perhaps not a true believer in missile defense, but rather viewed it as a means to the end of space colonies. Military programs to deploy large lasers in orbit could develop heavy-lift launch vehicles, space power supplies (such as nuclear reactors or solar-power satellites), mass drivers or solar sails and other expensive space technologies needed for colonizing space. As Henson put it, 'It's a long way from military space bunkers to what we are really interested in, the human habitation of space.' But the military angle justified the investment. 'The most significant advantage of this project is that it gives us a toehold in space. Once we have a steady supply line for extraterrestrial materials [...] we will have made the step from being visitors in a hostile environment to being homesteaders in the promised land.'[15]

Gregory Benford (1941–), a physicist and science-fiction author, including of the popular time-travel novel *Timequake* of 1980, similarly sought to ride military-space coat-tails to space colonies (and similarly invoked frontier rhetoric) in an *Omni* article titled 'Zeus in Orbit,' urging the deployment of space-based beam weapons. Benford allowed that military space stations were 'a far cry from the peaceful scenarios of space industrialization, or from the even less likely vision of farmers in the sky so beloved of the L-5 Society.' However, he continued, 'we should face the fact that putting a cap on our nuclear madness is more important.' The military had the money, and once it had established a toehold in outer space, colonies could follow.[16]

Ben Bova (1932–), a former marketing manager for defense firm Avco-Everett (an early producer of high-power lasers) turned *Omni* editor, reversed this approach. In a 1981 book titled *The High Road*, Bova argued, by this time familiarly, that 'all the disasters we face, from nuclear war to ecological collapse to the tide of irrationality, have one factor in common; population pressure.' But that assumed that humans were confined to earth. Humanity could tap the vast resources of the solar system (energy, minerals and so on) to relieve pressure on earth – or, better yet, send humans themselves out into space. Bova also noted that both the United States and Soviets were pursuing space-based beam weapons for missile defense. But this was *not* 'a step away from nuclear holocaust,' because a defensive shield would encourage the nation that had it to launch a devastating first strike, safe from retaliation. In other words, as SDI critics would soon argue, missile defense was destabilizing. The only way to avoid 'a new and deadly space race' was to push the private, commercial development of outer space. Space resources, Bova argued, would ease tensions on earth and civil space systems would crowd out the military in outer space. Military programs need not justify development of civil-space technologies; rather, civilian space programs would prevent the threat of a militarized space.[17]

II   Strange bedfellows

Missile defense, thanks especially to Reagan and SDI, is often seen as an issue exclusive to conservative politicians. In the 1970s, however, elements of the left-leaning counterculture found themselves, sometimes to their surprise, supporting missile defense alongside more conservative space enthusiasts, in particular libertarians. These 1970s debates over space programs made for strange ideological and cultural bedfellows, from Barry Goldwater to Jerry Brown and Timothy Leary.

One historian of the American space program has stated that by the late 1970s, O'Neill's 'vision to move billions of Earthlings to environmentally correct space colonies looked like a dreamy hangover from the radicalism of the 1960s.'[18] The 1960s influence was not just a dream. LSD guru Timothy Leary (1920–96) supported the L5 Society with his SMI^{2}LE (Space Migration, Intelligence Increase, Life Extension) movement; *Future Life* articles criticized creationism and the Moral Majority, while its letters to the editor included calls for legalizing marijuana. L5's president, Carolyn Henson, in particular imparted an environmentalist, feminist sensibility to the group and borrowed techniques from the anti-war and other protest movements for L5's political activism.[19] Such countercultural currents help explain the vocal response to the military-space issue, with many L5 members lamenting the militarization of outer space.

Carolyn Henson herself wondered how she had gone from protesting against the Vietnam War to supporting space lasers for missile defense. 'Had we repented of our fuzzy-brained leftist credos? Were we victims of the creeping conservatism of middle age? Perhaps we'd traded in our ideals for the promise of more hardware in space? Come now. I'd never admit to imperfection. Let me explain why we veterans of the war against war should back space laser ABM.' Her reasons: space lasers for missile defense were surgical; they were elegant; and, above all, they promised a solution to the threat of nuclear war.[20]

An exemplar of this cultural overlap appeared in 1978 in a book by J. Peter Vajk, titled *Doomsday Has Been Cancelled*. Vajk's publisher, Peace Press, was founded by activists from Students for a Democratic Society (SDS) in the 1960s to publish anti-war literature, and for the next two decades it provided a consistent home for progressive politics and alternative culture, publishing books by Timothy Leary, marijuana-growing guides, *The Art of Zen Meditation* and *Mysterious Herbs and Roots*. Vajk's book urged space colonization to dispel the era of limits and restore optimism in the human future. Amid discussion of space stations, space mining and space industrialization, Vajk included a section on space-based lasers for missile defense, perhaps powered by solar-power satellites. With such a system 'we may well see Armageddon becoming far less likely by the late 1990s'; humanity could 'escape from the risk of nuclear holocaust.'[21]

Vajk's book mixed references to C.S. Lewis and Charles Lindbergh with ones to Alvin Toffler, Carl Sagan, Fritjof Capra, E.F. Schumacher, Gail Sheehy and Stewart Brand. Vajk's background did not suggest a spaced-out

1960s guru; after getting his PhD in physics from Princeton in 1968, he joined the Theoretical Physics Division at the Livermore weapons lab in California. At Livermore he worked on astrophysics and cosmology, and developed an interest in space colonization, which he then pursued at defense contractor Science Applications International – and as one of the early activists building the L5 Society. The image of a former Livermore physicist smoothly mixing discussion of the Gaia hypothesis and biofeedback with a positive review of space-based missile defense seems jarring only in retrospect.

Vajk was not alone at Livermore in his eclectic interests. L5 board member Lowell Wood, a protégé of Edward Teller, led Livermore's O Group, a freewheeling outfit that developed advanced weapon concepts including the X-ray laser, which would later be the public centerpiece of SDI. Several O-Group members shared Wood's outer-space interests. Peter Hagelstein (1954–) was one of the main developers of the X-ray laser; his dissertation at MIT in 1981, based on his Livermore research, gave a nod to the 'vivid imaginations of the technological dreamers and the wondrous and exciting universes of science-fiction writers,' especially for speculations about beam weapons based on the X-ray laser, and included citations to Larry Niven's *Ringworld*, a popular science-fiction work on space colonies.[22]

O Group also included Rod Hyde, who had been inspired by science fiction while in high school, went to MIT to study aeronautical engineering, and then to Livermore as a summer intern in 1972, where he designed a rocket using laser-fusion propulsion. Hyde returned to Livermore while working on his PhD under Wood.[23] A technological optimist, Hyde proved a fertile source of solutions to obstacles on the X-ray laser, especially the optics for overcoming beam diffraction, and hence contributed to Wood's (and Teller's) sanguine projections on its potential for SDI.[24] Like Wood, Hyde was an early member of the L5 Society. Later in the 1980s Lowell Wood also hired Stewart Nozette into O Group, after Nozette received his PhD in planetary science from MIT; while in grad school, Nozette had written articles for *L5 News* and *Future Life* about possible mining expeditions in the solar system as a way to dispense with 'gloomy scenarios for our collective futures.'[25]

L5 fellow-travelers included the original Dr. Strangelove. Herman Kahn (1922–83) earned an image as a cold if not bloodthirsty nuclear strategist for his book *On Thermonuclear War* (1960), a product of his work for the Rand Corporation. He had gone on to help found the Hudson Institute, a futurism think tank, and contributed to a Hudson study backing missile defense.[26] In 1976 Kahn and Hudson colleagues published a lengthy report, *The Next 200 Years*, that explicitly challenged pessimistic talk of limits. The study took what Kahn called an 'Earth-centered perspective,' since looking to outer space as a solution seemed like 'a cop-out.' However, Kahn declared, 'It turned out that if you can't solve the problems on Earth, you can solve them in space, as far as we can see.'[27]

Kahn addressed this comment to California Governor Jerry Brown, noted Democrat, and Amory Lovins, a leading proponent of 'soft' or 'appropriate' technology, in a remarkable meeting in Brown's office in 1976. The conversation between 'Governor Moonbeam' and Kahn exemplified the flexible political alignments of this period. Brown had identified the 'era of limits' soon after his inauguration in 1975, and had earlier opposed large, centralized technologies as symptomatic of the ills of modern society. But he now embraced outer space with a typically visionary approach. Shortly after this meeting, the *LA Times* commented on the conversion of 'our new, spaced-out governor': 'Gov. Brown is blasting into space. But to achieve lift-off he has had to jettison much of his old rhetorical baggage. He no longer speaks of an "era of limits." His new high is the "era of possibilities." Nor is small always beautiful. "In space," he exults, "big is better".'[28] For his part, Kahn, who later enjoyed a stint as the Barry Goldwater Professor of American Institutions at Arizona State, described himself at the meeting as a neoconservative: 'Sometimes my friends ask, what do I mean by neoconservative? Well, I'm conservative, but you can't trust me.'[29]

Kahn won the admiration of Stewart Brand (1938–), the countercultural, LSD-dropping founder of *Whole Earth Catalog*. Brand published a chapter from *The Next 200 Years* in his *CoEvolution Quarterly*, with an approving editorial comment that urged more consideration of opposing views, singling out the intolerance of space-colony critics.[30] He also published the transcript to the Kahn-Brown-Lovins meeting, with more editorial notes expressing his admiration for Kahn: 'Herman Kahn is to American political intellectuals what Tom Wolfe is to the journalists [...] they are more open, less predictable, and far more informative than your standard ideologue. When it comes to political, social, military, or economic analysis there is no view I'm more interested in than Kahn's.' Kahn himself had earlier declared, 'I like the hippies. I've been to Esalen. I've had LSD a couple times. In some ways I'd like to join them.'[31]

A year after Kahn's meeting with Governor Brown and Lovins, he put out a report for NASA with William Brown, a Hudson Institute colleague, on the space program. Like Harry Stine, Kahn and William Brown saw a third watershed in human history underway, after the agricultural revolution of 10,000 BC and the industrial revolution of the nineteenth century. The current shift to post-industrial society (or what they called 'super-industrial' society) did not require a space program, but it would surely help: 'It is almost certain that space exploration will lead to great benefits, and possibly to an extraordinary economic and technological impact' – and perhaps 'vast treasures which have not yet been dreamed of.' Outer space was 'in any case a true frontier.' Although the NASA-aimed report mostly skipped over military space, it did mention space-based beam weapons for attacking satellites and also presented, among possible future scenarios, the 'Great Space War of 2019,' a one-day conflagration where several spacefaring countries destroyed each other's satellites and manned space stations. Since the conflict was confined to outer space, the space war was only the 'moderate' scenario; the really 'pessimistic' scenario was a back-to-nature 'triumph of the garden' and rejection of technology. In short, while the report did not see outer space as critical to human future, since humanity could solve its problems with earth-based technologies, it claimed that adding space made things even easier.[32]

III   Star wars

On 23 March 1983, President Ronald Reagan announced a crash program for what became known as the SDI. SDI would consume some $26 billion over the next ten years; as a major shift in nuclear strategy from deterrence to defense, SDI was a centerpiece of American military strategy in the late Cold War, a lightning rod for public debate, and a focal point of diplomatic negotiations with the Soviet Union. SDI was often derided as science fiction, as the sort of fantasy technology – death rays, laser guns and space battle stations – common in science-fiction books and films (Figure 12.2). The popular name for the project, 'Star Wars,' derived from the George Lucas film trilogy – released in 1977, 1980 and 1983 – that exemplified the genre. Several writers have suggested the contributions of science fiction to SDI, usually through cultural influence.[33] Some observers, however, perceived a more direct policy influence from science fiction, via the L5 Society. In 1999 science-fiction author Norman Spinrad (1940–) made waves by claiming that his science-fiction colleague Jerry Pournelle (1933–2017) had conceived the crash program for missile defense, sold the Reagan Administration on the idea and even written Reagan's speech. Pournelle replied that he did not write the speech,

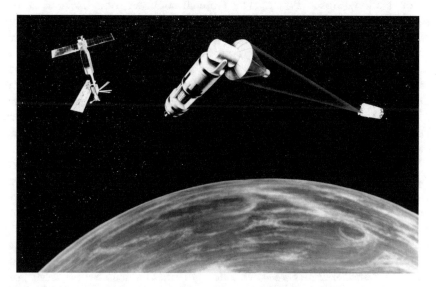

Figure 12.2   A 1984 artist's concept of a space-based laser for SDI.
Source: Courtesy of US Air Force.

but otherwise acknowledged that a group he had led, an offshoot of the L5 Society, was indeed the source of SDI.[34]

Pournelle was a space-colony enthusiast and an L5 board member in the late 1970s. He had advanced degrees in statistics and systems engineering and PhDs in psychology and political science; he had worked for Boeing and North American-Rockwell before establishing himself as a leading science-fiction author in the 1970s, with tales that combined a strong libertarian bent with technological enthusiasm. His book *A Step Farther Out* of 1979 was an explicit response to the Club of Rome, 'an effort to show that we are not faced with doom,' thanks in part to the potential of space colonization.[35] Earlier, Pournelle had coauthored, with his mentor Stefan Possony (1913–95), *The Strategy of Technology*, which held that the US competitive advantage in the Cold War lay in high technology; one of its chapters, titled 'Assured Survival,' argued for a shift from nuclear deterrence to defense. Pournelle was also one of the science-fiction writers cited as an influence by Peter Hagelstein at Livermore.

In November 1980, Harry Stine was thinking of ways to get the newly elected Ronald Reagan to support an ambitious space-exploration program. Meanwhile, Reagan transition-team staff had asked General Bernard Schriever (1910–2005) to help prepare transition papers on space policy for the incoming Administration. Schriever enlisted a former member of his planning staff, Colonel Francis Kane, and Stine and Kane separately contacted Pournelle, who thought to combine the efforts in a sweeping framework for civilian and military space programs. Pournelle, with support from the L5 Society and the American Astronautical Society, formed a 'Citizens Advisory Council on National Space Policy.' The Council convened one weekend at the end of January 1981, soon after Reagan's inauguration, at the spacious Tarzana home of science-fiction author Larry Niven, Pournelle's frequent collaborator. The Council counted about 50 members: eight of them from the L5 board; science-fiction authors Robert Heinlein and Poul Anderson; scientists Freeman Dyson, Marvin Minsky, John McCarthy and Eric Drexler; former astronauts Wally Schirra and Gordon Cooper; weapon designers Lowell Wood and Rod Hyde from Livermore; and futurist writers Barbara Marx Hubbard, Harry Stine and Peter Vajk. Thirty members attended this first meeting, with Pournelle as chairman.[36]

These initial meetings led to a Council report, in spring 1981, titled 'Space: The Crucial Frontier.' Its second and third sentences declared: 'Progress is possible. We do not have to accept limits to growth.' Among chapters recommending space stations, space solar power satellites, space commerce and a manned lunar colony by 2000, the report included a chapter urging space-based laser defenses.[37] According to Pournelle, the Council had a direct line to Reagan's National Security Adviser, Richard V. Allen (1936–), a protégé of Council member Possony; Allen got the Council's report from Possony and brought its message to Reagan.[38] Heinlein supported this claim in a 1985 letter to Barney Oliver, SETI scientist and SDI skeptic, that also expressed hardening ideological positions: 'It was endless effort by a mere handful of us that got the matter to Mr. Reagan's attention and resulted in his "Star Wars" speech.' Heinlein viewed the issue in the starkest light: 'Please understand that this is not just a friendly debate, this is not the world series; this is grim death, the very survival of our country.' He concluded with a query to Oliver: 'Which side are you on, Doctor? "Better Red than Dead?" Or "Better Dead than Red?"'[39]

Spinrad's 1999 article reintroduced the old coat-tails argument, attributing to Pournelle the strategy of supporting SDI as a clever way to attain a long-range space colony.[40] If so, that would add an interesting twist to SDI. Some Reagan advisers have claimed that SDI was a clever sting operation, to force the Soviets to spend themselves into bankruptcy building high-tech space systems.[41] If Spinrad was right, SDI was indeed a sting operation – against America's own military, convincing it to spend money on a chimerical military program that would eventually help humanity populate outer space. Pournelle, however, was a true believer and a strong anti-communist, evident in his 1970 work with Possony on strategic defense and his authoring of the Council report chapter on space-based missile defense; he also wrote a popular article for *Omni* in 1981 warning about Soviet efforts in space-based beam weapons and urging the United States to prepare for space war.[42]

These suggestive links do not mean that the Council created Reagan's policy. There is no contemporary evidence that they played a direct policy role in the Reagan Administration's shift from strategic deterrence to defense.[43] The actual seeds for SDI came rather from technological developments including directed-energy weapons, such as the X-ray laser, but far more so in microelectronics, phased-array radars and infrared focal-plane arrays; from strategic debates over MX missile deployment and ballistic-missile submarine vulnerability; and moral and political concerns about the nuclear arms race, including the Freeze movement and the Catholic bishops' letter on nuclear weapons.

The L5 Society and space enthusiasts in general, however, were among the groups that prepared the ground for SDI through public arguments for space-based missile defense. And the L5 Society was one of the first, starting with Keith Henson's article of 1976, and prominent L5 members became important SDI advocates, in particular Lowell Wood, who made the X-ray laser a centerpiece of the public debate over SDI. The L5 Society also provided more concrete connections thanks to institutional links between L5 and the High Frontier organization, created by General Daniel Graham (1925–95) in 1981 as a political lobby for missile defense. Graham hired several administrative staff members from the L5 Society, including its secretary, who brought with them the Society's political-action experience and, more important, its mailing lists. Pournelle later claimed that 'Dan Graham essentially took the remnants of the L5 Society and built his High Frontier out of it.'[44]

The High Frontier group supplemented the grassroots, L5-style public approach with kitchen-cabinet lobbying by Reagan's friends, such as Karl Bendetsen, Jaquelin Hume and Joseph Coors, that helped bring missile defense more to the Administration's attention.[45] Early L5 activists Pournelle and Vajk also contributed to Graham's *High Frontier* report, for which Heinlein wrote the foreword. The report echoed not only the title of Gerard K. O'Neill's original opus on space colonies, but also the original name of what became the L5 Society. Another contributor to Graham's report, John Bonsma, soon afterward married L5 founder Carolyn Meinel, who had divorced Keith Henson. Meanwhile, Pournelle's next book, *Mutual Assured Survival*, written with Dean Ing and based on the Council's report, boasted a blurb from Reagan.[46]

The space-colony and science-fiction influence appears also in the rhetoric deployed in Reagan's announcement of SDI. One of the most striking lines in his March 1983 speech called on the scientific community, the creators of nuclear weapons, 'to give us the means of rendering these weapons impotent and obsolete.' In 1978 Vajk's book on space colonization had declared that space-based lasers 'would render ICBMs totally obsolete,'[47] and Benford's 1981 *Omni* article, 'Zeus in Orbit,' carried the subtitle, 'Particle-beam weapons threaten to scuttle ballistic missiles and make the MX obsolete.'[48]

IV   Transcending limits

SDI usually appears strictly as a product of the Cold War. This chapter broadens the context for SDI to consider not only the strategic developments of the late 1970s and early 1980s, but also the crisis earlier in the 1970s sparked by *Limits to Growth* and other studies. But this chapter also shows that the era of limits included the threat of nuclear Armageddon, and even in utopian space forums such as the L5 Society the Cold War crept in.

The L5 Society connected the Club of Rome to SDI and the romantic counterculture to military space programs. These connections forced some L5 members to overcome ambivalence regarding the militarization of space, but they also provided SDI with one of its key characteristics. L5 was among the first groups to publicly argue for space-based missile defense, but that was not all. L5 injected its utopianism into the missile defense issue, and thus helped change ideas about missile defense itself. L5's space cadets urged a shift in missile defense from the existing ground-based, incremental approach to a vastly more ambitious, space-based, beam-weapon, Buck-Rogers program. Instead of methodically climbing the ladder from familiar ground-based rocket interceptors, the L5 crowd wanted to shoot straight into space, at warp speed and with phasers set to kill. That is indeed what Reagan proposed with SDI, and this breathtaking ambition was what set SDI apart. That ambition was spurred by groups like L5 in the 1970s, who saw outer space as a way to make strategic weapons 'obsolete.' And this ambition, in turn, was energized by the belief, popularized by the Club of Rome, that there was an existential crisis facing humankind.

SDI and missile defense continued to shape nominally civilian space programs, from reusable launch vehicles to planetary spacecraft to infrared astronomy; Lowell Wood and Rod Hyde kept alive the space-colony dream in the early 1990s with proposals for inflatable space stations using SDI concepts.[49] And military space interests continued to engage the L5 Society; the society itself fractured in the mid-1980s along political lines, thanks in part to SDI, and L5 eventually merged with the National Space Society, which was more closely aligned with the aerospace industry.[50] But space-colony supporters continued to argue that space-weapon programs 'can ensure the growth of infrastructure and enable the establishment of human settlements in space.' Odd political alignments also persisted, though more recently on the other side of the issue: whereas in the 1970s liberals and libertarians could agree on the merits of space-based missile defense, in the 2000s a proposed ban on space weapons was supported from the left (the Berkeley city council) and the far right (militia movements challenging space-based surveillance systems as threats to individual liberty).[51]

The L5 story sheds light on the context of the 1970s, when far-out ideas – asteroid mining, biofeedback or space-based missile defense – could receive a hearing at either end of the ideological spectrum. It also demonstrates how the far left and far right could, like a closed universe, bend around and meet at particular points.[52] In 1983, when L5 had already grown more conservative, an unpublished survey of its members found 31 percent identified themselves as Republicans, 25 percent as Democrats, 8 percent as Libertarians and 33 percent as none of the above. Addressing an L5 conference the next year, Congressman George E. Brown (1920–99) declared rather that L5 members were '5 percent Democrat, 5 percent Republican, and 90 percent anarchist.'[53]

This episode occurred amid an evolving ideological landscape for the general American polity and for the space program in particular. Andrew Butrica has detailed the emergence of a conservative space agenda in the 1980s under Reagan, but its roots reach deeper than Reagan's election.[54] The 1970s marked a conservative ideological turn in the United States, as the New Right merged evangelical moralists, Goldwaterite populists, free-market economists and neoconservatives.[55] Conservatives, traditionally skeptical of the Enlightenment faith in progress and improvability of humankind, became technological enthusiasts. The period also coincided with an ideological sea change for the US space program. As the highest expression of socially directed technical innovation, the early space program received its main support from politicians on the left. But political liberals in the 1960s began directing federal spending toward social problems instead of technoscientific extravaganzas that seemed only to benefit a few scientists and aerospace corporations. As liberals drifted down to earth, conservatives were abandoning fiscal austerity and their traditional aversion to Enlightenment faith in human progress. Conservatives became technological enthusiasts and embraced the vision of space as new frontier: the space program could rekindle the old pioneer spirit, inspiring noble achievements and opening up a new realm for commerce. Some liberal commentators for their part came to view the frontier myth as an emblem of imperial conquest, environmental damage, selective government subsidies and corporate profiteering.[56]

The 1970s provided a moment of flux before new alignments had solidified, when alternative connections could be explored. Thus, for example, despite the countercultural past of some of the L5 Society's leaders, the group embraced the rhetoric of colonization. Outer space was a frontier not just to explore, but also to settle and exploit; hence their references to 'farmers in the sky' and 'homesteaders in the promised land.' The latter echoed the eschatological imagery of the Club of Rome itself, which to some observers seemed to claim that the world was going to hell, figuratively if not literally, and the solution lay in more virtuous living. That was the view of its critics: Kahn viewed the Club of Rome as 'an exercise in moralistic judgments,' a 'package of New Testament virtues and preindustrial simplicity,' with whiffs of 'hellfire and sin.'[57] L5 critics in turn detected a revivalist, religious bent to the space-colony movement, with O'Neill as 'prophet.'[58] One L5 News reader wrote in protest of a 'pseudo-evangelical/fundamental editorial viewpoint,' claiming that '"Space Salesmen" seem very much the harbingers of a new religion. A religion promising to wipe away the problems of war and peace, of population, of nuclear war and wastes, of diminishing resources.'[59] Space colonists, however, countered the virtuous moralism of the 'era of limits' with faith in science and technology. When L5 enthusiasts talked about rising up to reside in the heavens, they meant it literally.

The quasi-spiritual dimension to these debates perhaps reflects the countercultural influence, which injected the romantic, utopian idealism of the 1960s with technological enthusiasm. How could former 1960s feminists, environmentalists and anti-war protesters support missile defense? The 1960s counterculture is supposed to represent a backlash against science and technology, which had become associated with the military, environmental pollution and technocratic challenges to democracy. Technology was the source of society's problems, not the solution. But science and technology were not necessarily opposed to the counterculture. Science and technology reflect their social context, and for the 1960s and 1970s that included the counterculture, as historians are beginning to recognize.[60]

This combination of cosmic idealism and technological enthusiasm was not new, nor confined to the United States. Consider the ostensible target of missile defense, the Soviet Union, whose active space program at the time aroused envy among some American space enthusiasts. The L5 News provided regular updates on Soviet efforts, reprinting articles on Soviet programs and noting that 'scientific bigthink is a fully approved literary genre in the Soviet Union.'[61] In 1981, American space writer James E. Oberg (1944–) predicted that by 2000 the Soviets would have dozens of 'space outposts' with hundreds of residents, 'the first true "space colonists".'[62] Soviet interest likewise had deep utopian roots. Although era-of-limits rhetoric certainly did not jibe with Soviet ideology, Russian and Soviet space enthusiasts had long engaged Malthusian concerns. In the late nineteenth century Nikolay Fyodorov (1829–1903) urged space colonization to relieve population pressure. Fyodorov's disciple Konstantin Tsiolkovsky (1857–1935), who provided the technical and popular foundations of Soviet astronautics and popularized the notion of spaceflight, helped form the Anarcho-Biocosmists (or Cosmists, for short) in the 1920s.[63] Asif Siddiqi has shown how the Cosmists, along with science-fiction authors such as Aleksey Tolstoy (1882–1945) inspired several amateur space societies in the Soviet Union in the 1920s and 1930s that combined mystical and spiritual idealism with technological enthusiasm.[64] To the Cosmists, faith and reason were not mutually exclusive; utopian mysticism and romantic idealism resonated with science and technology.

The space-colony advocates in the 1970s US similarly did not reject science or push for decentralized, small-scale, 'appropriate' technology; on the contrary, they openly embraced big, high-tech systems in their idealistic quest to save the planet. The common denominator of the L5 Society, Gerard K. O'Neill, Herman Kahn, Livermore's O Group, Stewart Brand and Jerry Brown was technological enthusiasm. They quelled the doom-and-gloom talk of the 'era of limits' with their belief that technology could solve existential threats, including that posed by nuclear weapons. The 1970s space cadets helped revive faith in high technology – in this case, very high technology.

Notes

1. California governor Jerry Brown popularized the 'era of limits' term in campaign speeches in 1976. See, for example, William Endicott, 'Brown, on TV, Talks Issues, Concedes Carter is Strong but Retains Delegates,' *Los Angeles Times* (26 June 1976), A5, and Orville Schell, 'Jerry Brown: Economics in an Era of Limits,' *Los Angeles Times* (5 August 1979), G1; see also Dominic Sandbrook, *Mad as Hell: The Crisis of the 1970s and the Rise of the Populist Right*, New York: Anchor Books, 2012, 275–6. The Club of Rome's original report was Donella H. Meadows, Dennis L. Meadows, Jørgen Randers, William W. Behrens III and Club of Rome, *The Limits to Growth: A Report for the Club of Rome's Project on the Predicament of Mankind*, New York: Universe Books, 1972. I thank Patrick McCray for many conversations and generous sharing of sources. Archived issues of *L5 News* are available at www.nss.org/settlement/L5news (accessed 1 October 2017).

2. Francis FitzGerald, *Way Out There in the Blue: Reagan, Star Wars, and the End of the Cold War*, New York: Touchstone, 2000; Paul Lettow, *Ronald Reagan and his Quest to Abolish Nuclear Weapons*, New York: Random House, 2005; Nigel Hey, *The Star Wars Enigma: Behind the Scenes of the Cold War Race for Missile Defense*, Washington, DC: Potomac Books, 2006. Histories of SDI based more heavily on archival research are Donald R. Baucom, *The Origins of SDI, 1944–1983*, Lawrence: University Press of Kansas, 1992, and Andrew Butrica, *Single Stage to Orbit: Politics, Space Technology, and the Quest for Reusable Rocketry*, Baltimore: Johns Hopkins University Press, 2003.

3. Peter J. Westwick, '"Space-Strike Weapons" and the Soviet Response to SDI,' *Diplomatic History* 32.5 (November 2008), 955–79.

4. Idem, 'The International History of the Strategic Defense Initiative: American Influence and Economic Competition in the Late Cold War,' *Centaurus* 52.4 (Fall 2010), 338–51.

5. Paul R. Ehrlich, *The Population Bomb*, New York: Ballantine Books, 1968; Roberto Vacca, *The Coming Dark Age*, Garden City: Doubleday, 1973.

6. On O'Neill, see 'Future Vacation Spot: The L5 Libration Point,' *New York Times* (19 May 1974), 6; and W. Patrick McCray, *The Visioneers: How a Group of Elite Scientists Pursued Space Colonies, Nanotechnologies, and a Limitless Future*, Princeton: Princeton University Press, 2012. On appropriate technology, see Ernst Friedrich Schumacher, *Small is Beautiful: Economics as if People Mattered*, New York: Harper Perennial, 1989, 190–201; and Carroll Pursell, 'The Rise and Fall of the Appropriate Technology Movement in the United States, 1965–1985,' *Technology and Culture* 34.3 (July 1993), 629–37. See also Amory B. Lovins, *Soft Energy Paths: Toward a Durable Peace*, San Francisco: HarperCollins, 1977.

7. G. Harry Stine, *The Third Industrial Revolution*, New York: Putnam, 1975, 14, 8.

8. On the Moon Treaty and other examples of L5 activism, see Michael A.G. Michaud, *Reaching for the High Frontier: The American Pro-Space Movement, 1972–84*, New York: Praeger, 1986, esp. 90–4, and Luca Follis's contribution, Chapter 8 in this volume.

9. Stine, *Third Industrial Revolution*, 14; see also Robert Poole's contribution, Chapter 5 in this volume.

10. Keith Henson, 'Military Aspects of SSPS Power,' *L5 News* 1.9 (May 1976), 2.

11. John Holt, 'Space Research and the Military,' *L5 News* 2.2 (February 1977), 3–4; 'Space Warfare Replies' [quote], *L5 News* 2.7 (July 1977), 6–7; see also letters to editor, *L5 News* 2.4 (April 1977), 14–15.

12. Michael Mautner to editor, *L5 News* 1.11 (July 1976), 5, and editor's response; James E. Oberg, 'Facts and Fallacies of Space Warfare,' *L5 News* 3.10 (October 1978), 9–11, and idem, 'Space Wars,' *L5 News* 4.4 (April 1979), 10–11.

13. Keith Henson, 'Space Forts: Or "Where Are You Obi Wan Kenobi?",' *L5 News* 4.6 (June 1979), 1–3; Keith Henson, 'Space Foxholes: Or Beetle Bailey in Orbit,' *L5 News*, 4.10 (October 79), 3–4.

14. Carolyn Henson, 'Lase the Nukes!,' *Future Life* 26 (May 1981), 21, 42.

15. Henson, 'Space Foxholes.'

16. Gregory Benford, 'Zeus in Orbit,' *Omni* 3.12 (September 1981), 53–5, 115–16.

17. Ben Bova, *The High Road*, Boston: Houghton Mifflin, 1981, 13, 206–18.

18. Howard E. McCurdy, *Space and the American Imagination*, Washington, DC: Smithsonian Institution Press, 1997, 154.

19. Harlan Ellison, 'An Edge in My Voice,' *Future Life* 27 (June 1981), 54; on Leary and on Carolyn Henson, see Michaud, *Reaching for the High Frontier*, 86–9.

20. Henson, 'Lase the Nukes!,' 21, 42. Carolyn's father, Aden Meinel, had followed a similar path from anti-war protester to missile-defense supporter.

21. J. Peter Vajk, *Doomsday Has Been Cancelled*, Culver City: Peace Press, 1978, 160.

22. Peter L. Hagelstein, *Physics of Short Wavelength Laser Design*, PhD thesis, Lawrence Livermore National Laboratory, 1981.

23. William J. Broad, *Star Warriors: A Penetrating Look into the Lives of the Young Scientists Behind Our Space Age Weaponry*, New York: Touchstone, 1985, 119–20, 131.

24. George Chapline interview, 19 April 2007; Robin Staffin interview, 6 November 2007, both with the author.

25. Stewart Nozette quote from 'Prospecting in the Solar System,' *Future Life* 21 (September 1980), 40–3; idem, 'Intrinsically Valuable Materials in Space,' *L5 News* 3.6 (June 1978), 5–6; idem, 'The Asteroid/Meteorite Connection,' *L5 News* 4.10 (October 1979), 6–7. Nozette recently made headlines on charges of attempted espionage for Israel; see Peter Grier, 'Stewart Nozette: The American Who Wanted to Spy for Israel,' *Christian Science Monitor* (21 October 2009), 2.

26. Johan J. Holst and William Schneider Jr., eds, *Why ABM?*, New York: Pergamon Press, 1969.

27. Herman Kahn, Jerry Brown and Amory Lovins, 'The New Class,' *CoEvolution Quarterly* 13 (Spring 1977), 8–39.

28. 'Our New, Spaced-out Governor,' *LA Times* (5 August 1977), D6. Joseph Lelyveld, 'Jerry Brown's Space Program,' *New York Times Magazine* (17 July 1977), 182. In a 1982 interview Newt Gingrich (1943–) noted his curious alignment with Brown, but suggested that 'both Jerry Brown and I are trying to find a way to re-create the spirit of optimism and vision of hope that once characterized America'; see Miles Beller, 'Q&A with Gingrich,' *LA Herald Examiner* (13 April 1982), 2, 4.

29. Kahn et al., 'The New Class.'

30. Herman Kahn, William Brown and Leon Martel, 'From Present to Future: The Problems of Transition to a Postindustrial Society,' *CoEvolution Quarterly* 11 (Fall 1976), 4–17.

31. Kahn et al., 'The New Class,' 8 [Brand], 10, and 24 [Kahn]; Kahn quote on hippies from 1968, in Fred Turner, *From Counterculture to Cyberculture: Stewart Brand, the Whole Earth Network, and the Rise of Digital Utopianism*, Chicago: University of Chicago Press, 2006, 186. Kahn also said, 'I stylize myself as a Stoic. My wife says, "You can't be a 300-pound Stoic," so I'm a neo-Stoic.' Quoted in Kahn et al., 'The New Class,' 36. On Kahn, see also Sharon Ghamari-Tabrizi, *The Worlds of Herman Kahn: The Intuitive Science of Nuclear War*, Cambridge, MA: Harvard University Press, 2005.

32. William M. Brown and Herman Kahn, 'Long-Term Prospects for Developments in Space (A Scenario Approach),' Hudson Institute report HI-2638-RR, NASA report CR-156837, Croton-on-Hudson: Hudson Institute, 1977, 77, 218, 223–4, 259.

33. On science-fiction links to SDI, see William J. Broad, 'Science Fiction Authors Choose Sides in "Star Wars,"' *New York Times* (26 February 1985), C1. Thomas M. Disch, 'The Road to Heaven: Science Fiction and the Militarization of Space,' *The Nation* (10 May 1986), 650–6; H. Bruce Franklin, *War Stars: The Superweapon and the American Imagination*, Oxford: Oxford University Press, 1988, 199–201; Chris Hables Gray, '"There Will Be War!": Future War Fantasies and Militaristic Science Fiction in the 1980s,' *Science Fiction Studies* 21.3 (1994), 315–36; David Seed, *American Science Fiction and the Cold War*, Edinburgh: Edinburgh University Press, 1999, 181–93. Michael P. Rogin, *Ronald Reagan, The Movie: And Other Exercises in Political Demonology*, Berkeley: University of California, 1987, was one of the first to argue that SDI was a case of life imitating art, through Reagan's own acting career, especially *Murder in the Air*, the 1940 film that featured an 'inertia projector' similar to SDI-era beam weapons. Rebecca

Slayton has examined how popular images of lasers shaped their role in SDI; see idem, 'From Death Rays to Light Sabers: Making Lasers Surgically Precise,' *Technology and Culture* 52.1 (January 2011), 45–74.

34. Norman Spinrad, 'Too High the Moon: From Jules Verne to Star Wars,' *Le Monde Diplomatique* (July 1999), posted at mondediplo.com/1999/07/14star (accessed 1 October 2017); Jerry Pournelle, 'Chaos Manor debates,' 25 February 2007; available at www.jerrypournelle.com/debates/nasa-sdi.html#respond1 (accessed 1 October 2017).

35. Idem, *A Step Farther Out*, New York: Ace Books, 1979, 3.

36. The *L5 News* from April 1981 has photos and a summary of the meeting. See also 'Greg Bear Interview Part 2: The Bullet You Don't Hear,' *PBS Wired Science interview* (22 December 2007); available at http://www.pbs.org/kcet/wiredscience/blogs/2007/12/ (accessed 28 May 2014).

37. Citizens Advisory Council on National Space Policy, 'Space: The Crucial Frontier,' Spring 1981; available at www.nss.org/settlement/L5news/1981-council.htm (accessed 1 October 2017).

38. Pournelle interview by Patrick McCray and Peter Westwick, 12 October 2007.

39. Heinlein to Barney Oliver, 22 February 1985, Robert Heinlein papers, UC Santa Cruz.

40. Spinrad, 'Too High the Moon.'

41. For a short summary of the sting theory, see Hendrik Hertzberg, 'Laser Show,' *New Yorker* 76.11 (15 May 2000), 92–6. The theory is presented, for instance, in Robert C. McFarlane, *Special Trust*, New York: Cadell and Davies, 1994, and Peter Schweizer, *Victory: The Reagan Administration's Secret Strategy that Hastened the Collapse of the Soviet Union*, New York: Atlantic Monthly Press, 1994.

42. Jerry Pournelle, 'Space: The Decisive Frontier,' *Omni* 4.2 (November 1981), 30, 138–9.

43. One could also argue that SDI did not produce a major shift in policy, and that deterrence continued as the underlying principle of US nuclear doctrine: see Janne E. Nolan, *Guardians of the Arsenal: The Politics of Nuclear Strategy*, New York: Basic Books, 1989.

44. Pournelle interview; on L5's political activity, see Michaud, *Reaching for the High Frontier*, 85–98.

45. Bendetsen to Joseph Coors, 25 October 1982; Coors to Bendetsen, 18 October 1982; Weinberger to Graham, 24 November 1982; Clark to Graham, 6 December 1982; Teller to Bendetsen; 29 December 1982; Bendetsen to Keyworth, 27 December 1982; and Bendetsen et al., 'Proposal for Inclusion in the President's State of the Union Address,' all in Edward Teller papers, box 273/folder 'Bendetsen,' Hoover Institution, Stanford University. Joseph Coors interview by Donald Baucom, 31 July 1987, and Jaquelin Hume interview by Donald Baucom, 29 October 1987, SDI Interview Collection, Ronald Reagan Presidential Library, Simi Valley, CA.

46. Michaud, *Reaching for the High Frontier*, 228–9. Jerry Pournelle and Dean Ing, *Mutual Assured Survival*, New York: Baen Enterprises, 1984.

47. Vajk, *Doomsday*, 160. Vajk may have picked it up from the Hitchcock film *Torn Curtain* of 1966, in which an American agent played by Paul Newman says the line, 'We will produce a defensive weapon that will make all nuclear weapons obsolete.' FitzGerald, *Way Out There in the Blue*, 23.

48. Gregory Benford, 'Zeus in Orbit,' *Omni* 3.12 (September 1981), 53–5, 115–16. Reagan first used the term after the Joint Chiefs meeting in February 1983 that reconsidered missile defense, described by Reagan in his diary: 'Out of it came a super idea […] a defensive weapon that could make nuclear weapons obsolete.' Ronald Reagan, *The Reagan Diaries*, ed. Douglas Brinkley, New York: HarperCollins, 2007, s.v. 11 February 1983. Pournelle has stated another famous line from Reagan's speech, 'wouldn't it be better to save lives than to avenge them,' was suggested by Jim Baen, editor of the science-fiction magazine *Galaxy*, at the Citizens Council meeting in 1981. The quote does not appear in the Council's report.

49. Butrica, *Single Stage*; Peter J. Westwick, *Into the Black: JPL and the American Space Program, 1976–2004*, New Haven: Yale University Press, 2007. See also Felicity Mellor, 'Colliding Worlds: Asteroid Research and the Legitimization of War in Space,' *Social Studies of Science* 37.4 (August 2007), 499–531.

50. Michaud, *Reaching for the High Frontier*, 243–4; Pournelle interview by McCray and Westwick.

51. John Carter McKnight, 'Let's Weaponize Space,' *Space Daily* (30 January 2003); available at www.spacedaily.com/news/oped-03d.html (accessed 1 October 2017).

52. John Judis, 'Libertarianism: Where the Left Meets the Right,' *Progressive* 44 (September 1980), 36–8.

53. Michaud, *Reaching for the High Frontier*, 120, 97.

54. Butrica, *Single Stage*.

55. Laura Kalman, *Right Star Rising: A New Politics, 1974–1980*, New York: Norton, 2010; Sandbrook, *Mad as Hell*; on the 1970s in general, see also Niall Ferguson, Charles S. Maier, Erez Manela and Daniel J. Sargent, eds, *The Shock of the Global: The 1970s in Perspective*, Cambridge, MA: Harvard University Press, 2010. The extensive literature on neocons includes Jacob Heilbrunn, *They Knew They Were Right: The Rise of the Neocons*, New York: Doubleday, 2008; James Mann, *Rise of the Vulcans: The History of Bush's War Cabinet*, New York: Penguin, 2004; Murray Friedman, *The Neoconservative Revolution: Jewish Intellectuals and the Shaping of Public Policy*, Cambridge: Cambridge University Press, 2005; and Anne Norton, *Leo Strauss and the Politics of American Empire*, New Haven: Yale University Press, 2005. Some commentators argue that no right turn occurred; see, for example, Thomas Ferguson and Joel Rogers, 'The Myth of America's Turn to the Right,' *Atlantic Monthly* 257.5 (May 1986), 43–53.

56. Roger D. Launius and Howard E. McCurdy, 'Epilogue: Beyond NASA Exceptionalism,' in eidem, eds, *Spaceflight and the Myth of Presidential Leadership*, Urbana: University of Illinois Press, 1997, 221–50, here 234–40.

57. James Reston Jr., 'The Wrath of Kahn,' *Omni* 5.12 (September 1983), 69–76. Literal biblical millennialism appeared at the time in the 1970 bestseller, *The Late Great Planet Earth*, and Reagan himself embraced end-times eschatology; see Richard N. Ostling, Michael P. Harris and James Castelli, 'Armageddon and the End Times,' *Time* 124.19 (5 November 1984), 73. Frances FitzGerald has argued that Reagan used SDI as an expression of America's civic religion, following G. Simon Harak's earlier identification of the 'soteriology of SDI.' See FitzGerald, *Way Out There in the Blue*; and Harak, 'One Nation, under God: The Soteriology of SDI,' *Journal of the American Academy of Religion* 56.3 (1988), 497–527.

58. Michaud, *Reaching for the High Frontier*, 98.
59. Gary Goodman, letter to editor, *L5 News* 2.7 (July 1977), 6–7.
60. David Kaiser, *How the Hippies Saved Physics: Science, Counterculture, and the Quantum Revival*, New York: Norton, 2012; Michael D. Gordin, *The Pseudoscience Wars: Immanuel Velikovsky and the Birth of the Modern Fringe*, Chicago: University of Chicago Press, 2012.
61. Iosif Zorich, 'Energy from Space,' reprinted in *L5 News* 4.6 (June 1979), 5–6; Dick Fredericksen, 'When the Soviets Let Their Hair Down,' ibid., 4.
62. James E. Oberg, *Red Star in Orbit*, New York: Random House, 1981, 235–6. Soviet writers criticized American references to 'space colonization' as another example of American imperialism. Frank H. Tucker, 'Soviet Science Fiction: Recent Developments and Outlook,' *Russian Review* 33.2 (1974), 189–200, here 199.
63. Paul Avrich, 'Russian Anarchists and the Civil War,' *Russian Review* 27.3 (1968), 296–306; James T. Andrews, *Science for the Masses: The Bolshevik State, Public Science, and the Popular Imagination in Soviet Russia, 1917–1934*, College Station: Texas A&M University Press, 2003, 85–6; Daniel Gerould, 'On Soviet Science Fiction,' *Science Fiction Studies* 10.3 (November 1983), 341–3. Fyodorov and the Cosmists also speculated that space travel could lead to immortality, another idea revived in the Soviet Union in the 1970s – at the same time as an American revival; see McCray, *Visioneers*.
64. Asif A. Siddiqi, 'Imagining the Cosmos: Utopians, Mystics, and the Popular Culture of Spaceflight in Revolutionary Russia,' *Osiris* 23.1 (2008), 260–88; Aleksey Tolstoy, *The Garin Death Ray*, Moscow: Foreign Languages Publishing House, 1955. Tolstoy entranced generations of Soviet scientists and engineers with his tales of space travel and, in his science-fiction novel *The Garin Death Ray* of 1927, laser-like beam weapons; Peter Zarubin interview by Peter Westwick, 5 October 2006.

Epilogue

CHAPTER 13

Final Frontiers? Envisioning Utopia in the Era of Limits

David A. Kirby

In the summer of 2011, NASA ended their Space Shuttle program after 135 flights with the last mission landing back on earth on 21 July. The Space Shuttles themselves were not just being put out of commission by the agency but were joining other symbols of technological obsolescence, like steam engines and biplanes, by literally becoming museum pieces. So, on 17 April 2012 the Space Shuttle *Discovery* was transported to Washington, DC to be put on display at the Smithsonian National Air and Space Museum. I watched news coverage of this event on BBC television where the newscasters waxed poetically about the Shuttle and bemoaned the state of a species that was no longer reaching for the stars. The segment's tone of melancholic nostalgia was summed up at the very end as they showed the transport vehicle touch down with a newscaster claiming 'and with that landing, we have the end of the age of discovery.'

When I heard the newscaster utter this sentence I thought, *Really? The end of the age of discovery?* I understand hyperbole in news casting. *But, really?* If only these newscasters had looked around they would have found plenty of social, cultural and economic evidence for the continuing existence and appeal of space travel. These BBC newscasters could have looked at the United Kingdom's former colony India or to China for evidence of governments that still prioritize space travel in the twenty-first century. They could have also looked beyond governments at private industry where space exploration

David A. Kirby (✉)
University of Manchester, Manchester, UK
e-mail: david.kirby@manchester.ac.uk

© The Author(s) 2018
Alexander C.T. Geppert (ed.), *Limiting Outer Space*
European Astroculture, vol. 2
https://doi.org/10.1057/978-1-137-36916-1_13

companies including SpaceX and Virgin Galactic have been operating since 2002 and 2004 respectively. The newscasters should have also been aware that the top-grossing movie of all time was a 2009 film featuring space travel to the utopian planet Pandora. Pandora was a planet so wonderful that a supposedly new psychological condition emerged called Post-*Avatar* depression syndrome that manifested through adults depressed about the fact that they would not be able to visit this planet in real life.[1] The subsequent critical and box office success of films like *Gravity* (2013) and *Interstellar* (2014) shows that our desire for space exploration is stronger than ever – even after the retirement of the Space Shuttle program.[2] To paraphrase Mark Twain: Reports concerning the death of the age of discovery have been greatly exaggerated.

The eleven chapters in this volume, together with the introduction and this epilogue, explore an earlier time period when many commentators also believed the age of space exploration to be over. The authors show how outer space was still a central component of our collective imagination in the 1970s. The complex social, political and cultural discussions surrounding space exploration during that decade still reverberate today. The launch of the first Space Shuttle *Columbia* on 12 April 1981 was just a highly visible symbol of continued space activity throughout a decade that was supposed to have witnessed the death of space travel on 19 December 1972 when Apollo 17 returned to earth. Space programs were still operating in the United States and the Soviet Union in the 1970s. A consortium of European countries had developed an independent space agency during the decade – the European Space Agency – in order to get out from under their dependence on the United States and the Soviet Union launching European satellites.[3] The chapters in this volume address new configurations of outer space utopias in the face of the very real limits placed by the 1970s and in the shadow of old utopian ideas developed in the 1950s and 1960s.

It is a long-standing joke amongst non-German speakers that the Germans seem to have specific, very long words to describe every unique situation, idea or thing. Leave it to the Germans, then, to have the perfect word to capture why astroculture still thrived in the 1970s and continues to thrive today – *Aufbruchstimmung*.[4] Not only does this word mean the 'expectation of wonderful things,' it is also a word that is often associated with traveling or new ventures.[5] Aufbruchstimmung is what makes space exploration unique amongst our technoscientific activities. By its very nature, human space travel is a story about severe limitations. Early rocket scientists and engineers were well aware of the physical limitations that kept humanity earthbound.[6] Without the expectation of wonderful things cultivated by astroculture we might never have made any attempt to transcend these limits.[7]

There is another very long, specific German word that explains why it became increasingly difficult for Aufbruchstimmung to generate public excitement for space exploration in the 1970s – *Katerstimmung*.[8] Katerstimmung is the perfect word to capture the post-Apollo climate that space travel faced in the 1970s. Not only does the word mean hangover, it also refers specifically to a post-celebration hangover. Certainly the celebration of space accomplishments in the 1960s gave way in the 1970s to a growing awareness of the problems facing humanity on earth and the inability of space exploration to solve these problems. There were certainly still attempts by the proponents of space travel to reignite a general feeling of Aufbruchstimmung, and there were some notable successes during the decade such as the space stations Skylab and Salyut 1. But the feeling of Katerstimmung created a melancholic atmosphere in the 1970s that overwhelmed these efforts as it became more difficult to generate any public enthusiasm for space projects.

I   Optimistic past meets bitter present

The 1970s were a period of transition in how we perceived human spaceflight. Space exploration may be about the expectation of wonderful things in the future, but societal attention in the 1970s shifted explicitly to the problems the world was facing in the present. New limits to the goal of human spaceflight emerged that were not imposed by nature like gravity or the vacuum of space, but instead grew out of humanity's increasing concern about how it had treated the natural world. The first Earth Day celebration took place in 1970 as a way to call attention to environmental problems. The 1972 book *The Limits to Growth* commissioned by the Club of Rome reinforced fears of environmental activists that the earth could not sustain the level of human growth.[9] Several contributors to this volume agree that there are essential links between space programs and the development of environmentalism. Images of the earth from space, like the famous Earthrise picture, played a significant role in the growing environmental movement by showing how fragile the planet looked and how limited our resources really were.[10] Space travel in the 1960s showed us why we needed to care about the earth, but for some people space travel in the 1970s also offered a solution to these environmental problems.

Narratives about space exploration in the 1950s and 1960s were also about looking backward to move forward. Before 1970, stories concerning space travel's 'new frontiers' sought to replicate the feelings of adventure and conquest associated – at least in popular culture – with the old frontiers of Europe's colonial past. However, these narratives of conquest were not appropriate for the environmentally conscious 1970s. Discussions about exploring new frontiers had too many connotations of the environmentally destructive conquest of the 'new world' by Europeans.[11] So advocates for space exploration put forward a new narrative based on the utopian vision of space exploration as an antidote to the dystopian novel come to life that was the 1970s. The blackness of outer space had become a green promised land. For technological optimists such as the Committee for the Future, Gerard K. O'Neill, the L5 Society and the Association of Space Explorers, space colonies and space stations fit easily into a rhetoric about outer space as a refuge from an ecologically destroyed earth.[12] These narratives of technological salvation were also presented as a means to prevent nuclear war in the 1970s and 1980s.[13] Groups like the L5 Society portrayed space colonies simultaneously as the means by which humanity could alleviate the population problem on earth and as a technology that would be the first step toward developing a world-saving missile deflection shield such as the Strategic Defense Initiative. Their convictions mirrored the same transcendent promises for outer space found in the rhetoric of space travel's earliest proponents. It was an ideology built upon a vision of salvation – and the heavens – if humanity chose the right technological path for its future. But, the golden age promised by space boosters never arrived. The failure of NASA and ESA to build on the successes of the 1960s meant that the 1970s were littered with the dashed hopes of the true believers.[14]

The promise of a technological utopia certainly appealed to some members of the counterculture such as Stewart Brand and The Point Foundation whose miniscule funding actually propelled O'Neill into his celebrity career.[15] The move toward the post-industrial world of the 1960s was a story of technocratic optimism, but the changes in late capitalism during this period led to the rise of the individual as a site of power in the 1970s.[16] The balance between claims of rampant powerful global forces shaping the world and the rise of individual agency in the decade changed the sense of what this technological optimism should be for. For some, satellites were a truly transnational development whereby global dissemination of ideas could turn a locally produced countercultural product like Brand's *Whole Earth Catalog* into a literal world view.[17]

But, what had once been a mainstream conviction in the utopian technological promise of outer space had become a minority view in the 1970s. It was a remnant of a pre-Apollo past that had turned into a countercultural belief mainly limited to leftists on the West Coast of America, who saw technology as part of the growing New Age movement. Most environmental or social activists did not share these narratives of technological salvation through space exploration. For them, space travel was not considered a solution to environmental issues; rather, they saw continued funding of space activities as contributing to the problems facing society on earth.[18] The oil and economic crises of the 1970s confirmed their fears that humanity did not have the resources to combat the earth's environmental destruction. For most environmentalists, space programs were not only a colossal waste of money, they were literally stealing resources from the earth. Various groups like Friends of the Earth and Greenpeace (established in 1969 and 1971, respectively) complained about space, but the attitude toward the space program coming out of the environmental movement is summed up more appropriately through one of the most notorious and radical groups to emerge out of the 1970s: Earth First! Their name was a direct rebuke to anyone who would think about spending resources on something as frivolous as going into outer space.[19]

Space enthusiasts like the British Interplanetary Society and the L5 Society may have believed that space promised a refuge from an ecologically destroyed earth, but it was that kind of thinking that made them villains to environmentalists who argued that they were guilty of irresponsible daydreaming. In the 1970s, it seemed preposterous to environmentalists that space stations or colonies on Mars justified the resources that could be devoted to other more pressing problems on earth. Most environmentalists would have taken O'Neill's notion of the 'humanization of space' as more of a menace than a promise of salvation. Environmentalists saw something sinister in the fusion of the Space Age and the Nuclear Age.[20] It may well be the case that once we establish colonies in space, humanity will become both nuclear war-proof and able to survive environmental destruction on earth. But this was a pretty bleak message to convey to people. The grimness of this solution is evident in the 1972 film *Silent Running*. Images of the Valley Forge and the other spaceships traveling through the blackness of space provide a gloomy juxtaposition of dirty impersonal machines with tiny pods of green that represent the last remaining earth plants in the universe (Figure 13.1). Sure, space travel may allow us to save our natural resources, but it is a pretty sad solution to consider a space station in outer space to be the last existing 'wilderness' habitat.

Figure 13.1 This still from the 1972 film *Silent Running* depicts a grim technological solution to environmental problems with the last remaining forests existing only in spaceships.
Source: Courtesy of Universal Pictures.

II Success begets boredom

Whether economic resources were shifted to environmental priorities or not, it is clear that financial support was being taken away from space programs in the 1970s as governments decided to spend their money on what they saw as more pressing, or worse, more *interesting* projects. Space exploration programs were the victims of their own success.[21] What do you do when you have built up the Aufbruchstimmung, this expectation of wonderful things, to hyperbolic levels with visions of cities on the moon and holidays on Mars, and then your real successes begin to seem pedestrian and disappointing rather than astonishing? The 1970s were, as Martin Collins calls it, an 'in-between decade' where a gap was emerging between space agencies and the public regarding their enthusiasm about the prospect of further space exploration.[22] Agencies like NASA and ESA still considered outer space to be a place of adventure, but for the public space exploration had begun to feel like an unexciting endeavor. This public perception developed in spite of the fact that there was still significant uncertainty and danger involved in space travel as best evidenced by the near-disaster of Apollo 13. Apollo 17's 1972 mission might have involved significant risk, but it was the sixth moon landing. To the public, this mission did not appear substantially different from the previous moon landings and this gave the impression that space travel was now just a matter of routine for NASA.

The perception of human spaceflight as mundane rather than glamorous was reflected in the cinema of the 1970s. Movies like *Solaris* (1972) and *Eolomea* (1972) showed a more gritty notion of space travel with spaceships depicted as heavily inhabited spaces where the astronauts are more like dock workers than fighter pilots. Images of the *Nostromo*'s common area in *Alien* (1979), for example, demonstrate the unglamorous nature of space travel by highlighting the messiness of the shared space, the disheveled nature of the crew's clothing and the crew's casual body language such as a crewman putting his feet up on the table (Figure 13.2). The excitement of space travel was replaced by the monotony of hanging out on a space station. Being an astronaut on Skylab conveyed all the thrill of being a toll booth operator.[23] Homer Simpson's response in an episode of the *Simpsons* to accidently stumbling across television coverage of a Space Shuttle launch – 'Boring!' – echoed by Bart's shout of 'No, not another boring space launch!' perfectly sums up what most people had begun to feel about real-life space missions in the 1970s.[24] This *Simpsons* episode also hits upon why this lack of public enthusiasm was an issue for NASA with the fictional NASA administrator telling his colleagues: 'People, we're in danger of losing our funding. America isn't interested in space exploration anymore.' One solution for space programs in the 1970s could have been to go even bigger: to devote sufficient resources to space exploration to undertake even larger-scale projects, like massive space stations, so that space exploration would become interesting again. However, NASA in particular was a conservative organization that was not ambitious

Figure 13.2 This still from the 1979 film *Alien* shows the spaceship crew looking and acting more like blue collar workers than heroic astronauts.
Source: Courtesy of Twentieth-Century Fox.

enough to justify that kind of monetary expenditure.[25] NASA had become incapable of conjuring the visions of utopia they had once projected in the 1960s.

In 1970s Britain it was just as difficult maintaining public enthusiasm for space travel because space exploration was also competing with its more successful terrestrial space science cousins in radio astronomy. The Jodrell Bank observatory was a national British treasure that was far cheaper than space travel and its images of the universe actually exceeded the expectations of amateur astronomers. The British Interplanetary Society's activities had always been dependent on stoking people's imaginations about the possibility of space travel, but by the 1970s they could no longer stir the public's imaginations the same way given the reality of space travel and its disappointments. *Dan Dare* comics may have been the perfect vision of the future in the 1950s, but when space travel failed to live up to this utopian fantasy it was a bitter disappointment in the 1970s. The United Kingdom's first, and only, successful satellite launch through the Black Arrow rocket program – Prospero – came only after the program had already been canceled. This incident provides a good representation of attitudes toward space travel in the 1970s: a resounding success followed by a shrug from the public.[26]

III   Searching for meaning in an age of limits

Yet, the chapters in this volume challenge the notion that the 1970s were the end of utopia concerning human spaceflight. The ideals and visions for space exploration emerging out of the 1950 and 1960s had changed, but new utopian visions emerged to replace the old utopian notions even for traditional space programs like NASA. Of course, O'Neill and his space colonies fit within this new conception of utopia, but the changes in utopian thinking went beyond ideas for new space technologies. In the 1960s the meanings of space exploration were closely tied to Cold War political and military motivations – 'cosmopolitics' – which were hardly a solid foundation for utopian narratives. By the 1970s, though, the meanings of outer space had shifted toward a counterculture-inspired utopian narrative of transcending earthly politics where the earth's lack of visible borders when seen from space would lead to transnationalism. Like the multinational, multiracial, mixed-gender crew of the USS Enterprise on *Star Trek*, there was hope that space exploration would inspire humanity to build a unified earth civilization, then a galactic civilization and, finally, a universal civilization. Thus, the utopian visions of space exploration in the 1970s were often about collaborations. Certainly, the joint US/USSR 1975 Apollo-Soyuz Test Project fits into this move toward collaboration. But there was also a shifting away from a focus on the US/USSR as Europeans began to feel like second-class citizens in any collaborative projects. European nations seemed to always be 'bumming rides' for their satellites and payloads on American or Soviet rockets. When your supposed collaborators continually treat you like a junior partner, then it makes true collaboration seem like a utopian ideal.[27]

Along these lines, many governments envisioned spaceflight as the means by which to transcend the limits of Cold War divisions. The vision of a Space Age utopia that existed in the 1950s and 1960s emerged within a specific Western European, American and Soviet perspective. Even the 1967 Outer Space Treaty on peaceful uses of outer space was the product of Cold War thinking based on East versus West. The mere fact of discussing the moon as a resource already established it as an entity that could be owned, which was a concept that was in the best interests of developed nations even if developing countries hoped to share in this resource at some point. As the 1970s progressed it became clear that economic issues and the energy crisis would highlight a North-South divide rather than old Cold War divisions. Supposed third-world countries had their own conceptions of a Space Age utopia that significantly differed from the utopian narratives of the developed world.[28] The 1979 Moon Agreement promised a share of the moon pie for third-world countries. But by the end of the 1970s, third-world countries began to see that language about the benefit of all humanity was really about the benefit of those that could actually get to the moon and that they would be left out. Why would countries that would not even share their ability to launch telecommunications satellites be willing to share in the moon's resources? Collaboration might have been a utopian theme in the 1970s, but collaboration was still on the terms of those with technology. This is why nations that possessed the ability to launch their own spacecraft and groups like the L5 Society ultimately helped kill the 1979 Moon Agreement.[29]

This perceived lack of a common purpose was one reason why the public heavily questioned the integrity of the scientific community in the 1970s. Critiques in this decade differed from previously expressed concerns about science because detractors often aimed their critiques at scientists who were nominally working on projects designed for peaceful purposes like space exploration. Critics of devoted-space proponents like O'Neill, Barbara Marx Hubbard and G. Harry Stine were not calling into question their priorities based on scientific grounds; rather they viewed these space enthusiasts as immoral for prioritizing technological solutions to what they saw as a problem of human behavior.[30] Likewise, the field of space exploration took a public-relations hit in the 1970s as the public began to realize that a 'peaceful' scientific agency like NASA was not only involved in research designed to visualize outer space but also in research intended to reveal human targets to the US military in Vietnam. Astronauts themselves were also beginning to see their image tarnished in the media. The formation of the Association of Space Explorers in the early 1980s represented less a collaborative endeavor than a symbol that astronauts, cosmonauts or space explorers – who could not even agree on a name – were a fractured fraternity.[31] And, importantly, it was an unquestionably male fraternity well into the 1980s. Criticism of NASA forced the organization to shift from its white male version of the future toward the utopian multiracial, multigender crew future depicted in *Star Trek*. Even LEGO colorized their astronauts in order to ensure the inclusion of black astronaut minifigures in the late 1970s.[32] However, as with the crew in *Star Trek*, it was still white males who were in charge of space travel at the start of the 1980s.

Although astronauts had become a bit tarnished in the 1970s, as a group they are still regarded as heroes today. De Witt Douglas Kilgore has identified the belief that space travel was the key to a future of unlimited human progress as 'astrofuturism.'[33] By the 1970s, it was certainly hard to ignore the religious connotations astrofuturists had imparted to space exploration through their worship of astronauts and the spiritual dimension ascribed to space travel.[34] Leo Tolstoy once claimed that 'science is meaningless because it gives no answer to our question, the only question important for us: "What shall we do and how shall we live?"'[35] Many people would agree with his assessment, but proponents of space exploration believe that science *can* answer this question and that the continuing development of human spaceflight is the key to that answer. Much of the astroculture in the 1970s was grappling with the idea that spaceflight was not a spiritually transcendent activity, but rather it was a way of exposing our limitations. Screenwriter and director Joss Whedon, who wrote many space-themed movies including *Titan A.E.* (2000), *Serenity* (2005) and *Avengers Assemble* (2012), drew this message from 1979's *Close Encounters of the Third Kind*:

> More than anything, seeing *Close Encounters of the Third Kind* was a germ that opened my mind: the idea that Roy was going to leave Earth and travel through space, and that when he came back it would be several decades later and everybody he had known would be dead, hit home the reality of being human. It made me consider what we are, what we can be,

what our limitations are. That blew the brains out of my head and I wore them on my shoulders as epaulettes. When I got back to school I told my best friend what had happened and he handed me a copy of *Nausea* by Jean-Paul Sartre. Basically, the film had made me an existentialist.[36]

In the late 1970s *Close Encounters of the Third Kind* made Whedon consider: Is space travel worth it? What do we get at the end of the day for investing in this endeavor? Everyday hero Roy certainly gets to experience the wonders of the universe, but he loses what he had on earth because he is limited as a human being.

The limitation of being human is a sentiment that was already evident at the apex of the Space Age in the film *2001: A Space Odyssey* with its quasi-religious allegory showing a tension between our optimistic secular notions of technological progress trapped within an ultimately pessimistic narrative about humanity's inherent failings as a species. According to the film, human behavior itself becomes a limit to space travel. Humanity is capable of creating such wonderful technologies like spaceships, but humans are ultimately flawed animals whose behavioral imperfections get in the way of their own salvation. The substantial social, political and environmental problems that emerged in the 1970s reinforced the film's message that human behavior was still rooted in our evolutionary past. No amount of technological development could save humanity unless there were fundamental changes in human nature. *2001* made it clear that it was an illusion to think that the Space Age could save us from the self-created dangers of the Nuclear Age. Ultimately, the film implies that unless some benevolent aliens are willing to help us change ourselves, then utopia is out of reach.[37]

Increased concern about the state of humanity in the 1970s prompted changes in broader scientific research priorities. The rise in primatology, as well as the development of molecular biology, shifted the scientific focus from outer space to the internal spaces of human nature. Part of this shift from looking outwards to looking inwards was also due to discoveries made about outer space itself. Before space travel, outer space was full of mystery and the promise of exciting new worlds to explore, but all we encountered when we finally left the planet in the 1960s was the emptiness of outer space and a barren celestial body. The realization that humans are likely alone in the universe was the catalyst for many people to not only look more closely at our planet but also to try and understand ourselves better as a species. This shift toward greater self-reflection in the sciences and in society matched the general trend in the 1970s toward a more individualized society.[38] This cosmologically driven self-reflection became a staple of astroculture in the 1970s, including a new literary genre dubbed 'inner-space fiction.' Novels in this genre critiqued the 'utopian anthropocentrism' of the Space Race by using a medium that novelists from non-spacefaring nations, including Doris Lessing, A.S. Byatt and John Banville, argued was perfectly suited to the task of looking inwards rather than outwards. If our experiences with outer space had left humanity feeling alone and searching for meaning, then inner space fiction novels allowed humans to do what they do best – make it all about themselves.[39]

IV   Transcending limits and reaching for outer space in a post-Apollo world

This volume is titled *Limiting Outer Space*, so it is worth concluding by examining how these chapters have engaged with the notion of limits or, more appropriately, how they have explored how space advocates and others still envisioned utopias in the face of limits in the 1970s. One of the primary themes that emerges in this volume is that humanity has a very difficult time envisioning limits. Humans are outstanding at creating all sorts of limits, but they also have the capacity for envisioning the ways we could transcend the limits, whether physical, economic, psychological, social, cultural or spiritual that keep us earthbound. This volume highlights the role that well-developed imagined futures play in actually making these futures become realities. More so than in any other technoscientific activity, storytelling is essential in keeping dreams of space travel alive and in inspiring people to want to see these astrofutures come to pass. It is hard to convey the necessity of space travel to the public without telling stories about how wonderful the future will be if we could only travel to the stars.

Space exploration's success as a technoscientific endeavor was – and is – dependent on the 'persuasive fictions' that its proponents create.[40] It would have been difficult to excite the public about the space stations that emerged in the 1970s without good storytelling. Orbital space stations became a way to remain in outer space, but they needed to be just as enticing for the public as the original spaceships of the 1960s. Proponents needed to craft stories that would convince the public that space stations represented the pathway to a new and exciting future, when they were essentially just talking about what amounted to depressingly bland tin cans orbiting the earth.[41] Science-fiction authors like Jerry Pournelle also utilized their storytelling skills to convince the Soviets that they needed to spend funds to compete with a technology – the Strategic Defense Initiative – that *only* existed in stories.[42] The need for storytelling about outer space is why ESA and NASA became involved with the Legoland Space toy line in the 1970s. Real-life space exploration had become limited, mundane and no longer able to excite the public. If NASA and ESA could not provide children with the wonderful futures promised by previous astroculture, then LEGO allowed for 'unlimited play in a world of limits' so that children could create their own utopias. Perhaps the best way to make sure that our expectations for wonderful things are met is to create those wonderful things ourselves through the telling of our own personal stories about space exploration.[43]

By its very nature, human space travel is a story about limits and our attempts to transcend these limits. This brings us back to Aufbruchstimmung and the expectation that space exploration will lead to wonderful things. This volume shows that it is impossible for human cultures to ever meet the wonderful expectations that space boosters, space agencies, governments and our collective astroculture have set for space travel. One of the central motivations for human space travel is to escape the issues facing society here on earth. Yet, the failure of space travel to alleviate social ills and conflicts on earth was one factor contributing to the Katerstimmung of the 1970s. The problems facing the earth were not only still existent but had multiplied to include concerns about overpopulation, land ownership and exploitation of resources. The fact is that wherever humans go we will bring these problems with us. One of the major ideas this volume brings to light is that, from a human perspective, there is no outer space, there are just the spaces humans inhabit.

Notes

1. See Jo Piazza, 'Audiences experience *Avatar* blues,' *CNN Online*, 11 January 2011, available at http://www.edition.cnn.com/2010/SHOWBIZ/Movies/ 01/11/avatar.movie.blues/ (accessed 1 October 2017).
2. Each of these films had significant input from space scientists during production. See Daniel Clery, 'The Theoretical Physicist Behind *Interstellar*,' *Science* 346 (November 2014), 800–1; and Paul Smaglik, 'Media Consulting: Entertaining Science,' *Nature* 511 (July 2014), 113–15.
3. For a general history of the European Space Agency, see, John Krige, Arturo Russo and Lorenza Sebesta, *A History of the European Space Agency, 1958–1987*, 2 vols, Noordwijk: ESA, 2000; see also Tilmann Siebeneichner's contribution, Chapter 11 in this volume.
4. On the concept of astroculture, see Alexander C.T. Geppert, 'Rethinking the Space Age: Astroculture and Technoscience,' *History and Technology* 28.3 (September 2012), 219–23.
5. Thanks to Ralf Bülow for bringing the word *Aufbruchstimmung* to my attention.
6. For a general history of early rocketry, see Frank H. Winter, *Rockets Into Space*, Cambridge, MA: Harvard University Press, 1990.
7. There are a number of works that explore astroculture's role in creating a public desire for space exploration. For example, see Howard E. McCurdy, *Space and the American Imagination*, Washington, DC: Smithsonian Institution Press, 1997; David A. Kirby, *Lab Coats in Hollywood: Science, Scientists, and Cinema*, Cambridge, MA: MIT Press, 2011; and Alexander C.T. Geppert, ed., *Imagining Outer Space: European Astroculture in the Twentieth Century*, Basingstoke: Palgrave, 2012 (= *European Astroculture*, vol. 1).
8. Thanks to Alexander Geppert for bringing the word *Katerstimmung* to my attention; see also his contribution, Chapter 1 in this volume.
9. For a history of the development of the environmental movement in the 1970s, see John McCormick, *Reclaiming Paradise: The Global Environmental Movement*, Bloomington: Indiana University Press, 1989.
10. A discussion about the relationship between the Earthrise photographs and the environmental movement can be found in Robert Poole, *Earthrise: How Man First Saw the Earth*, New Haven: Yale University Press, 2008.
11. See Luca Follis's contribution, Chapter 8 in this volume.

12. See the contributions by Andrew Jenks (Chapter 9), Roger Launius (3) and Peter Westwick (12).
13. See Peter Westwick's contribution, Chapter 12 in this volume.
14. See Roger Launius's contribution, Chapter 3 in this volume.
15. See the contributions by Martin Collins (Chapter 2) and Peter Westwick (12).
16. See Florian Kläger's contribution, Chapter 6 in this volume.
17. See Martin Collins's contribution, Chapter 2 in this volume.
18. See the contributions by Roger Launius (Chapter 3) and Peter Westwick (12).
19. See the contributions by Andrew Jenks (Chapter 9) and Peter Westwick (12).
20. See Peter Westwick's contribution, Chapter 12 in this volume.
21. See the contributions by Andrew Jenks (Chapter 9), Roger Launius (3), Doug Millard (4) and Regina Peldszus (10).
22. See Martin Collins's contribution, Chapter 2 in this volume.
23. See Regina Peldszus's contribution, Chapter 10 in this volume.
24. *The Simpsons*, Episode 1F13, 'Deep Space Homer,' 24 February 1994.
25. See Roger Launius's contribution, Chapter 3 in this volume.
26. See Doug Millard's contribution, Chapter 4 in this volume.
27. See the contributions by Andrew Jenks (Chapter 12), Regina Peldszus (10) and Tilmann Siebeneichner (11).
28. See Luca Follis's contribution, Chapter 8 in this volume.
29. See Peter Westwick's contribution, Chapter 12 in this volume.
30. See the contributions by Andrew Jenks (Chapter 9), Roger Launius (3) and Peter Westwick (12).
31. See Andrew Jenks's contribution, Chapter 9 in this volume.
32. See Thore Bjørnvig's contribution, Chapter 7 in this volume.
33. De Witt Douglas Kilgore, *Astrofuturism: Science, Race, and Visions of Utopia in Space*, Philadelphia: University of Pennsylvania Press, 2003.
34. See Roger D. Launius, 'Escaping Earth: Human Spaceflight as Religion,' *Astropolitics* 11.1–2 (2013), 45–64.
35. Max Weber, 'Wissenschaft als Beruf,' in idem, *Wissenschaft als Beruf, 1917/1919; Politik als Beruf, 1919*, Wolfgang J. Mommsen and Wolfgang Schluchter, eds, Tübingen: Mohr, 1992, 72–111, here 93; quoted after Max Weber, 'Science as a Vocation,' in Hans Heinrich Gerth and C. Wright Mills, eds, *From Max Weber: Essays in Sociology*, New York: Oxford University Press, 1946, 129–56, here 143.
36. Quoted in Gemma Kappala-Ramsamy, 'Joss Whedon: The Film that Changed My Life,' *Guardian* (15 April 2012), available at www.theguardian.com/film/2012/apr/15/joss-whedon-film-changed-spielberg (accessed 1 October 2017).
37. See Robert Poole's contribution, Chapter 5 in this volume.
38. See Martin Collins's contribution, Chapter 2 in this volume.
39. See Florian Kläger's contribution, Chapter 6 in this volume.
40. Thanks to Debbora Battaglia for bringing the phrase 'persuasive fictions' to my attention.
41. See Regina Peldszus's contribution, Chapter 10 in this volume.
42. On SDI, see the contributions by Tilmann Siebeneichner (Chapter 11) and Peter Westwick (12).
43. See Thore Bjørnvig's contribution, Chapter 7 in this volume.

BIBLIOGRAPHY

Filmography

The Invaders, television series created by Larry Cohen, 43 episodes, USA 1967–68 (Quinn Martin Productions).

Was sucht der Mensch im Weltraum? (What Are Men Looking for in Space?), television series created by Heinz Haber, 13 episodes, BRD 1968 (ZDF).

2001: A Space Odyssey, directed by Stanley Kubrick, USA 1968 (Metro-Goldwyn-Mayer).

Marooned, directed by John Sturges, USA 1969 (Columbia Pictures).

Erinnerungen an die Zukunft, directed by Harald Reinl, BRD 1970 (Terra Filmkunst).

Die Delegation (The Delegation), directed by Rainer Erler, BRD 1970 (Bavaria Film).

Signale: Ein Weltraumabenteuer (Signals: A Space Adventure), directed by Gottfried Kolditz, DDR/PL 1970 (VEB DEFA-Studio für Spielfilme/DEFA Gruppe Roter Kreis/Film Polski/Iluzjon).

U.F.O. – S.H.A.D.O., television series created by Gerry Anderson and Sylvia Anderson, 26 episodes, UK 1970–73 (ITC Films).

Earth II, directed by Tom Gries, USA 1971 (Metro-Goldwyn-Mayer).

Eolomea, directed by Herrmann Zschoche, DDR/USSR/BUL 1972 (VEB DEFA-Studio für Spielfilme/Künstlerische Arbeitsgruppe Berlin Mosfilm/Boyana).

Silent Running, directed by Douglas Trumbull, USA 1972 (Universal Pictures).

Solaris, directed by Andrey Tarkovsky, USSR 1972 (Creative Unit of Writers & Cinema Workers/Mosfilm/Unit Four).

Ukroshcheniye ognya (Taming of the Fire), directed by Daniil Khrabrovitsky, USSR 1972 (Mosfilm).

La planète sauvage (Fantastic Planet), directed by René Laloux, FR 1973 (Argos Films).

Moskva – Kassiopeya (Moscow – Cassiopeia), directed by Richard Viktorov, USSR 1973 (Gorky Studio).

Sleeper, directed by Woody Allen, USA 1973 (Rollins-Joffe Productions).

Soylent Green, directed by Richard Fleischer, USA 1973 (Metro-Goldwyn-Mayer).

© The Editor(s) (if applicable) and The Author(s) 2018
Alexander C.T. Geppert (ed.), *Limiting Outer Space*
European Astroculture, vol. 2
https://doi.org/10.1057/978-1-137-36916-1

Welt am Draht (World on a Wire), directed by Rainer Werner Fassbinder, BRD 1973 (Westdeutscher Rundfunk).

Dark Star, directed by John Carpenter, USA 1974 (Jack H. Harris Enterprises).

Otroki vo vselennoy (Teens in the Universe), directed by Richard Viktorov, USSR 1974 (Gorky Studio).

Space is the Place, directed by John Coney, USA 1974 (North American Star System).

Zardoz, directed by John Boorman, USA 1974 (Twentieth Century Fox).

Space: 1999, television series, 48 episodes, UK 1975–77 (ITC Films/RAI Radiotelevisione Italiana).

Cinderella 2000, directed by Al Adamson, USA 1976 (Independent-International Pictures).

Im Staub der Sterne (In the Dust of the Stars), directed by Gottfried Kolditz, DDR 1976 (VEB DEFA-Studio für Spielfilme, Künstlerische Arbeitsgruppe 'Futurum').

Logan's Run, directed by Michael Anderson, USA 1976 (Metro-Goldwyn-Mayer).

Star Maidens, television series, 13 episodes, UK/BRD 1976 (Jost Grad von Hardenberg).

The Man Who Fell to Earth, directed by Nicolas Roeg, UK 1976 (British Lion Films).

Close Encounters of the Third Kind, directed by Steven Spielberg, USA 1977 (Columbia Pictures).

Operation Ganymed, directed by Rainer Erler, BRD 1977 (Pentagramma Filmproduktion).

Star Wars, directed by George Lucas, USA. Episode IV: *A New Hope* (1977); Episode V: *The Empire Strikes Back* (1980); Episode VI: *Return of the Jedi* (1983) (Lucasfilm).

Capricorn One, directed by Peter Hyams, USA 1978 (ITC Entertainment).

Alien, directed by Ridley Scott, UK/USA 1979 (Brandywine/Twentieth Century Fox).

The Black Hole, directed by Gary Nelson, USA 1979 (Walt Disney Productions).

Moonraker (James Bond), directed by Lewis Gilbert, USA/UK/FR 1979 (United Artists).

Literature

Abelson, Philip H., 'Apollo and Post-Apollo,' *Science* 166 (10 October 1969), 171.

Agar, Jon, *The Science of Spectacle: The Work of Jodrell Bank in Post-War British Culture*, Amsterdam: Harwood Academic, 1998.

———, *Science in the Twentieth Century and Beyond*, Cambridge: Polity Press, 2012.

Agel, Jerome, ed., *The Making of Kubrick's 2001*, New York: Signet, 1970.

Allen, Michael, *Live from the Moon: Film, Television and the Space Race*, London: I.B. Tauris, 2009.

Anders, Günther, *Der Blick vom Mond: Reflexionen über Weltraumflüge*, Munich: C.H. Beck, 1970 (2nd edn 1994).

Anderson, Benedict, *Imagined Communities: Reflections on the Origin and Spread of Nationalism*, New York: Verso, 1991.

Andrews, James T. and Asif A. Siddiqi, eds, *Into the Cosmos: Space Exploration and Soviet Culture*, Pittsburgh: University of Pittsburgh Press, 2011.

Anker, Peder, 'The Ecological Colonization of Space,' *Environmental History* 10.2 (April 2005), 239–68.

————, 'Buckminster Fuller as Captain of Spaceship Earth,' *Minerva* 45.4 (December 2007), 417–34.

'Apollo: "Mann, ist der Berg groß,"' *Der Spiegel* 26.53 (25 December 1972), 76–8.

Arendt, Hannah, *The Human Condition*, Chicago: University of Chicago Press, 1958.

————, 'The Conquest of Space and the Stature of Man' [1963], in idem, *Between Past and Future: Eight Exercises in Political Thought*, New York: Penguin, 2006, 260–74.

Armstrong, Neil, Michael Collins and Edwin E. Aldrin Jr., eds, *First on the Moon: A Voyage*, Boston: Little, Brown, 1970.

Asimov, Isaac, 'After Apollo: A Colony on the Moon,' *New York Times Magazine* (28 May 1967), 30–6.

————, 'Plädoyer für Science-fiction,' *Der Spiegel* 26.11 (6 March 1972), 138–9.

————, 'The Next Frontier?,' *National Geographic* 150 (July–December 1976), 76–89.

————, '20 Ways the World Could End,' *Popular Mechanics* (March 1977), 86–9, 162–4.

Asimov, Isaac and Robert McCall, *Our World in Space*, Greenwich: New York Graphic Society, 1974.

Atwill, William D., *Fire and Power: The American Space Program as Postmodern Narrative*, Athens: University of Georgia Press, 1994.

Auden, W.H., 'Moon Landing,' *New Yorker* (6 September 1969), 38.

Augustine, Dolores L., *Red Prometheus: Engineering and Dictatorship in East Germany, 1945–1990*, Cambridge, MA: MIT Press, 2007.

Bailey, Beth and David Farber, eds, *America in the Seventies*, Lawrence: University Press of Kansas, 2004.

Bainbridge, William Sims, *The Spaceflight Revolution: A Sociological Study*, New York: John Wiley, 1976.

————, *The Meaning and Value of Spaceflight: Public Perceptions*, Heidelberg: Springer, 2015.

Ballard, J.G., 'Which Way to Inner Space?,' *New Worlds Science Fiction* 118 (May 1962), 2–3, 116–18.

————, *Memories of the Space Age*, Sauk City: Arkham House, 1988.

————, *A User's Guide to the Millennium: Essays and Reviews*, London: HarperCollins, 1996.

————, *Extreme Metaphors: Selected Interviews, 1967–2008*, London: Fourth Estate, 2012.

Ballard, J.G. and Lynn Barber, 'Sci-Fi Seer,' *Penthouse* [UK] 5.5 (May 1970), 26–30.

Ballard, J.G. and Christopher Evans, 'The Space Age Is Over,' *Penthouse* [UK] 14.1 (January 1979), 39–42, 102, 106.

Basalla, George, *Civilized Life in the Universe: Scientists on Intelligent Extraterrestrials*, Oxford: Oxford University Press, 2006.

Battaglia, Debbora, ed., *E.T. Culture: Anthropology in Outerspaces*, Durham: Duke University Press, 2005.

Baucom, Donald R., *The Origins of SDI, 1944–1983*, Lawrence: University Press of Kansas, 1992.

Bauman, Zygmunt, *The Individualized Society*, Cambridge: Polity, 2001.

Baxter, John, *Stanley Kubrick: A Biography*, London: HarperCollins, 1997.

Baz, Peter, Eckart Elsner and Götz Niederau, *Zur Frage einer deutschen Beteiligung am Post-Apollo-Programm: Weltraumprogramm-Alternativen für die siebziger Jahre*, Berlin: Technische Universität Berlin, 1970.

Beck, Ulrich, *Risikogesellschaft: Auf dem Weg in eine andere Moderne*, Frankfurt am Main: Suhrkamp, 1986 (Eng. *Risk Society: Towards a New Modernity*, London: Sage, 1992).

Beckett, Andy, *When the Lights Went Out: Britain in the Seventies*, London: Faber & Faber, 2009.

Beebe, Barton 'Law's Empire and the Final Frontier: Legalizing the Future in the Early *Corpus Juris Spatialis*,' *The Yale Law Journal* 108.7 (1999), 1737–73.

Bell, Daniel, *The Coming of Post-Industrial Society: A Venture in Social Forecasting*, New York: Basic Books, 1973.

Bell, David and Martin Parker, eds, *Space Travel and Culture: From Apollo to Space Tourism*, Chichester: Wiley-Blackwell, 2009 (= *Sociological Review* 57.s1).

Benjamin, Marina, *Rocket Dreams: How the Space Age Shaped Our Vision of a World Beyond*, New York: Free Press, 2003.

Bergaust, Erik, *The Next 50 Years on the Moon*, New York: Putnam, 1974.

———, *Colonizing the Planets*, New York: Putnam, 1975.

———, *Colonizing Space*, New York: Putnam, 1978.

Berkowitz, Edward D., *Something Happened: A Political and Cultural Overview of the Seventies*, New York: Columbia University Press, 2006.

'Beyond Apollo: Where? A Prospectus by the Space Science Board, National Academy of Sciences,' *Bulletin of the Atomic Scientists* 25.7 (September 1969), 68.

Bizony, Piers, *2001: Filming the Future*, London: Aurum, 1994.

———, *The Making of Stanley Kubrick's 2001: A Space Odyssey*, Cologne: Taschen, 2014.

Bjørnvig, Thore, 'Outer Space Religion and the Overview Effect: A Critical Inquiry into a Classic of the Pro-Space Movement,' *Astropolitics* 11.1–2 (2013), 4–24.

Black, Jeremy, *Europe Since the Seventies*, London: Reaktion Books, 2009.

Black, Lawrence, 'An Enlightening Decade? New Histories of 1970s' Britain,' *International Labor and Working-Class History* 82 (Fall 2012), 174–86.

Black, Lawrence, Hugh Pemberton and Pat Thane, eds, *Reassessing 1970s Britain*, Manchester: Manchester University Press, 2013.

Bloomfield, Lincoln P., ed., *Outer Space: Prospects for Man and Society*, Englewood Cliffs: Prentice-Hall, 1962.

Blumenberg, Hans, *Die Genesis der kopernikanischen Welt*, Frankfurt am Main: Suhrkamp, 1981 (Eng. *The Genesis of the Copernican World* [1975], Cambridge, MA: MIT Press, 1987).

———, *Die Vollzähligkeit der Sterne*, Frankfurt am Main: Suhrkamp, 1997.

Bond, Alan, ed., *Project Daedalus: The Final Report on the BIS Starship Study*, London: British Interplanetary Society, 1978 (= *Journal of the British Interplanetary Society Supplement*).

Bonnett, Roger M. and Vittorio Manno, *International Cooperation in Space: The Example of the European Space Agency*, Cambridge, MA: Harvard University Press, 1994.

Borstelmann, Thomas, *The 1970s: A New Global History from Civil Rights to Economic Inequality*, Princeton: Princeton University Press, 2012.

Bova, Ben, *The High Road*, Boston: Houghton Mifflin, 1981.

Boyer, Paul, *By the Bomb's Early Light: American Thought and Culture at the Dawn of the Atomic Age*, Chapel Hill: University of North Carolina Press, 1985.

Boym, Svetlana, *The Future of Nostalgia*, New York: Basic Books, 2001.

———, 'Kosmos: Remembrances of the Future,' in Adam Bartos and Svetlana Boym, *Kosmos: A Portrait of the Russian Space Age*, New York: Princeton Architectural Press, 2001, 82–99.

Bracewell, Ronald, *The Galactic Club: Intelligent Life in Outer Space*, San Francisco: Freeman, 1975.

Bradbury, Ray, Arthur C. Clarke, Bruce Murray and Walter Sullivan, *Mars and the Mind of Man*, New York: Harper & Row, 1973.

Brauch, Hans-Günther, ed., *Star Wars and European Defence: Implications for Europe. Perceptions and Assessments*, Basingstoke: Macmillan, 1987.

Braun, Wernher von, *Das Marsprojekt: Studie einer interplanetarischen Expedition*, Frankfurt am Main: Umschau, 1952.

———, 'NASA's Plans for the 70s,' in Michael Cutler, ed., *International Cooperation in Space Operations and Exploration: Proceedings of the American Astronautical Society's 9th Goddard Memorial Symposium, 10–11 March 1971, Washington, DC*, San Diego: Univelt, 7–20.

Brick, Howard, 'Optimism of the Mind: Imagining Postindustrial Society in the 1960s and 1970s,' *American Quarterly* 44.3 (September 1992), 348–80.

Brode, Douglas and Leah Deyneka, eds, *Myth, Media, and Culture in* Star Wars: *An Anthology*, Lanham: Scarecrow Press, 2012.

———, eds, *Sex, Politics, and Religion in* Star Wars: *An Anthology*, Lanham: Scarecrow Press, 2012.

Brooks, Courtney G., Roland W. Newkirk and Ivan D. Ertel, *Skylab Chronology: The Story of the Planning, Development, and Implementation of America's First Manned Space Station*, Washington, DC: NASA, 1977.

Brown, Allan H., 'The Post-Apollo Era: Decisions Facing Nasa,' *Bulletin of the Atomic Scientists* 23.4 (April 1967), 11–16.

Büdeler, Werner, *Raumfahrt in Deutschland: Forschung, Entwicklung, Ziele*, Frankfurt am Main: Ullstein, 1978.

Burrows, William E., *This New Ocean: The Story of the First Space Age*, New York: Random House, 1998.

Butrica, Andrew, *Single Stage to Orbit: Politics, Space Technology, and the Quest for Reusable Rocketry*, Baltimore: Johns Hopkins University Press, 2003.

Calic, Marie-Janine, Dietmar Neutatz and Julia Obertreis, eds, *The Crisis of Socialist Modernity: The Soviet Union and Yugoslavia in the 1970s*, Göttingen: Vandenhoeck & Ruprecht, 2011.

Carroll, Peter N., *It Seemed Like Nothing Happened: The Tragedy and Promise of America in the 1970s*, New York: Holt, Rinehart and Winston, 1982.

Chaikin, Andrew L., *A Man on the Moon: The Voyages of the Apollo Astronauts*, New York: Viking, 1994.

Chassaigne, Philippe, *Les années 1970: Fin d'un monde et origine de notre modernité*, Paris: Armand Colin, 2008.

Clarke, Arthur C., 'Extra-Terrestrial Relays: Can Rocket Stations Give World-wide Radio Coverage?,' *Wireless World* 51.10 (October 1945), 305–8.

———, 'The Rocket and the Future of Warfare,' *RAF Quarterly* 17.2 (March 1946), 61–9.

———, 'The Challenge of the Spaceship: Astronautics and its Impact upon Human Society,' *Journal of the British Interplanetary Society* 6.3 (December 1946), 66–81.

———, 'The Sentinel,' in idem, ed., *Expedition to Earth* [1953], New York: Ballantine Books, 1965, 155–65.

———, *The Challenge of the Spaceship*, London: Frederick Muller, 1960.

———, *Profiles of the Future: An Enquiry into the Limits of the Possible*, London: Gollancz, 1962.

———, 'Man and Space,' in Eric Burgess, ed., *Voyage to the Planets: Proceedings of the American Astronautical Society's 5th Goddard Memorial Symposium, 14–15 March 1967, Washington, DC*, Washington, DC: American Astronautical Society, 1968, 9–22.

———, *2001: A Space Odyssey*, London: Hutchinson, 1968.

———, *The Promise of Space*, London: Hodder & Stoughton, 1968.

———, 'Beyond Apollo,' in Neil Armstrong, Michael Collins and Edwin E. Aldrin Jr., eds, *First on the Moon: A Voyage*, Boston: Little, Brown, 1970, 371–419.

———, 'At the Interface: Technology and Mysticism,' *Playboy* 19.1 (January 1972), 94–6, 130, 256–74.

———, *Report on Planet Three and Other Speculations*, London: Gollancz, 1972.

———, *Greetings, Carbon-Based Bipeds! Collected Essays, 1934–1998*, London: HarperCollins, 1999.

Cleaver, Arthur V., 'European Space Activities Since the War: A Personal View,' *Spaceflight* 16.6 (June 1974), 220–38.

———, 'On the Realisation of Projects: With Special Reference to O'Neill Space Colonies and the Like,' *Journal of the British Interplanetary Society* 30.8 (August 1977), 282–8.

Cockroft, John D., ed., *Atomic Challenge: A Symposium*, London: Winchester, 1948.

Collins, Guy, *Europe in Space*, Basingstoke: Macmillan, 1990.

Collins, Martin, 'One World... One Telephone: Iridium, One Look at the Making of a Global Age,' *History and Technology* 21.3 (September 2005), 301–24.

Collins, Martin, ed., *After Sputnik: 50 Years of the Space Age*, New York: Smithsonian Books, 2007.

Compton, William David and Charles D. Benson, *Living and Working in Space: A History of Skylab*, Washington, DC: NASA, 1983.

Cooke, Hereward Lester and James D. Dean, eds, *Eyewitness to Space: Paintings and Drawings Related to the Apollo Mission to the Moon. Selected, with a few Exceptions, from the Art Program of the National Aeronautics and Space Administration (1963 to 1969)*, New York: Abrams, 1971.

Cooper, Henry S.F., Jr., *A House in Space: The First True Account of the Skylab Experience*, London: Granada, 1977.

Coopey, Richard and Nicholas Woodward, eds, *Britain in the 1970s: The Troubled Economy*, London: UCL Press, 1996.

Corn, Joseph J., *The Winged Gospel: America's Romance with Aviation*, Oxford: Oxford University Press, 1983.

Corn, Joseph J., ed., *Imagining Tomorrow: History, Technology, and the American Future*, Cambridge, MA: MIT Press, 1986.

Corn, Joseph J. and Brian Horrigan, eds, *Yesterday's Tomorrows: Past Visions of the American Future*, Baltimore: Johns Hopkins University Press, 1996.

Cosgrove, Denis E., 'Contested Global Visions: One-World, Whole-Earth, and the Apollo Space Photographs,' *Annals of the Association of American Geographers* 84.2 (June 1994), 270–94.

———, *Apollo's Eye: A Cartographic Genealogy of the Earth in the Western Imagination*, Baltimore: Johns Hopkins University Press, 2001.

Cowen, Robert C., 'The Post-Apollo Era,' *Christian Science Monitor* (3 March 1981), 12–3.

Cowie, Jefferson, *Stayin' Alive: The 1970s and the Last Days of the Working Class*, New York: New Press, 2010.

Däniken, Erich von, *Erinnerungen an die Zukunft: Ungelöste Rätsel der Vergangenheit*, Düsseldorf: Econ, 1968 (Eng. *Chariots of the Gods? Unsolved Mysteries of the Past*, London: Souvenir, 1969).

————, *Erscheinungen: Phänomene, die die Welt erregen*, Düsseldorf: Econ, 1974.

Day, Dwayne A., 'The Von Braun Paradigm,' *Space Times* 33 (November–December 1994), 12–15.

De Grazia, Victoria, *Irresistible Empire: America's Advance through Twentieth-century Europe*, Cambridge, MA: Harvard University Press, 2005.

Diamond, Edwin, *The Rise and Fall of the Space Age*, Garden City: Doubleday, 1964.

Dick, Steven J., 'Anthropology and the Search for Extraterrestrial Intelligence: An Historical View,' *Anthropology Today* 22.2 (April 2006), 3–7.

Dick, Steven J., ed., *Remembering the Space Age: Proceedings of the Fiftieth Anniversary Conference*, Washington, DC: NASA, 2008.

Dick, Steven J. and Roger D. Launius, eds, *Critical Issues in the History of Spaceflight*, Washington, DC: NASA, 2006.

————, eds, *Societal Impact of Spaceflight*, Washington, DC: NASA, 2007.

Dickens, Peter and James S. Ormrod, eds, *The Palgrave Handbook of Society, Culture and Outer Space*, Basingstoke: Palgrave Macmillan, 2016.

Diederichsen, Diedrich and Anselm Franke, eds, *The Whole Earth: California and the Disappearance of the Outside*, Berlin: Sternberg, 2013.

Doering-Manteuffel, Anselm and Lutz Raphael, *Nach dem Boom: Perspektiven auf die Zeitgeschichte seit 1970*, Göttingen: Vandenhoeck & Ruprecht, 2008 (3rd edn 2012).

Doering-Manteuffel, Anselm, Lutz Raphael and Thomas Schlemmer, eds, *Vorgeschichte der Gegenwart: Dimensionen des Strukturbruchs nach dem Boom*, Göttingen: Vandenhoeck & Ruprecht, 2015.

Dolezol, Theodor and Georg M. Januszewski, *Planet des Menschen: Entwicklung und Zukunft der Erde*, Vienna: Ueberreuter, 1975.

Dunnett, Oliver, *The British Interplanetary Society and Cultures of Outer Space*, PhD thesis, University of Nottingham, 2011.

————, 'Patrick Moore, Arthur C. Clarke and "British Outer Space" in the Mid 20th Century,' *Cultural Geographies* 19.4 (October 2012), 505–22.

Dyson, Freeman, 'Human Consequences of the Exploration of Space,' *Bulletin of the Atomic Scientists* 25.7 (September 1969), 8–13.

Ebling, F.J.G. and G.W. Heath, eds, *The Future of Man: Proceedings of a Symposium held at the Royal Geographical Society London, on 1 April, 1971*, London: Academic Press for the Institute of Biology, 1972.

Ehricke, Krafft A., 'The Anthropology of Astronautics,' *Astronautics* 2.4 (November 1957), 26–7, 65–8.

Engel, Johannes and Hermann Schreiber, 'Rücksturz zur Erde: *Spiegel*-Gespräch mit Nasa-Planungschef Wernher von Braun über die US-Raumfahrt der siebziger Jahre,' *Der Spiegel* 25.7 (8 February 1971), 137–44.

Engell, Lorenz, 'Das Mondprogramm: Wie das Fernsehen das größte Ereignis aller Zeiten erzeugte und wieder auflöste, um zu seiner Geschichte zu finden,' in Friedrich Lenger and Ansgar Nünning, eds, *Medienereignisse der Moderne*, Darmstadt: Wissenschaftliche Buchgesellschaft, 2008, 150–71.

Erler, Rainer, *Die Delegation: Ein Bericht*, Frankfurt am Main: Fischer, 1973.

Europe in Space: A Survey Prepared by the European Space Research Organisation (ESRO), Paris: ESRO, 1974.

Evans, Ben, *At Home in Space: The Late Seventies into the Eighties*, New York: Springer, 2012.

Ezell, Edward Clinton and Linda Neuman Ezell, *The Partnership: A History of the Apollo-Soyuz Test Project*, Washington, DC: NASA, 1978.

Fallaci, Oriana, *Se il sole muore*, Milano: Rizzoli, 1965 (Eng. *If the Sun Dies*, New York: Atheneum, 1966).

Farry, James and David A. Kirby, 'The Universe Will Be Televised: Space, Science, Satellites and British Television Production, 1946–1969,' *History and Technology* 28.3 (September 2012), 311–33.

Faulstich, Werner, ed., *Die Kultur der siebziger Jahre*, Munich: Fink, 2004.

Ferguson, Niall, Charles S. Maier, Erez Manela and Daniel J. Sargent, eds, *The Shock of the Global: The 1970s in Perspective*, Cambridge, MA: Harvard University Press, 2010.

Fink, Carole and Bernd Schäfer, eds, *Ostpolitik, 1969–1974: European and Global Responses*, Cambridge: Cambridge University Press, 2009.

Fischer, Joachim and Dierk Spreen, eds, *Soziologie der Weltraumfahrt*, Bielefeld: transcript, 2014.

FitzGerald, Francis, *Way Out There in the Blue: Reagan, Star Wars, and the End of the Cold War*, New York: Touchstone, 2000.

Forman, Paul, 'The Primacy of Science in Modernity, of Technology in Postmodernity, and of Ideology in the History of Technology,' *History and Technology* 23.1–2 (March 2007), 1–152.

Forster, Laurel and Sue Harper, eds, *British Culture and Society in the 1970s: The Lost Decade*, Newcastle-upon-Tyne: Cambridge Scholars, 2010.

Fourastié, Jean, *Les trente glorieuses: Ou, la révolution invisible de 1946 à 1975*, Paris: Fayard, 1979.

Franklin, H. Bruce, 'Don't Look Where We're Going: Visions of the Future in Science-Fiction Films, 1970–82,' *Science Fiction Studies* 10.1 (March 1983), 70–80.

———, *War Stars: The Superweapon and the American Imagination*, Oxford: Oxford University Press, 1988.

Frayling, Christopher, *The 2001 File: Harry Lange and the Design of the Landmark Science Fiction Film*, London: Reel Art Press, 2015.

Frewin, Antony, *Are We Alone? The Stanley Kubrick Extraterrestrial Intelligence Interviews*, London: Elliott & Thompson, 2005.

Freytag, Nils, 'Eine Bombe im Taschenbuchformat? Die "Grenzen des Wachstums" und die öffentliche Resonanz,' *Zeithistorische Forschungen* 3.3 (2006), 465–9.

'From the '60s to the '70s: Dissent and Discovery,' *Time* 94.25 (19 December 1969), 20–6.

Fuchs, Walter R., *Leben unter fernen Sonnen? Wissenschaft und Spekulation*, Munich: Droemer Knaur, 1973.

Fuller, R. Buckminster, *Operating Manual for Spaceship Earth* [1969], Baden: Lars Müller Publishers, 2008.

Gatland, Kenneth W., ed., *Spaceflight Today*, London: Liffe Books, 1963.

———, 'Europe and Post-Apollo: A Chaotic Position,' *Spaceflight* 14.9 (September 1972), 322–4.

Geis, Larry and Fabrice Florin, eds, *Worlds Beyond: The Everlasting Frontier*, Berkeley: And/Or Press, 1978.

Geppert, Alexander C.T., 'Flights of Fancy: Outer Space and the European Imagination, 1923–1969,' in Steven J. Dick and Roger D. Launius, eds, *Societal Impact of Spaceflight*, Washington, DC: NASA, 2007, 585–99.

———, 'Space *Personae*: Cosmopolitan Networks of Peripheral Knowledge, 1927–1957,' *Journal of Modern European History* 6.2 (2008), 262–86.

___, 'Rethinking the Space Age: Astroculture and Technoscience,' *History and Technology* 28.3 (September 2012), 219–23.

———, 'Extraterrestrial Encounters: UFOs, Science and the Quest for Transcendence, 1947–72,' *History and Technology* 28.3 (September 2012), 335–62.

———, 'Infrastrukturen der Weltraumimagination: Außenstationen im 20. Jahrhundert,' in Kunst- und Ausstellungshalle der Bundesrepublik Deutschland, ed., *Outer Space: Faszination Weltraum*, Bonn: Nicolai, 2014, 124–7.

———, 'Die Zeit des Weltraumzeitalters, 1942–1972,' in idem and Till Kössler, eds, *Obsession der Gegenwart: Zeit im 20. Jahrhundert*, Göttingen: Vandenhoeck & Ruprecht, 2015, 218–50 (= *Geschichte und Gesellschaft*. Sonderheft 25).

Geppert, Alexander C.T., ed., *Imagining Outer Space: European Astroculture in the Twentieth Century*, Basingstoke: Palgrave Macmillan, 2012; 2nd edn, London: Palgrave Macmillan, 2018 (= *European Astroculture*, vol. 1).

———, ed., *Astroculture and Technoscience*, London: Routledge, 2012 (= *History and Technology* 28.3).

Geppert, Alexander C.T., Daniel Brandau and Tilmann Siebeneichner, eds, *Militarizing Outer Space: Astroculture, Dystopia and the Cold War*, London: Palgrave Macmillan, forthcoming (= *European Astroculture*, vol. 3).

Geyer, Martin H., 'Die Gegenwart der Vergangenheit: Die Sozialstaatsdebatten der 1970er Jahre und die umstrittenen Entwürfe der Moderne,' *Archiv für Sozialgeschichte* 47 (2007), 47–93.

———, 'Auf der Suche nach der Gegenwart: Neue Arbeiten zur Geschichte der 1970er und 1980er Jahre,' *Archiv für Sozialgeschichte* 50 (2010), 643–69.

———, '"Gaps" and the (Re-)Invention of the Future: Social and Demographic Policy in Germany During the 1970s and 1980s,' *Social Science History* 39.1 (April 2015), 39–61.

Gilcher-Holtey, Ingrid, Rainer Eckert, Etienne François, Christoph Kleßmann and Krzysztof Ruchniewicz, 'Die 1970er-Jahre in Geschichte und Gegenwart,' *Zeithistorische Forschungen* 3.3 (2006), 422–38.

Godwin, Matthew, *The Skylark Rocket: British Space Science and the European Space Research Organisation 1957–1972*, Paris: Beauchesne, 2007.

Golden, Frederic and Kiyoaki Komoda, *Colonies in Space: The Next Giant Step*, New York: Harcourt Brace Jovanovich, 1977.

Goldsen, Joseph M., ed., *Outer Space in World Politics*, New York: Praeger, 1963.

Goldstein, Laurence, ed., *The Moon Landing and its Aftermath*, Ann Arbor: University of Michigan, 1979 (= *Michigan Quarterly Review* 18.2, 153–363).

Goodrich, Malinda K., Alice R. Buchalter and Patrick Miller, *Toward a History of the Space Shuttle: An Annotated Bibliography. Part II: 1992–2011*, Washington, DC: NASA, 2012.

Gorman, Alice and Beth O'Leary, 'An Ideological Vacuum: The Cold War in Outer Space,' in John Schofield and Wayne D. Cocroft, eds, *A Fearsome Heritage: Diverse Legacies of the Cold War*, Tucson: Left Coast Press, 2007, 73–92.

Gray, Colin and John Sheldon, 'Space Power and the Revolution in Military Affairs: A Glass Half-Full?,' *Airpower Journal* 13.3 (1999), 23–38.

'The Greening of the Astronauts,' *Time* (11 December 1972), 57.

Grosvenor, Gilbert M., 'Summing up Mankind's Greatest Adventure,' *National Geographic* 144 (July–December 1973), 289–331.

Grosvenor, Gilbert M., Isaac Asimov, Richard F. Babcock, Edmund N. Bacon, Buckminster Fuller and Gerard Piel, 'Five Noted Thinkers Explore the Future,' *National Geographic* 150 (July 1976), 68–74.

Hall, Rex D. and David J. Shayler, *Soyuz: A Universal Spacecraft*, London: Springer, 2003.

Hall, Simon, 'Protest Movements in the 1970s: The Long 1960s,' *Journal of Contemporary History* 43 (October 2008), 655–72.

Halle, Louis J., 'A Hopeful Future for Mankind,' *Foreign Affairs* 58.5 (Summer 1980), 1129–36.

Haller, Fritz, *Umweltgestaltung einer prototypischen Raumkolonie*, Karlsruhe: Universität Karlsruhe, 1980.

Handberg, Roger, *Seeking New World Vistas: The Militarization of Space*, Westport: Praeger, 2000.

Harak, G. Simon, 'One Nation, Under God: The Soteriology of SDI,' *Journal of the American Academy of Religion* 56.3 (1988), 497–527.

Hart, Michael H., 'An Explanation for the Absence of Extraterrestrials on Earth,' *Quarterly Journal of the Royal Astronomical Society* 16.2 (June 1975), 128–35.

Harvey, David, *The Condition of Postmodernity: An Enquiry into the Origins of Cultural Change*, Oxford: Blackwell, 1989.

Hasenkamp, Andreas, *Raumfahrtpolitik in Westeuropa und die Rolle Frankreichs: Macht – Nutzen – Reformdruck*, Münster: Lit, 1996.

Helmreich, Stefan, 'From Spaceship Earth to Google Ocean: Planetary Icons, Indexes, and Infrastructures,' *Social Research* 78.4 (Winter 2011), 1211–42.

Heppenheimer, Thomas A., *Colonies in Space*, Harrisburg: Stackpole Books, 1977.

———, *Toward Distant Suns*, Harrisburg: Stackpole Books, 1979.

———, *History of the Space Shuttle*, vol. 1: *Space Shuttle Decision, 1965–1972*, vol. 2: *Development of the Space Shuttle, 1972–1981*, Washington, DC: Smithsonian Institution Press, 2002.

Hersch, Matthew H., *Inventing the American Astronaut*, Basingstoke: Palgrave Macmillan, 2012.

Hill, C.N., *A Vertical Empire: History of the British Rocketry Programme*, London: Imperial College Press, 2001 (2nd edn 2012).

Hoch, David G., 'Mythic Patterns in *2001: A Space Odyssey*,' *Journal of Popular Culture* 4.4 (Spring 1971), 961–5.

Höhler, Sabine, 'The Environment as a Life Support System: The Case of Biosphere 2,' *History and Technology* 26.1 (March 2010), 39–58.

———, *Spaceship Earth in the Environmental Age, 1960–1990*, London: Pickering & Chatto, 2015.

Hoeveler, J. David, *The Postmodernist Turn: American Thought and Culture in the 1970s*, New York: Twayne, 1996.

Horrigan, Brian, 'Popular Culture and Visions of the Future in Space, 1901–2001,' in Bruce Sinclair, ed., *New Perspectives on Technology and American Culture*, Philadelphia: American Philosophical Society, 1986, 49–67.

Horst, Ernst, 'Da stellen wir uns mal ganz klug: Sind die siebziger Jahre schon ein Thema für Historiker?,' *Frankfurter Allgemeine Zeitung* (27 April 2007), 40.

Huntley, Wade L., Joseph G. Bock and Miranda Weingartner, 'Planning the Unplannable: Scenarios on the Future of Space,' *Space Policy* 26.1 (February 2010), 25–38.

Hurup, Elsebeth, ed., *The Lost Decade: America in the Seventies*, Aarhus: Aarhus University Press, 1996.

Ivanovich, Grujica S., *Salyut – The First Space Station: Triumph and Tragedy*, New York: Springer, 2008.

Jameson, Fredric, *Postmodernism, or, the Cultural Logic of Late Capitalism*, Durham: Duke University Press, 1991.

———, *Archaeologies of the Future: The Desire Called Utopia and Other Science Fictions*, New York: Verso, 2005.

Jarausch, Konrad H., 'Krise oder Aufbruch? Historische Annäherungen an die 1970er-Jahre,' *Zeithistorische Forschungen* 3.3 (2006), 334–41.

———, *Out of Ashes: A New History of Europe in the Twentieth Century*, Princeton: Princeton University Press, 2015.

Jarausch, Konrad H., ed., *Die 1970er-Jahre: Inventur einer Umbruchzeit*, Göttingen: Vandenhoeck & Ruprecht, 2006 (= *Zeithistorische Forschungen* 3.3).

———, ed., *Das Ende der Zuversicht? Die siebziger Jahre als Geschichte*, Göttingen: Vandenhoeck & Ruprecht, 2008.

Jasanoff, Sheila, 'Heaven and Earth: The Politics of Environmental Images,' in idem and Marybeth Long Martello, eds, *Earthly Politics: Local and Global in Environmental Governance*, Cambridge, MA: MIT Press, 2004, 31–52.

Johnson, Richard D. and Charles Holbrow, eds, *Space Settlements: A Design Study in Colonization*, Washington, DC: NASA, 1977.

Judt, Tony, *Postwar: A History of Europe Since 1945*, New York: Penguin, 2005.

Kaelble, Hartmut, 'Vers une histoire sociale et culturelle de l'Europe pendant les années de l'"après-prospérité,"' *Vingtième Siècle* 84.4 (2004), 169–79.

———, *The 1970s in Europe: A Period of Disillusionment or Promise?*, London: German Historical Institute, 2010.

Kaiser, David, *How the Hippies Saved Physics: Science, Counterculture, and the Quantum Revival*, New York: Norton, 2012.

Kalic, Sean N., 'Reagan's SDI Announcement and the European Reaction: Diplomacy in the Last Decade of the Cold War,' in Leopoldo Nuti, ed., *The Crisis of Détente in Europe: From Helsinki to Gorbachev, 1975–1985*, London: Routledge, 2009, 99–110.

Kelley, Kevin W., ed., *The Home Planet*, London: Queen Anne Press, 1988.

Kelsey, Robin, 'Reverse Shot: *Earthrise* and *Blue Marble* in the American Imagination,' in Gareth Doherty and El Hadi Jazairy, eds, *Scales of the Earth*, Cambridge, MA: Harvard University Press, 2011, 10–16.

Keys, Barbara, Jack Davies and Elliott Bannan, 'The Post-Traumatic Decade: New Histories of the 1970s,' *Australasian Journal of American Studies* 33.1 (July 2014), 1–17.

Kilgore, De Witt Douglas, *Astrofuturism: Science, Race, and Visions of Utopia in Space*, Philadelphia: University of Pennsylvania Press, 2003.

———, 'Exploring Astroculture,' *Science Fiction Studies* 41.2 (July 2014), 447–50.

King, Geoff and Tanya Krzywinska, *Science Fiction Cinema: From Outerspace to Cyberspace*, New York: Wallflower, 2000 (2nd edn 2006).

Kirk, Andrew, 'Appropriating Technology: *The Whole Earth Catalog* and Counterculture Environmental Politics,' *Environmental History* 6.3 (July 2001), 374–94.

Kirby, David A., *Lab Coats in Hollywood: Science, Scientists and Cinema*, Cambridge, MA: MIT Press, 2011.

Kolker, Robert Phillip, ed., *Stanley Kubrick's* 2001: A Space Odyssey: *New Essays*, Oxford: Oxford University Press, 2006.

Kopal, Zdeněk, *The Moon in the Post-Apollo Era*, Dordrecht: Reidel, 1974.

Krämer, Peter, *2001: A Space Odyssey*, Basingstoke: Palgrave Macmillan, 2010.

Kraemer, Robert S., *Beyond the Moon: A Golden Age of Planetary Exploration, 1971–1978*, Washington, DC: Smithsonian Institution Press, 2000.

Krige, John and Arturo Russo, *Europe in Space 1960–1973*, Noordwijk: ESA, 1994.

———, *A History of the European Space Agency 1958–1987*, 2 vols, Noordwijk: ESA, 2000.

Krige, John, Angelina Long Callahan and Ashok Maharaj, *NASA in the World: Fifty Years of International Collaboration in Space*, Basingstoke: Palgrave Macmillan, 2013.

Krugman, Herbert E., 'Public Attitudes Toward the Apollo Space Program, 1965–1975,' *Journal of Communication* 27.4 (Fall 1977), 87–93.

Kubbig, Bernd W., ed., *Die militärische Eroberung des Weltraums*, 2 vols, Frankfurt am Main: Suhrkamp, 1990.

Kuberski, Philip, 'Kubrick's *Odyssey*: Myth, Technology, Gnosis,' *Arizona Quarterly* 64.3 (Fall 2008), 51–73.

Kuchenbuch, David, '"Eine Welt": Globales Interdependenzbewusstsein und die Moralisierung des Alltags in den 1970er und 1980er Jahren,' *Geschichte und Gesellschaft* 38.1 (2012), 158–84.

Kupper, Patrick, 'Die "1970er Diagnose": Grundsätzliche Überlegungen zu einem Wendepunkt der Umweltgeschichte,' *Archiv für Sozialgeschichte* 43 (2003), 325–48.

Lapp, Ralph E., *Man and Space: The Next Decade*, London: Secker & Warburg, 1961.

Larabee, Ann, '"Nothing Ends Here": Managing the *Challenger* Disaster,' in Steven Biel, ed., *American Disasters*, New York: New York University Press, 2001, 197–220.

Larson, Jerrord, 'Limited Imagination: Depictions of Computers in Science Fiction Film,' *Futures* 40.3 (April 2008), 293–9.

Launius, Roger D., 'NASA and the Decision to Build the Space Shuttle, 1969–72,' *The Historian* 57 (Fall 1994), 17–34.

———, 'Public Opinion Polls and Perceptions of US Human Spaceflight,' *Space Policy* 19.3 (August 2003), 163–75.

———, 'Perfect Worlds, Perfect Socities: The Persistent Goal of Utopia in Human Spaceflight,' *Journal of the British Interplanetary Society* 56.5 (September/October 2003), 338–49.

———, 'Perceptions of Apollo: Myth, Nostalgia, Memory or All of the Above?,' *Space Policy* 21.2 (May 2005), 129–39.

———, 'Interpreting the Moon Landings: Project Apollo and the Historians,' *History and Technology* 22.3 (September 2006), 225–55.

———, 'Planning the Post-Apollo Space Program: Are There Lessons for the Present?,' *Space Policy* 28.1 (2012), 38–44.

Lauwaert, Maaike, 'Playing Outside the Box: On LEGO Toys and the Changing World of Construction Play,' *History and Technology* 24.3 (September 2008), 221–37.

Lavery, David, *Late for the Sky: The Mentality of the Space Age*, Carbondale: Southern Illinois University Press, 1992.

Lazier, Benjamin, 'Earthrise; or, The Globalization of the World Picture,' *American Historical Review* 116.3 (June 2011), 602–30.

Leavitt, William, 'Post-Apollo Policy: A Look Into the 1970s,' *Bulletin of the Atomic Scientists* 25.7 (September 1969), 41–4.

Lente, Dick van, ed., *The Nuclear Age in Popular Media: A Transnational History, 1945–1965*, Basingstoke: Palgrave Macmillan, 2012.

Lewis, Richard S., ed., *Men on the Moon: An Assessment*, Chicago: Educational Foundation for Nuclear Science, 1969 (= *Bulletin of the Atomic Scientists* 25.7).

Listner, Michael, 'The Moon Treaty: Failed International Law or Waiting in the Shadows?,' *The Space Review* (24 October 2011), available at http://www.thespacereview.com/article/1954/1 (accessed 1 October 2017).

Logsdon, John M., *The Decision to Go to the Moon: Project Apollo and the National Interest*, Chicago: University of Chicago Press, 1970.

———, 'The Decision to Develop the Space Shuttle,' *Space Policy* 2.2 (May 1986), 103–19.

———, 'The Space Shuttle Program: A Policy Failure?,' *Science* 232 (30 May 1986), 1099–105.

———, 'Evaluating Apollo,' *Space Policy* 5.3 (August 1989), 188–92.

———, *John F. Kennedy and the Race to the Moon*, Basingstoke: Palgrave Macmillan, 2010.

———, *After Apollo? Richard Nixon and the American Space Program*, Basingstoke: Palgrave Macmillan, 2015.

Lord, Douglas R., *Spacelab: An International Success Story*, Washington: NASA, 1987.

Lowe, Rodney, 'Life Begins in the Seventies? Writing and Rewriting the History of Postwar Britain,' *Journal of Contemporary History* 42.1 (January 2007), 161–9.

Lowman, Paul D., Jr., 'The Apollo Program: Was it Worth it?,' *World Resources* 49 (August 1975), 291–302.

———, 'T Plus Twenty-Five Years: A Defense of the Apollo Program,' *Journal of the British Interplanetary Society* 49.2 (February 1996), 71–9.

Lunan, Duncan, *Man and the Stars: Contact and Communication with Other Intelligence*, London: Corgi, 1978.

MacDonald, Fraser, 'Anti-*Astropolitik*: Outer Space and the Orbit of Geography,' *Progress in Human Geography* 31.5 (October 2007), 592–615.

———, 'Space and the Atom: On the Popular Geopolitics of Cold War Rocketry,' *Geopolitics* 13.4 (Winter 2008), 611–34.

Mack, Pamela E., *Viewing the Earth: The Social Construction of the Landsat Satellite System*, Cambridge, MA: MIT Press, 1990.

MacLeish, Archibald, 'A Reflection: Riders on Earth Together, Brothers in Eternal Cold,' *New York Times* (25 December 1968), 1.

Maier, Charles S., 'Consigning the Twentieth Century to History: Alternative Narratives for the Modern Era,' *American Historical Review* 105.3 (June 2000), 807–31.

———, 'Two Sorts of Crisis? The "Long" 1970s in the West and the East,' in Hans Günter Hockerts, ed., *Koordinaten deutscher Geschichte in der Epoche des Ost-West-Konflikts*, Munich: Oldenbourg, 2004, 49–62.

———, *Among Empires: American Ascendancy and Its Predecessors*, Cambridge, MA: Harvard University Press, 2006.

Maier, Charles S., ed., *The Cold War in Europe: Era of a Divided Continent*, 3rd edn, New York: Markus Wiener, 1996.

Mailer, Norman, *Of a Fire on the Moon*, Boston: Little, Brown, 1970.

Malkin, Myron S., 'The Space Shuttle,' *American Scientist* 66.6 (November–December 1978), 718–23.

Malzberg, Barry N., *Beyond Apollo*, New York: Random House, 1972.

Marotta, Michael E., *Space Colonization: An Annotated Bibliography*, Mason: Loompanics, 1979.

Maruyama, Magoroh and Arthur Harkins, eds, *Cultures Beyond the Earth: The Role of Anthropology in Outer Space*, New York: Vintage Books, 1975.

Massey, Harrie and Malcolm O. Robins, *History of British Space Science*, Cambridge: Cambridge University Press, 1986.

Maurer, Eva, Julia Richers, Monica Rüthers and Carmen Scheide, eds, *Soviet Space Culture: Cosmic Enthusiasm in Socialist Societies*, Basingstoke: Palgrave Macmillan, 2011.

McAleer, Neil, *Odyssey: The Authorized Biography of Arthur C. Clarke*, London: Gollancz, 1992 (2nd edn 2013).

McCormick, John, *Reclaiming Paradise: The Global Environmental Movement*, Bloomington: Indiana University Press, 1989.

McCray, W. Patrick, *The Visioneers: How a Group of Elite Scientists Pursued Space Colonies, Nanotechnologies, and a Limitless Future*, Princeton: Princeton University Press, 2012.

McCurdy, Howard E., *The Space Station Decision: Incremental Politics and Technological Choice*, Baltimore: Johns Hopkins University Press, 1990.

———, *Space and the American Imagination*, Washington DC: Smithsonian Institution Press, 1997 (2nd edn, Baltimore: Johns Hopkins University Press, 2011).

McDonald, Ian, 'All Aboard the Last Apollo: End of the Great American Space Adventure,' *Times* (6 December 1972), 16.

McDougall, Walter A., 'Space-Age Europe: Gaullism, Euro-Gaullism, and the American Dilemma,' *Technology and Culture* 26.2 (April 1985), 179–203.

———, *...The Heavens and the Earth: A Political History of the Space Age* [1985], Baltimore: Johns Hopkins University Press, 1997.

McLean, Alasdair W.M., *Western European Military Space Policy*, Aldershot: Dartmouth, 1992.

McLuhan, Marshall, *Understanding Media: The Extensions of Man*, New York: McGraw-Hill, 1964.

McMillen, Ryan J., *Space Rapture: Extraterrestrial Millennialism and the Cultural Construction of Space Colonization*, PhD thesis, University of Texas, Austin, 2004.

McQuaid, Kim, 'Selling the Space Age: NASA and Earth's Environment, 1958–1990,' *Environment and History* 12.2 (May 2006), 127–63.

———, 'Earthly Environmentalism and the Space Exploration Movement, 1960–1990: A Study in Irresolution,' *Space Policy* 26.3 (August 2010), 163–73.

Meadows, Donella H., Dennis L. Meadows, Jørgen Randers, William W. Behrens III and Club of Rome, *The Limits to Growth: A Report for the Club of Rome's Project on the Predicament of Mankind*, New York: Universe Books, 1972.

Mehl, Friederike, 'Conference Report "Envisioning Limits: Outer Space and the End of Utopia. 19.–21. April 2012," Berlin,' *H-Soz-u-Kult* (9 July 2012).

———, 'Berlin Symposium on Outer Space and the End of Utopia in the 1970s,' *NASA History News & Notes* 29.2–3 (2012), 1–5.

Michaud, Michael A.G., 'Spaceflight, Colonization, and Independence: A Synthesis,' *Journal of the British Interplanetary Society* 30.3/6/9 (March/June/September 1977), 83–95, 203–12, 323–31.

———, *Reaching for the High Frontier: The American Pro-Space Movement, 1972–84*, New York: Praeger, 1986.

Millard, Douglas, *The Black Arrow Rocket: A History of a Satellite Launch Vehicle and its Engines*, London: Science Museum, 2001.

Miller, Ron, *The Art of Space: The History of Space Art, from the Earliest Visions to the Graphics of the Modern Era*, Minneapolis: Zenith, 2014.

Mindell, David A., *Digital Apollo: Human and Machine in Spaceflight*, Cambridge, MA: MIT Press, 2008.

Moltz, James C., *The Politics of Space Security: Strategic Restraint and the Pursuit of National Interests*, Stanford: Stanford University Press, 2008.

'The Moon Age,' *Newsweek* (7 July 1969), 26–52.

Moore, Patrick, *Space in the Sixties*, Harmondsworth: Penguin, 1963.

———, *Can You Speak Venusian? A Guide to the Independent Thinkers*, Newton Abbot: David and Charles, 1972.

Moore, Patrick and David A. Hardy, *The New Challenge of the Stars*, London: Mitchell Beazley, 1972.

Moy, Tim, 'Culture, Technology, and the Cult of Tech in the 1970s,' in Beth Bailey and David Farber, eds, *America in the Seventies*, Lawrence: University Press of Kansas, 2004, 208–27.

Mueller, George E., 'Apollo Actions in Preparation for the Next Manned Flight,' *Astronautics and Aeronautics* 5.8 (August 1967), 28–33.

———, 'In the Next Decade: A Lunar Base, Space Laboratories and a Shuttle Service,' *New York Times* (21 July 1969), 14.

———, 'Post-Apollo Revisited,' *Astronautics and Aeronautics* 17.7–8 (July–August 1979), 24–31.

Muir-Harmony, Teasel, 'Selling Space Capsules, Moon Rocks, and America: Spaceflight in U.S. Public Diplomacy, 1961–1979,' in Hallvard Notaker, Giles Scott-Smith and David J. Snyder, eds, *Reasserting America in the 1970s: U.S. Public Diplomacy and the Rebuilding of America's Image Abroad*, Manchester: Manchester University Press, 2016, 127–42.

Mumford, Lewis, *The Myth of the Machine*, 2 vols, New York: Harcourt Brace Jovanovich, 1967/1970.

Murphy, Douglas, *Last Futures: Nature, Technology and the End of Architecture*, London: Verso, 2016.

Nardon, Laurence, 'Cold War Space Policy and Observation Satellites,' *Astropolitics* 5.1 (August 2007), 29–62.

Neal, Valerie, 'Bumped from the Shuttle Fleet: Why Didn't *Enterprise* Fly in Space?,' *History and Technology* 18.3 (September 2002), 181–202.

———, *Spaceflight in the Shuttle Era and Beyond: Redefining Humanity's Purpose in Space*, New Haven: Yale University Press, 2017.

Neufeld, Michael J., *The Rocket and the Reich: Peenemünde and the Coming of the Ballistic Missile Era*, New York: Free Press, 1995.

———, *Von Braun: Dreamer of Space, Engineer of War*, New York: Alfred A. Knopf, 2007.

Neufeld, Michael J., ed., *Spacefarers: Images of Astronauts and Cosmonauts in the Heroic Era of Spaceflight*, Washington, DC: Smithsonian Institution Press, 2013.

Nicholson, Tony and Geoffrey Pardoe, 'After Apollo: Why Man Must Stay in Space,' *Times* (20 December 1972), 12.

Nicolson, Iain, *The Road to the Stars*, Abbot: Westbridge Books, 1978.

Nixon, Richard, 'Statement Announcing Decision to Proceed with Development of the Space Shuttle,' 5 January 1972, *The American Presidency Project*, available at http://www.presidency.ucsb.edu/ws/?pid=3574 (accessed 1 October 2017).

Noble, David F., *The Religion of Technology: The Divinity of Man and the Spirit of Invention*, New York: Alfred A. Knopf, 1997.

Nordern, Eric, 'Interview: Stanley Kubrick. A Candid Conversation with the Pioneering Creator of "2001: A Space Odyssey," "Dr. Strangelove" and "Lolita,"' *Playboy* 15.9 (September 1968), 85–96, 180–95.

Nowotny, Helga, 'Vergangene Zukunft: Ein Blick zurück auf die "Grenzen des Wachstums,"' in Michael Globig, ed., *Impulse geben – Wissen stiften: 40 Jahre VolkswagenStiftung*, Göttingen: Vandenhoeck & Ruprecht, 2002, 655–94.

Nuti, Leopoldo, ed., *The Crisis of Détente in Europe: From Helsinki to Gorbachev, 1975–1985*, London: Routledge, 2009.

Nye, David E., *American Technological Sublime*, Cambridge, MA: MIT Press, 1994.

———, 'Don't Fly Us to the Moon: The American Public and the Apollo Space Program,' *Foundation* 66 (Spring 1996), 69–81.

———, 'The Energy Crisis of the 1970s as a Cultural Crisis,' in Cristina Giorcelli and Peter G. Boyle, eds, *Living with America, 1946–1996*, Amsterdam: VU University Press, 1997, 82–102.

Oliver, Kelly, *Earth and World: Philosophy After the Apollo Missions*, New York: Columbia University Press, 2015.

Oliver, Kendrick, *To Touch the Face of God: The Sacred, the Profane, and the American Space Program, 1957–1975*, Baltimore: Johns Hopkins University Press, 2013.

O'Neill, Gerard K., 'A Lagrangian Community?,' *Nature* 250 (23 August 1974), 636.

———, 'The Colonization of Space,' *Physics Today* 27.9 (September 1974), 32–40.

———, 'Space Colonies and Energy Supply to the Earth,' *Science* 190 (5 December 1975), 943–7.

———, 'Colonies in Orbit,' *New York Times Magazine* (18 January 1976), 11, 25–9.

———, *The High Frontier: Human Colonies in Space*, New York: William Morrow, 1976.

Ordway, Frederick I., '*2001: A Space Odyssey* in Retrospect,' in Eugene M. Emme, ed., *Science Fiction and Space Futures: Past and Present*, San Diego: American Astronautical Society, 1982, 47–105.

Osaka, Takuro, 'Art in the Space Age: Exploring the Relationship between Outer Space and Earth Space,' *Leonardo* 37.4 (2004), 273–6.

Osborne, Catherine R., 'From Sputnik to Spaceship Earth: American Catholics and the Space Age,' *Religion and American Culture* 25.2 (Summer 2015), 218–63.

Owen, Kenneth, 'Europe's Future in Space,' *Flight* (6 July 1961), 5–9.

———, 'Unparalleled Achievement of Apollo Programme,' *Times* (8 December 1972), 25.

Parkinson, Bob, ed., *Interplanetary: A History of the British Interplanetary Society*, London: British Interplanetary Society, 2008.

Parks, Lisa, 'Technology in the Twilight: A Cultural History of the First Earth Satellite,' *Humanities and Technology Review* 16 (Fall 1997), 3–20.

———, *Cultures in Orbit: Satellites and the Televisual*, Durham: Duke University Press, 2005.

Pelton, Joseph N., 'The Space Shuttle: Evaluating an American Icon,' *Space Policy* 26 (November 2010), 246–8.

Peoples, Columba, *Justifying Ballistic Missile Defense: Technology, Security and Culture*, Cambridge: Cambridge University Press, 2010.

Phillips, Gene D., ed., *Stanley Kubrick Interviews*, Jackson: University of Mississippi Press, 2001.

Pias, Claus, 'Schöner wohnen: Weltraumkolonien als Wille und Vorstellung,' in Annett Zinsmeister, ed., *Welt[stadt]raum: Mediale Inszenierungen*, Bielefeld: transcript, 2008, 25–51.

Pierre, Arnauld, *Maternités cosmiques. La recherche des origines, de Kupka à Kubrik*, Paris: Hazan, 2010.

Ponnamperuma, Cyril and A.G.W. Cameron, eds, *Interstellar Communication: Scientific Perspectives*, Boston: Houghton Mifflin, 1974.

Poole, Robert, *Earthrise: How Man First Saw the Earth*, New Haven: Yale University Press, 2008.

———, 'The Challenge of the Spaceship: Arthur C. Clarke and the History of the Future, 1930–1970,' *History and Technology* 28.3 (September 2012), 255–80.

Poundstone, William, *Carl Sagan: A Life in the Cosmos*, New York: Henry Holt, 1999.

Proske, Rüdiger, *Zum Mond und weiter*, Bergisch Gladbach: Bastei, 1966.

Pursell, Carroll, 'The Rise and Fall of the Appropriate Technology Movement in the United States, 1965–1985,' *Technology and Culture* 34.3 (July 1993), 629–37.

Puttkamer, Jesco von, *Raumstationen: Laboratorien im All*, Weinheim: Verlag Chemie, 1971.

Pynchon, Thomas, *Gravity's Rainbow*, New York: Viking, 1973.

Rabinowitch, Eugene I. and Richard S. Lewis, eds, *Men in Space: The Impact on Science, Technology, and International Cooperation*, Chicago: Education Foundation for Nuclear Science, 1969.

Reinke, Niklas, *Geschichte der deutschen Raumfahrtpolitik: Konzepte, Einflussfaktoren und Interdependenzen 1923–2002*, Munich: Oldenbourg, 2002 (Eng. *The History of German Space Policy: Ideas, Influences, and Interdependence 1923–2002*, Paris: Beauchesne, 2007).

Ritter, Gerhard A., Margit Szöllösi-Janze and Helmuth Trischler, eds, *Antworten auf die amerikanische Herausforderung: Forschung in der Bundesrepublik und der DDR in den 'langen' siebziger Jahren*, Frankfurt am Main: Campus, 1999.

Roberts, Adam, *The History of Science Fiction*, Basingstoke: Palgrave Macmillan, 2005.

Rodgers, Daniel T., *Age of Fracture*, Cambridge, MA: Harvard University Press, 2011.

Roland, Alex, 'Barnstorming in Space: The Rise and Fall of the Romantic Era of Spaceflight, 1957–1986,' in Radford Byerly Jr., ed., *Space Policy Reconsidered*, Boulder: Westview Press, 1989, 33–52.

Rome, Adam, *The Genius of Earth Day: How a 1970 Teach-in Unexpectedly Made the First Green Generation*, New York: Hill and Wang, 2013.

Rosenberg, Emily S., 'Far Out: The Space Age in American Culture,' in Steven J. Dick, ed., *Remembering the Space Age: Proceedings of the Fiftieth Anniversary Conference*, Washington, DC: NASA, 2008, 157–84.

Roslansky, John D., ed., *Shaping the Future: A Discussion at the Nobel Conference*, Amsterdam: North-Holland, 1972.

Rudzinski, Kurt, 'Die Monderoberung verweist auf die Erde zurück: Zum Abschluß des Apollo-Programms,' *Frankfurter Allgemeine Zeitung* (20 December 1972), 2.

Russo, Arturo, 'Science in Space vs Space Science: The European Utilisation of Spacelab,' *History and Technology* 16.2 (October 1999), 137–78.

Sänger, Eugen, *Raumfahrt: heute – morgen – übermorgen*, Düsseldorf: Econ, 1963 (Eng. *Space Flight: Countdown for the Future*, New York: McGraw-Hill, 1965).

Sage, Daniel, *How Outer Space Made America: Geography, Organization and the Cosmic Sublime*, Farnham: Ashgate, 2014.

Salkeld, Robert, 'Space Colonization Now?,' *Astronautics and Aeronautics* 13.9 (September 1975), 30–4.

Schaber, Elisabeth, 'Der rote Weltraum: Die künstlerische Darstellung von Raumfahrt auf Briefmarken der DDR,' *Neue Kunstwissenschaftliche Forschungen* 1 (October 2014), 48–60.

Schmidt-Gernig, Alexander, 'Scenarios of Europe's Future: Western Future Studies of the Sixties and Seventies as an Example of a Transnational Public Sphere of Experts,' *Journal of European Integration History* 8.2 (2002), 69–90.

———, 'The Cybernetic Society: Western Future Studies of the 1960s and 1970s and Their Predictions for the Year 2000,' in Richard N. Cooper and Richard Layard, eds, *What the Future Holds: Insights from Social Science*, Cambridge, MA: MIT Press, 2002, 233–59.

Schoijet, Mauricio, '*Limits to Growth* and the Rise of Catastrophism,' *Environmental History* 4.4 (October 1999), 515–30.

Schreiber, Hermann, 'Dieser sogenannte Menschheitstraum,' *Der Spiegel* 25.33 (9 August 1971), 94.

Schulman, Bruce J., *The Seventies: The Great Shift in American Culture, Society and Politics*, New York: Free Press, 2001.

Schulman, Bruce J. and Julian E. Zelizer, eds, *Rightward Bound: Making America Conservative in the 1970s*, Cambridge, MA: Harvard University Press, 2008.

Schulte-Hillen, Jürgen, *Die Luft- und Raumfahrtpolitik der Bundesrepublik Deutschland: Forschungs- und Entwicklungsprogramme in der Kritik*, Göttingen: Schwartz, 1975.

Schwamm, Stephanie, *The Making of* 2001: A Space Odyssey, New York: Random House, 2000.

Scott, David Meerman and Richard Jurek, *Marketing the Moon: The Selling of the Apollo Lunar Program*, Cambridge, MA: MIT Press, 2014.

Sebesta, Lorenza, *Alleati competitivi: Origini e sviluppo della cooperazione spaziale fra Europa e Stati Uniti, 1957–1973*, Rome: Laterza, 2003.

Seed, David, *American Science Fiction and the Cold War*, Edinburgh: Edinburgh University Press, 1999.

Seefried, Elke, 'Towards *The Limits to Growth*? The Book and its Reception in West Germany and Britain, 1972–1973,' *Bulletin of the German Historical Institute* 33.1 (2011), 3–37.

———, *Zukünfte: Aufstieg und Krise der Zukunftsforschung, 1945–1980*, Berlin: de Gruyter, 2015.

Segal, Howard P., *Technological Utopianism in American Culture*, Chicago: University of Chicago Press, 1985.

———, *Technology and Utopia*, Washington, DC: American Historical Association, 2006.

———, *Utopias: A Brief History from Ancient Writings to Virtual Communities*, Chichester: Wiley-Blackwell, 2012.

Sheehan, Michael J., *The International Politics of Space*, London: Routledge, 2007.

Siebeneichner, Tilmann, 'Exploring the Heavens: Space Technology and Religious Implications,' *Technology and Culture* 55.4 (October 2014), 988–94.

———, 'Europas Griff nach den Sternen: Das Weltraumlabor Spacelab, 1973–1998,' *Themenportal Europäische Geschichte* (2015), available at http://www.europa.clio-online.de/2015/Article=739 (accessed 1 October 2017).

Siddiqi, Asif A., *Challenge to Apollo: The Soviet Union and the Space Race, 1945–1974*, Washington, DC: NASA, 2000.

———, 'Soviet Space Power During the Cold War,' in Paul G. Gillespie and Grant T. Weller, eds, *Harnessing the Heavens: National Defense Through Space*, Colorado Springs: United States Air Force Academy, 2008, 135–50.

———, 'Competing Technologies, National(ist) Narratives, and Universal Claims: Toward a Global History of Space Exploration,' *Technology and Culture* 51.2 (April 2010), 425–43.

Singer, S. Fred, 'Exploring Space in the Seventies,' *Bulletin of the Atomic Scientists* 26.9 (November 1970), 22–3.

Slack, Edward R., 'A Brief History of Satellite Communications,' *Pacific Telecommunications Review* 22.3 (2001), 7–20.

Sloterdijk, Peter, *Was geschah im 20. Jahrhundert?*, Berlin: Suhrkamp, 2016.

Slotten, Hugh R., 'Satellite Communications, Globalization, and the Cold War,' *Technology and Culture* 43.2 (April 2002), 315–50.

Smith, Michael L., 'Selling the Moon: The U.S. Manned Space Program and the Triumph of Commodity Scientism,' in Richard Wightman Fox and T.J. Jackson Lears, eds, *The Culture of Consumption: Critical Essays in American History, 1880–1980*, New York: Pantheon, 1983, 175–209.

Smith, Warren, 'To Infinity and Beyond?,' *Sociological Review* 57.1 (May 2009), 204–12.

Spangenburg, Ray and Diane Moser, *Carl Sagan: A Biography*, Amherst: Prometheus, 2004.

Stapledon, Olaf, 'Interplanetary Man?,' *Journal of the British Interplanetary Society* 7.6 (November 1948), 213–33.

Stine, G. Harry, *The Third Industrial Revolution*, New York: Putnam, 1975.

Stork, David G., ed., *HAL's Legacy: 2001's Computer as Dream and Reality*, Cambridge, MA: MIT Press, 1997.

Sutton, Jeff, *Beyond Apollo*, New York: Putnam, 1966.

Tassin, Jacques, *Vers l'Europe spatiale*, Paris: Denoël, 1970.

Taylor, Charles, *Modern Social Imaginaries*, Durham: Duke University Press, 2004.

Teilhard de Chardin, Pierre, 'Vie et planètes: Que se passe-t-il en ce moment sur la Terre?,' *Etudes: Revue de culture contemporaine* 248 (1946), 145–69 (Eng. 'Life and the Planets: What is Happening at this Moment on Earth?,' in idem, *The Future of Man*, London: William Collins, 1964, 97–123).

———, 'Un grand événement qui se dessine: La Planétisation humaine,' *Cahiers du Monde Nouveau* 2.7 (August–September 1946), 1–13 (Eng. 'A Great Event Foreshadowed: The Planetisation of Mankind,' in idem, *The Future of Man*, London: William Collins, 1964, 124–39).

———, *Le Phénomène humain*, Paris: Editions du Seuil, 1955 (Eng. *The Phenomenon of Man*, New York: Harper Perennial, 2008).

———, *L'Avenir de l'homme*, Paris: Editions du Seuil, 1959 (Eng. *The Future of Man*, London: William Collins, 1964).

Toffler, Alvin, *Future Shock*, New York: Random House, 1970.

Touraine, Alain, *The Post-industrial Society: Tomorrow's Social History. Classes, Conflicts and Culture in the Programmed Society*, New York: Random House, 1971.

Tribbe, Matthew D., *No Requiem for the Space Age: The Apollo Moon Landings and American Culture*, Oxford: Oxford University Press, 2014.

Trischler, Helmuth, *Luft- und Raumfahrtforschung in Deutschland 1900–1970: Politische Geschichte einer Wissenschaft*, Frankfurt am Main: Campus, 1992.

———, The 'Triple Helix' of Space: German Space Activities in a European Perspective, Noordwijk: ESA, 2002.

———, 'Contesting Europe in Space,' in idem and Martin Kohlrausch, Building Europe on Expertise: Innovators, Organizers, Networkers, Basingstoke: Palgrave Macmillan, 2014, 243–75.

Tuck, Stephen, 'Reconsidering the 1970s: The 1960s to a Disco Beat?,' Journal of Contemporary History 43.4 (October 2008), 617–20.

Tucker, Anthony, 'Europe Ditched in Space?,' Guardian (10 July 1972), 11.

Turchetti, Simone and Peder Roberts, eds, The Surveillance Imperative: The Rise of the Geosciences During the Cold War, Basingstoke: Palgrave Macmillan, 2014.

Turner, Alwyn W., Crisis? What Crisis? Britain in the 1970s, London: Aurum, 2008.

Turner, Fred, From Counterculture to Cyberculture: Stewart Brand, the Whole Earth Network, and the Rise of Digital Utopianism, Chicago: University of Chicago Press, 2006.

United States Congress, Committee on Aeronautical and Space Sciences, National Space Goals for the Post-Apollo Period: Hearings on Alternative Goals for the National Space Program Following the Manned Lunar Landing, Washington, DC: Government Printing Office, 1965.

United States Congress, Committee on Science and Technology, World-Wide Space Activities: National Programs Other than the United States and Soviet Union, International Participation in the U.S. Post-Apollo Program, International Cooperation in Space Science, Applications and Exploration, Organization, and Identification of Major Policy Issues, Washington, DC: Government Printing Office, 1977.

United States President's Science Advisory Committee, The Space-Program in the Post-Apollo Period, Washington, DC: Government Printing Office, 1967.

United States Space Task Group, The Post-Apollo Space Program: Directions for the Future, Washington, DC: Government Printing Office, 1969.

Valentine, Burl, 'Obstacles to Space Cooperation: Europe and the Post-Apollo Experience,' Research Policy 1.2 (April 1972), 104–21.

Varsori, Antonio, ed., Alle origini delle presente: L'Europa occidentale nella crisi degli anni settanta, Milan: Franco Angeli, 2007.

Waldrep, Shelton, ed., The Seventies: The Age of Glitter in Popular Culture, New York: Routledge, 2000.

Ward, Barbara, Spaceship Earth, New York: Columbia University Press, 1966.

Weart, Spencer R., Nuclear Fear: A History of Images, Cambridge, MA: Harvard University Press, 1988.

Weber, Ronald, Seeing Earth: Literary Responses to Space Exploration, Athens: Ohio University Press, 1985.

Welck, Stephan Freiherr von, 'Outer Space and Cosmopolitics,' Space Policy 2.3 (August 1986), 200–5.

Westwick, Peter J., Into the Black: JPL and the American Space Program, 1976–2004, New Haven: Yale University Press, 2007.

———, '"Space-Strike Weapons" and the Soviet Response to SDI,' Diplomatic History 32.5 (November 2008), 955–79.

———, 'The International History of the Strategic Defense Initiative: American Influence and Economic Competition in the Late Cold War,' Centaurus 52.4 (November 2010), 338–51.

Weyer, Johannes, 'European Star Wars: The Emergence of Space Technology through the Interaction of Military and Civilian Interest-Groups,' in Everett Mendelsohn, Meritt R. Smith and Peter Weingart, eds, *Science, Technology and the Military*, Dordrecht: Kluwer, 1988, 243–88.

Wheeler, Edd, 'The "Wow" Signal, Drake Equation and Exoplanet Considerations,' *Journal of the British Interplanetary Society* 67.11–12 (November/December 2014), 412–17.

White, Frank, *The Overview Effect: Space Exploration and Human Evolution*, Boston: Houghton Mifflin, 1987.

Wilford, John Noble, 'Last Apollo Wednesday: Scholars Assess Program,' *New York Times* (3 December 1972), 1, 68.

———, 'Meaning of Apollo: The Future Will Decide,' *New York Times* (21 December 1972), 21.

———, 'Space and the American Vision,' *New York Times Magazine* (5 April 1981), 53.

Winchester, Simon, 'Post-Apollo Theories Revised,' *Guardian* (10 April 1970), 6.

Wirsching, Andreas, Göran Therborn, Geoff Eley, Hartmut Kaelble and Philippe Chassaigne, 'The 1970s and 1980s as a Turning Point in European History?,' *Journal of Modern European History* 9.1 (2011), 8–26.

Wittner, Lawrence S., *One World or None? A History of the World Nuclear Disarmament Movement Through 1945*, Stanford: University of California Press, 1993.

Wolfe, Tom, 'The "Me" Decade and the Third Great Awakening,' *New York* (23 August 1976), 26–40.

———, 'One Giant Leap to Nowhere,' *New York Times Magazine* (19 July 2009), WK11.

'World Poll Finds Wide Belief in Life on Other Planets,' *New York Times* (13 June 1971), 20.

INDEX

Page numbers appearing in *italics* refer to illustrations. A page reference in the form 'n' indicates a note number; for example, 204n59 is note 59 on page 204. Index entries that begin with a number are indexed as if the number were spelled out; for example, *2001: A Space Odyssey* is alphabetized as 'Two thousand and one.'

© The Editor(s) (if applicable) and The Author(s) 2018
Alexander C.T. Geppert (ed.), *Limiting Outer Space*
European Astroculture, vol. 2
https://doi.org/10.1057/978-1-137-36916-1

Printed in the United States
by Baker & Taylor Publisher Services